Subpixel Mapping for Remote Sensing Images

This book provides readers with a complete overview of subpixel image processing methods, basic principles, and different subpixel mapping techniques based on single or multi-shift remote sensing images. Real-life applications are a great resource for understanding how and where to use subpixel mapping when dealing with different remote sensing imaging data.

FEATURES

- Provides the fundamentals of subpixel mapping technology and its applications.
- Discusses in detail the advantages of using different subpixel mapping techniques based on remote sensing data.
- Summarizes in a systematic way current subpixel mapping methods.
- Highlights authors' achievements in subpixel mapping technology.
- Includes case studies based on remote sensing data from the United States, Italy, China, and Cambodia.

This book will be of interest to undergraduate and graduate students majoring in remote sensing, surveying, mapping, and signal and information processing in universities and colleges, and it can also be used by professionals and researchers at different levels in related fields.

Subpixel Mapping for Remote Sensing Images

Peng Wang and Lei Zhang

CRC Press
Taylor & Francis Group
Boca Raton London New York

CRC Press is an imprint of the
Taylor & Francis Group, an **informa** business

First edition published 2023
by CRC Press
6000 Broken Sound Parkway NW, Suite 300, Boca Raton, FL 33487–2742

and by CRC Press
4 Park Square, Milton Park, Abingdon, Oxon, OX14 4RN

CRC Press is an imprint of Taylor & Francis Group, LLC

ISBN: 978-1-032-22938-6 (hbk)
ISBN: 978-1-032-24522-5 (pbk)
ISBN: 978-1-003-27908-2 (ebk)

DOI: 10.1201/9781003279082

Typeset in Times
by Apex CoVantage, LLC

Contents

Foreword

With the rapid development of the aerospace industry in recent years, remote sensing science, as a comprehensive technology in the aerospace field, has been widely and greatly developed in both theory and application. In particular, the spectral remote sensing developed in recent years can obtain the spatial information and spectral information of land cover classes at the same time, realize the real integration of maps, and provide strong technical support for the aerospace industry. Remote sensing technology has been successfully applied in military, civil, agricultural, marine, smart city, disaster reduction, and other fields.

As an important research direction of remote sensing technology, remote sensing image processing directly determines the accuracy and utilization of remote sensing information. One of the hot issues in remote sensing image processing technology is how to deal with mixed pixels to improve the spatial resolution of remote sensing images. Although spectral unmixing technology can obtain the proportion (abundance value) of each class in the mixed pixel, it cannot obtain the specific spatial distribution information of each class in the mixed pixel. Subpixel mapping (SPM) technology, as the subsequent processing of spectral unmixing technology, decomposes pixels into smaller subpixels and then obtains thematic mapping with distribution information of each class at the subpixel scale. SPM, as a potential technology to obtain land cover classes of spatial distribution information, has attracted more and more attention.

Combined with the authors' relatively new research achievements in this technology, this book makes a systematic arrangement and detailed explanation and contributes to readers' understanding, learning, and research of SPM technology. In this book, the authors compile the important theories and methods about the analysis and processing stage in the SPM process, involve the complex technical key scientific issues such as multi-source information fusion and artificial intelligence technology, and put forward many new theories and methods of remote sensing information. It is fascinating and refreshing to read. I would like to express my warm congratulations.

Professor Yongqi Xue
Academician of China Academy of Sciences

Preface

The main research contents of this book are supported by the Natural Science Foundation of Jiangsu Province (grant no. BK20221478), the Hong Kong Scholars Program (grant no. XJ2022043), and the National Natural Science Foundation of China (grant no. 61801211). The book is divided into seven chapters. Chapters 3 through 7 are written mainly by Peng Wang, covering five aspects: subpixel mapping based on single remote sensing image, subpixel mapping based on multi-shift remote sensing images, subpixel mapping of remote sensing image based on pansharpening technology, subpixel mapping of remote sensing image based on reconstruction then classification, and application of subpixel mapping technology in remote sensing image, which have been the research achievements of the authors for many years. The authors hope these contents can provide readers with some reference and inspiration in thought or method. Chapters 1 and 2 are written mainly by Lei Zhang, including an introduction to and basic principles of subpixel mapping, which are convenient for readers to have a comprehensive understanding of subpixel mapping technology of remote sensing images. Chapter 1 refers primarily to the relevant works of Qingxi Tong, Liguo Wang, Qunming Wang, and much academic literature at home and abroad. It introduces the research background and significance of subpixel mapping technology, the research status of subpixel mapping, the current problems of existing subpixel mapping technology, and the main research contents and chapter arrangement of this book. Chapter 2 introduces the basic principle of spectral unmixing technology, the basic principle of mainstream subpixel mapping, and the evaluation method of subpixel mapping accuracy. We sincerely hope that readers will critique, correct and comment, so we can further improve and revise in the follow-up work.

Peng Wang and Lei Zhang

Authors

Peng Wang earned his doctoral degree from the College of Information and Communications Engineering, Harbin Engineering University, Harbin, China, in 2018. He is currently an associate professor at the College of Electronic and Information Engineering, Nanjing University of Aeronautics and Astronautics, Jiangsu, China. He was a Hong Kong Scholar with the Institute of Space and Earth Information Science, the Chinese University of Hong Kong, Hong Kong SAR, China. His research interests include remote sensing imagery processing and machine learning. He has authored one book and more than fifty papers.

Lei Zhang earned his doctoral degree from the Graduate School of Chinese Academy of Sciences in 2008 and finished the postdoctoral program in Tsinghua University in 2010. From 2011 to 2012 he was an associate professor at the Chinese University of Hong Kong and from 2012 to 2015 at the Shanghai Institute of Technical Physics of Chinese Academy of Sciences. Currently, he is a professor in Tongji University. His research interests include intelligent information processing and spatiotemporal applications.

1 Introduction

This chapter first introduces the research background and significance of subpixel mapping, the research status of subpixel mapping, the problems that exist in existing subpixel mapping, and the main research content and chapter arrangement of the book in order to help readers understand the main content of the book.

1.1 BACKGROUND AND SIGNIFICANCE

1.1.1 BACKGROUND OF SUBPIXEL MAPPING

Remote sensing (RS) grew fast as a comprehensive earth observation technology in the 1960s. It mainly refers to the use of sensors on the platform to detect a target from a distance utilizing electromagnetic wave signals without having to come into direct touch with the study target or location. Meanwhile, the interaction mechanism between signals and surface items can be used to get useful data. RS has become one of the hottest scientific disciplines after decades of rapid development. The most significant benefit of RS technology is that a huge amount of data can be gathered in a short amount of time, and information may be expressed in both image and non-image formats [1].

RS images have gotten a lot of attention as a crucial aspect of RS technology. As demonstrated in Figure 1.1, RS images have evolved in recent years to include panchromatic images, color images, multispectral RS images, and hyperspectral RS images [2].

Because of the genuine combination of spectrum and space, hyperspectral RS images not only have been actively investigated by RS scholars in recent years but also have been widely utilized in other fields. The hyperspectral RS data is introduced in detail next since the RS experimental data utilized in this book is mostly hyperspectral RS data.

The schematic diagram of hyperspectral data is shown in Figure 1.2. Hyperspectral data are presented in the form of three-dimensional cube data, which contain two-dimensional spatial information and one-dimensional spectral information, where each image level represents a spectral band data, and the spectral dimension is composed of the image level according to the order composite [3]. At the same time, each pixel in spectral space represents a continuous spectral curve, and different spectral curves indicate varying radiation intensities for different substances. Hyperspectral RS images can acquire not just surface image information but also spectral information, allowing for "image and spectral integration." In comparison to other RS images, hyperspectral RS images not only enhance the information in RS images but also help to analyze and process spectral data more efficiently and effectively. As a result, RS images created at earlier phases of development cannot compare in terms of influence and development potential to hyperspectral RS images.

DOI: 10.1201/9781003279082-1

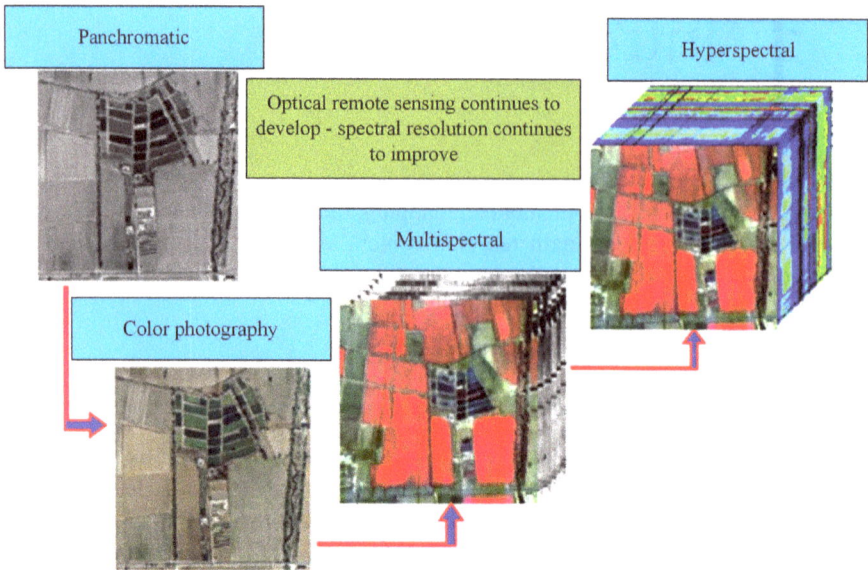

FIGURE 1.1   Remote sensing image development.

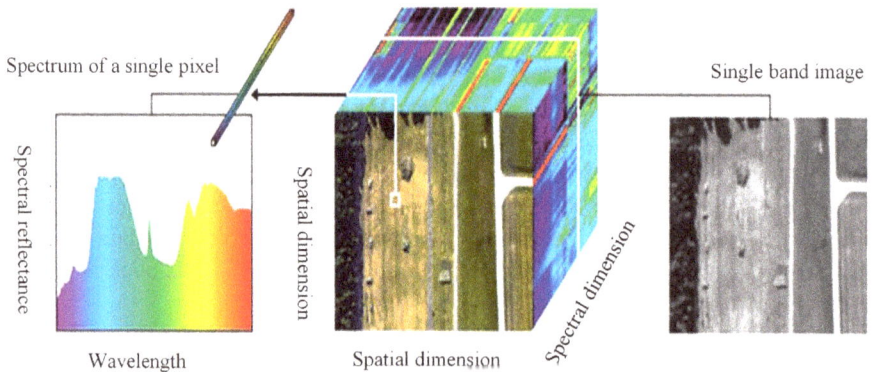

FIGURE 1.2   Hyperspectral data.

The imaging spectrometer installed on the acquisition platform acquires hyperspectral RS images by simultaneously imaging the target region with dozens to hundreds of continuous spectral bands. The ultraviolet, visible, near-infrared, mid-infrared, and thermal infrared bands of the electromagnetic spectrum are included in this group. The three most frequent hyperspectral RS image acquisition platforms are near-earth, aviation, and aerospace [4]. The major usage of the near-earth platform, which has been widely utilized in agriculture and laboratory areas, is in space with a height of little more than 50 meters. It is required to use the aviation platform when the altitude increases from 10 to 100 km. Small aircraft or unmanned aerial vehicles equipped with imaging spectrometers are referred to as aviation platforms. The use of this platform to collect hyperspectral RS data necessitates a high level of precision in terms of hardware and operation. Hyperspectral RS data are obtained at altitudes above 150 km using space platforms equipped with imaging spectrometers on space satellites; however, the expenses are enormous. As a result, the two most regularly employed platforms for collecting hyperspectral images are the ground platform and the aviation platform.

In addition to the selection of acquisition platform, the imaging spectrometer mounted on the acquisition platform also plays a key role in the quality of hyperspectral RS images. The spatial resolution and spectral resolution of hyperspectral images have been improved with the updating of imaging spectrometers. Throughout the history of foreign imaging spectrometers, the three generations of imaging spectrometers developed in the United States are very representative. The first generation of hyperspectral imaging spectrometer, aero imaging spectrometer (AIS), was designed by the jet propulsion laboratory of the National Aeronautics and Space Administration (NASA) in 1983. It mainly includes two kinds, AIS-1 and AIS-2, with spectral coverage range of 1.2 to 2.4 microns. AIS-1 is the world's first hyperspectral imaging spectrometer. The visible/infrared airborne hyperspectral imaging spectrometer was also developed by the jet propulsion laboratory of NASA in 1987. The airborne visible/infrared imaging spectrometer (AVIRIS) marks the advent of the second generation of hyperspectral imaging spectrometers. Since then, airborne hyperspectral imagers have improved in technology. The United States launched the Hyperspectral Digital Imagery Collection Experiment (HYDICE) and Spatially Enhanced Broadband Array Spectrograph System (SEBASS), and so on. During this period, other countries also successfully developed a variety of airborne hyperspectral imagers. For example, Fluorescence Line Imager (FLI), Compact Airborne Spectrographic Imager (CASI), and Shortwave Infrared Airborne Spectrographic Imager (SASI) were successfully developed in Canada. Thermal Airborne Spectrographic Imager (TASI) and Reflective Optics Imaging Spectrometer (ROSIS) were successfully developed in Germany. Australia developed Hyperspectral Mapper (HyMap). In the 21st century, hyperspectral imaging spectrometers gradually transition from airborne imaging spectrometers to spaceborne imaging spectrometers, which also marks the arrival of the third generation of hyperspectral imaging spectrometers. In 2000, Hyperion, the hyperspectral imager mounted on the EO-1 satellite launched by NASA, achieved a ground object resolution of 30 meters. The Coastal Ocean Imaging Spectrometer (COIS), carried by the Naval EarthMap Observer (NEMO) satellite launched by the United States in 2002, is used to detect the coastal zone and shallow sea environment, meeting both military and civilian needs. In addition, other countries, such as Germany, Finland, and Japan, have successfully launched satellites carrying imaging spectrometers.

Although hyperspectral imaging spectrometer technology started late in China, it has made rapid progress in recent years. In the mid to late 1980s, emphasis was placed on the development of autonomous airborne imaging spectrometers. The main airborne hyperspectral imaging spectrometers are pushbroom hyperspectral imaging (PHI) developed by Shanghai Institute of Physics, Chinese Academy of Sciences; the operational modular imaging spectrometer (OMIS); and the high-resolution imaging spectrometer C-HRIS (Changchun high-resolution imaging spectrometer) developed by the Changchun Optics Institute of the Chinese Academy of Sciences Imaging Spectrometer. At the same time, with the rapid development of China's space industry, the spaceborne imaging spectrometer has received more domestic RS scholars' attention and research. In 2002, the China Moderate Resolution Imaging Spectroradiometer (CMODIS), developed by the Shanghai Institute of Physics, Chinese Academy of Sciences, was successfully launched aboard the Shenzhou III spacecraft. And the first hyperspectral RS image of China was successfully obtained, which contains a total of 30 bands of near-infrared and 4 bands of mid-infrared to far-infrared, with a spatial resolution of 500 meters. The interferometric imaging spectrometer developed by Xi'an Institute of Optomechanics, Chinese Academy of Sciences, was carried on the Chang'e-1 satellite launched in 2007. The spectrometer is mainly used for collecting two-dimensional multispectral images of the lunar surface and high-resolution spectral images of the lunar surface. After this, China has launched a number of "remote sensing" and "Gaofen" series of RS satellites, which are equipped with different imaging spectrometers to collect RS image data.

It can be found that the demand for RS image data in various countries around the world is very large. These collected RS image data have been widely used in geographic information system, land management, resource investigation, environmental detection, and natural disaster resistance. However, in order to obtain better RS image data in some application fields, it is sometimes necessary to carry out some technical processing on the collected RS image data. RS image classification, endmember extraction, spectral unmixing and super-resolution restoration recovery, dimension reduction and compressing, anomaly detection, and subpixel mapping are common RS image processing techniques. Subpixel mapping is an important part of many RS image processing technologies. This RS image processing technology is studied in this text.

1.1.2 SIGNIFICANCE OF SUBPIXEL MAPPING

The spectral resolution of RS image data is constantly improving. For example, the spectral resolution of hyperspectral images can be less than 10 nanometers, but the spatial resolution of RS images is restricted. Among them, the spatial resolution of RS images is mainly affected by the mixed pixels produced by the diversity of feature categories and the limitations of the sensor's instantaneous field of view [4].

Figure 1.3 shows the two most common existing forms of mixed pixels [5]. Figure 1.3(a) shows the most common form of the existence of mixed pixels in RS images. Due to the spatial resolution of the terrain category below the sampling frequency spectral imager, even if the dimension of the multiple land cover classes is greater than the resolution of the RS image, the mixed pixels also will appear in their border mixed, producing a lot of mixed pixels. For example, the mixed pixels form at the boundary of many different kinds of crops in a field. Figure 1.3(b) shows the existence of another kind of mixed pixel, this is because the spatial resolution of the feature category is higher than the sampling frequency of the imager. That is to say, the size of a land cover class is smaller than the resolution of the RS image, and it occupies part of an isolated spatial region inside a pixel. For example, if a car exists

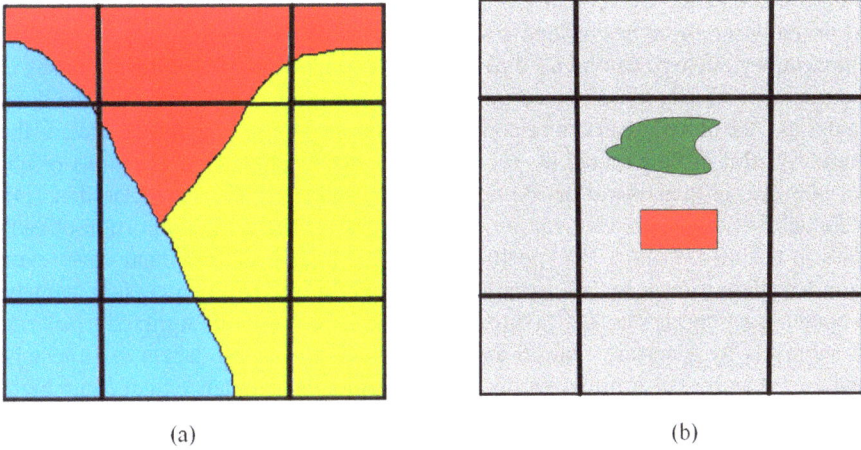

(a)   (b)

FIGURE 1.3   Two existing forms of mixed pixels: (a) border line and (b) small target.

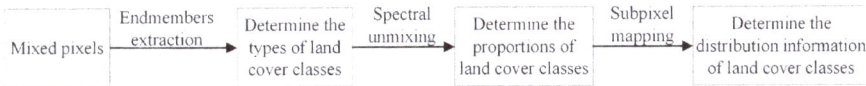

FIGURE 1.4   Mixed pixel processing process.

on a pixel in a MODIS RS image with a resolution of 250 m, the mixed pixel can be regarded as a small target type of mixed pixel. In addition to these two common mixed pixels, there are also compact mixed pixels and linear feature mixed pixels.

Thus, the mixed pixels widely exist in the RS images, and the mixed pixels affect the spatial resolution of the RS image. These mixed pixels have brought a lot of difficulty in extracting the accurate spatial distribution information, the accurate spatial distribution information has significance in the industry, agriculture, environment, and military sectors. Therefore, how to improve the spatial resolution of the RS image is one of the hot issues in RS research. In general, the spatial resolution of RS images can be improved through the following two approaches. One approach is to improve the precision of hardware equipment, such as improving sampling frequency, increasing pixel density, or reducing the size of photosensitive elements, but the high-precision equipment is often expensive, which limits the wide use of this approach. The other way is to improve the spatial resolution of the RS image by processing RS image, especially the processing of mixed pixels. Because this way is not limited by the hardware equipment itself, it has become the preferred way to improve the spatial resolution. Figure 1.4 is the current process for processing mixed pixels. The first step is endmembers extraction, which is used to determine the types of land cover classes existing in the mixed pixel. Spectral information of the land cover class in RS images is usually obtained by automatic acquisition or manual selection. Common endmember extraction methods include N-FinDR [3], pixel purity index

(PPI) [6], and iterative error analysis (IEA) [7]. The second step is spectral unmixing. After obtaining the types of land cover classes inside the mixed pixel, the spectral unmixing technology can be used to predict the proportion of land cover classes in the mixed pixel according to a certain spectral unmixing model, and then the basic analysis of the mixed pixel can be completed. Common unmixing models include the linear spectral mixing model [8–10], support vector machine [11], K-nearest neighbor classifier [12], artificial neural network [13], and fuzzy C-means classifier [14]. Although spectral unmixing can provide information about the proportion of each class in the mixed pixel, the specific spatial distribution of each land cover class cannot be determined. In this case, the third step of mixed pixel processing, namely, subpixel mapping, is needed. Subpixel mapping is to subdivide a mixed pixel into S^2 subpixels by a certain scale S and assign land cover class labels to subpixels, realizing the process of transforming the low-resolution abundance image into high-resolution classification thematic mapping.

In 1997, Atkinson first proposed the concept of subpixel mapping [15], whose purpose is to determine the specific distribution information of each land cover class in the mixed pixel, so as to obtain classification thematic mapping with finer spatial resolution. Existing literature is often named as subpixel mapping, super-resolution mapping, or downscaling. Since this technology is processed on a subpixel-level image, this paper uses subpixel mapping as the description of this technology [16]. As a follow-up step of spectral unmixing, subpixel mapping is of great scientific significance. It can extract subpixel-level land cover class distribution information, which can be applied to distinguish and invert ground object features from medium- and low-resolution RS data. Subpixel mapping has been successfully applied to land cover mapping [17–18], lake and coast boundary extraction [19–21], landscape index calculation [22–23], and change detection [24–25]. Therefore, subpixel mapping has been paid more and more attention in the field of mixed pixel processing.

1.2 RESEARCH STATUS OF SUBPIXEL MAPPING

Since the advent of subpixel mapping technology, domestic and overseas RS scholars have been widely concerned and studied a lot of research. The research results have been published in major RS geoscience journals, such as *Remote Sensing of Environment, IEEE Transactions on Geoscience and Remote Sensing, IEEE Geoscience and Remote Sensing Letters, Photogrammetric Engineering and Remote Sensing*, and *International Journal of Remote Sensing*.

In 1997, Atkinson first proposed the concept of subpixel mapping [15]. Subpixel mapping refers to the thematic mapping of the classification distribution of high-resolution images at the subpixel level by downscaling coarse RS image and conducting correlation analysis through subpixel spatial correlation or some spatial prior information models. The subpixel mapping is of great significance in both science and society, which is the follow-up processing step of spectral unmixing. The mapping accuracy of each class of coarse RS image can reach subpixel level through the subpixel mapping, and the mapping accuracy is greatly improved compared with the original hard classification technology. These advantages bring great benefits to the classification and inversion of land cover classes. When the low-resolution satellite RS data are used to estimate the land cover classification and land cover change detection, and so on, not only can subpixel mapping effectively alleviate the limitation of low spatial resolution of original RS image, but it can also overcome the misclassification phenomenon caused by serious mixed pixel problem. In addition, subpixel mapping provides a good idea for further classification and recognition by using spatial correlation. In practical social applications, subpixel mapping has been successfully applied to land classification thematic mapping, river and lake boundary extraction, coastline detection, landscape index calculation, change detection, and other related fields. Subpixel mapping has also been widely used in crop changes, investigation of drinking water quality, environmental detection, climate change, and analysis of natural disasters [17–25]. Therefore, in recent years, more and more scholars at home and abroad have paid attention to the related technologies of subpixel mapping, and the development of subpixel mapping has become a demand in the field of RS.

Wang et al., in "Allocating classes for soft-then-hard subpixel mapping methods in units of class" [26], summarized the existing mainstream subpixel mapping technologies and divided them into two main types. As shown in Figure 1.5, the first type is called initialization-then-optimization subpixel mapping, and the second type is called soft-then-hard subpixel mapping. The following is a systematic introduction to the current research status of subpixel mapping domestic and overseas.

1.2.1 INITIALIZE-THEN-OPTIMIZE SUBPIXEL MAPPING

In initialize-then-optimize subpixel mapping, land cover class labels are allocated randomly to subpixels, and the location of each subpixel is optimized to obtain the final mapping result. For a two-class land cover problem, the pixel-swapping algorithm (PSA) proposed by Atkinson [29] belongs to this type. In this method, the abundance image is randomly initialized, and the subpixels in each pixel are exchanged according to the attraction to obtain the optimal spatial distribution at the subpixel scale. To improve the defects of random initialization in the PSA method, Shen et al. used the spatial attraction model as the method to obtain the initialization image of the PSA method, producing the more accurate subpixel mapping results [30]. According to the influence of different spatial correlation weight parameters

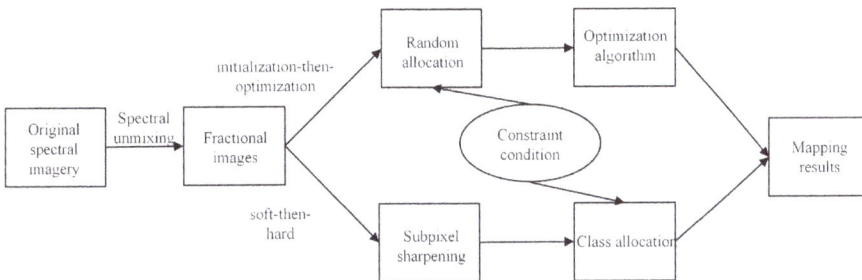

FIGURE 1.5   Two types of subpixel mapping.

on the mapping results of the PSA method, Makido made a detailed analysis in his doctoral thesis "Land Cover Mapping at Subpixel Scales" [31, 32]. At the same time, the other alternative models of the PSA method for multi-type land cover classes problems are also discussed [33].

In addition, as an optimization model, an artificial intelligence algorithm has been successfully applied in initialize-then-optimize subpixel mapping, such as particle swarm optimization algorithm [27, 28], simulated annealing algorithm [34–36], and genetic algorithm [37, 38]. At the same time, these optimization algorithms can also be combined with a two-point histogram [39, 40] or landscape structure information extracted from training images [41] to obtain subpixel mapping results with higher accuracy. It should be noted that in the optimization process, only the spatial position of subpixels changes, while the number of subpixels of each land cover class is fixed. Although the initialize-then-optimize subpixel mapping can yield the accurate subpixel mapping results, it often requires a complex optimization process, which makes the calculation burden of this type large.

1.2.2   Soft-Then-Hard Subpixel Mapping

The second type is soft-then-hard subpixel mapping that consists of two steps. Firstly, the abundance values of all subpixels in each class (probability membership information between 0 and 1) are estimated by some method. At this time, the low-resolution abundance images obtained by spectral unmixing technology will be transformed into high-resolution abundance images with subpixel abundance values. This step is called subpixel sharpening. Secondly, according to the abundance value of each subpixel and the fixed number of subpixels in each class, the class label is allocated to each subpixel, producing the final mapping result. This step is called class allocation.

Many existing methods, such as Hopfield neural network, spatial attraction model, backpropagation neural network, and some super-resolution algorithms, can be subpixel sharpening methods [26]. Tatem successively proposed a series of sub-pixel mapping technologies based on the Hopfield neural network (HNN) model [42–44]. In subpixel mapping based on the HNN method, each subpixel is regarded as a neuron, the spatial correlation between the central subpixel and its neighboring pixels is regarded as an energy function, the output neuron is solved according to the principle of minimizing the energy function, and the subpixel mapping result is finally obtained. Later, when the target size is smaller than the mixed pixel, Tatem used the semi-variance function to represent the prior information of spatial distribution and added this information into the HNN model [45], improving the accuracy of the subpixel mapping results. For the simultaneous existence of objects of different scales, Tatem proposed a multi-scale HNN model [46]. At the same time, Tatem successfully applied HNN-based subpixel mapping to land cover mapping in agricultural areas [47] and IKONOS images [48]. Muad et al. continued to discuss the ability of different parameter settings in the HNN method to locate targets of different scales [49].

The subpixel-pixel spatial attraction model (SPSAM) proposed by Mertens from Ghent University in Belgium [50] also belongs to the type of soft-then-hard subpixel mapping. In this method, the spatial attraction values between each subpixel in the center and the neighboring pixels are first calculated. The land cover class labels are assigned according to the attraction values, obtaining the final mapping results. In order to improve the mapping accuracy of this method, some scholars have successively proposed a spatial attraction optimization model that considers both subpixel-pixel and subpixel-subpixel at the same time [51, 52]. Meanwhile, Mertens also proposed another subpixel mapping method, named the backpropagation (BP) neural network, which belongs to the type of subpixel mapping soft-then-hard [53]. In the subpixel mapping based on the BP neural network, the corresponding relationship between the class attribute values of each subpixel and the abundance values of each pixel in the neighborhood are learned by training known high-resolution images. Then it is input into the trained network to realize the subpixel mapping results. Later, Mertens combined the network with wavelet transform to discover more details of spatial distribution, improving the mapping accuracy of this method [54]. To solve the potential problem of the edge of the subpixel mapping result generated by this method having severe serration, Zhang et al. proposed the BPMAP method by adding the observation model on the basis of the BP neural network [55]. Gu et al. added spatial-spectral information of the RS image into the BP neural network and obtained good subpixel mapping results [56, 57]. In addition, Xu et al. used the spatial autocorrelation function as a constraint condition to improve the mapping results.

Subpixel sharpening can be replaced by some appropriate super-resolution algorithms [58–63], that is to say, the soft-then-hard subpixel mapping method can effectively combine the super-resolution technology and subpixel mapping technology in the field of RS image processing. In particular, when simple interpolation algorithms are used, such as nearest interpolation, bilinear interpolation, and bicubic interpolation, the results of subpixel mapping can be obtained quickly, which significantly improves the operation speed of the subpixel mapping method. Subpixel mapping based on interpolation can be applied in environments where no prior structural information is required [64–65].

In addition to the subpixel sharpening method, the selection of class allocation method also has an important impact on the results of the soft-then-hard subpixel mapping. According to different class allocation principles, different class allocation methods can be obtained. The existing class allocation principles are mainly based on linear optimization technique (LOT) [59], the principle of units of subpixel (UOS) [66–68], the principle of highest soft attribute values first (HAVF) [69], the principle of units of class [70], the principle of hybrid constraints of pure and mixed pixels [71], and the principle of spatial distribution patterns of geographical objects [72]. As can be seen from the above description, the soft-then-hard subpixel mapping has the advantages of convenient application and fast computing speed, so it has attracted the attention and research of RS scholars in recent years and become the mainstream subpixel mapping type, which is also the focus of the introduction and research of this book.

1.2.3 OTHER TYPES OF SUBPIXEL MAPPING

In addition to the previous two main types of subpixel mapping, there is another subpixel mapping method that directly draws the class boundary [73–75] according to the component information of a coarse image. However, this method cannot protect the proportion information of different classes in the image and may not be able to restore the land cover class in a single coarse pixel.

The subpixel mapping based on the Markov random field (MRF) model is different from the two previous main types of subpixel mapping. The two previous main types of subpixel mapping are based on a low-resolution abundance image. However, the subpixel mapping based on MRF considers both the spatial constraints of land cover classes and the spectral constraints of mixed pixels to directly obtain the subpixel mapping result and is independent of the low-resolution abundance image after spectral unmixing [76]. Tolpekin et al. conducted a more in-depth study on this model, analyzed the uncertainty of subpixel mapping results based on the MRF model [77], discussed the influence of different spatial constraint and spectral constraint parameters on the mapping results [78], and studied the influence of spectral separability of land cover classes on mapping results [79]. At the same time, this model is successfully applied to the extraction of tree information from RS images [80]. In addition, Li et al. added the shape information of land cover class into the MRF model and proposed a subpixel mapping method based on the anisotropic MRF model [81]. Meanwhile, Li et al. also studied the setting of smoothing parameters in the MRF model [82]. Wang et al. applied multiple subpixel shifted images to the MRF model, thus providing spatial and spectral dual constraints [83]. However, the parameter selection of the constraint term in this method is not universal, and the optimal parameter selection always depends on the unique spatial pattern and spectral change of the RS image study area.

Since subpixel mapping is an ill-posed inverse problem [84], the inherent uncertainty limits the performance of many subpixel mapping methods. In recent years, various kinds of auxiliary data have been used to reduce these uncertainties and improve the accuracy of subpixel mapping results. For example, Foody used the finer spatial resolution images within the same scene to sharpen the low-resolution abundance images, thus providing more accurate land cover distribution information [85]. Aplin and Atkinson used auxiliary land linear vector boundary information to refine the land cover distribution within each polygon [86]. Based on the subpixel mapping method based on a two-point histogram, Atkinson added the restriction conditions of abundance values obtained from panchromatic images with medium resolution to the subjective function model of the method, improving the mapping accuracy of the method [40]. Nguyen et al. constructed a height function using light detection and ranging (LiDAR) data and added it into a subpixel mapping model based on HNN [87]. Ling et al. obtained the topographic information of land cover from the digital elevation model and modified the waterline mapping result based on this information [88]. Meanwhile, for the subpixel mapping of urban buildings, Ling et al. used prior shape information [89] as auxiliary data to improve the anisotropic model, making the subpixel mapping results of urban buildings better [90, 91]. In HNN-based subpixel mapping methods, the two methods proposed in the literature [92, 93] are provided the additional abundance limits on HNN energy functions to improve HNN-based subpixel mapping results. The specific method of these two methods is that Nguyen et al. obtained the abundance value of each land cover class in medium spatial resolution from panchromatic images [94] and regarded it as the abundance. A similar approach is that Nguyen fused the panchromatic image and multispectral image to get the multispectral image with the same spatial resolution as the panchromatic image. Nguyen then used spectral unmixing technology to the fused image, producing the abundance of value for each land cover class, which is the abundance limit [95]. In addition to the previous auxiliary information, multiple subpixel shifted images (MSIs) from the same scene can also be used as auxiliary data and have been widely used in subpixel mapping methods [96–100]. In particular, when MSIs are combined with simple and fast subpixel mapping methods, such as the spatial attraction model [101], bilinear interpolation model [102], and bicubic interpolation model [103], the final mapping results are significantly better than the subpixel mapping method based on a single image. Moreover, these methods are much faster than other subpixel mapping methods based on MSIs. In addition, these methods do not need any prior structure information. Deep learning has been successfully applied to subpixel mapping in recent years [104]. However, subpixel mapping based on deep learning usually needs a lot of fine priori training data to achieve the desired performance.

1.2.4 RESEARCH STATUS OF SUPER-RESOLUTION TECHNOLOGY

In subpixel mapping, some super-resolution algorithms can be used as subpixel sharpening methods, so there are some connections and differences between subpixel mapping technology and super-resolution technology. In addition, super-resolution technology is one of the important technical links of some proposed methods in the main research content of this book. Therefore, the research status of super-resolution technology is briefly introduced.

The super-resolution technology was first proposed by Harris and Goodman in the 1960s with the concept and method of single image restoration [104]. Subsequently, many people studied it and proposed various restoration methods successively, such as linear extrapolation method, long ellipsoid function method, and superimposed sine template method. Multivariate information fusion is a process of automatic detection, interconnection, correlation, estimation, and a combination of data and information from multiple sources, which is also a process of information enrichment. Wald et al. improved the spatial resolution of satellite RS images by using the high-frequency components calculated in multivariate data and by means of fusion, achieving good results [105]. Tsai and Huang first proposed the super-resolution reconstruction problem based on sequential images and proposed a reconstruction method based on frequency domain approximation [106]. Andrews and Hunt et al. proposed and developed many valuable super-resolution methods, such as convex set projection method [107], continuous energy reduction and subtraction method [108], and Bayes analysis method [109]. In addition, when the original information of the image is limited, the resolution of the image can be improved by increasing the number of pixels in the output image by the interpolation method. At present, there are many classical interpolation methods, among which the most commonly used are the nearest interpolation, bilinear interpolation, and cubic spline interpolation [110]. Schultz and Stevenson provided a brief summary of the current complex interpolation methods, including improved algorithms for cubic spline interpolation, rule-based methods, edge-preserving methods, and Bayes methods [111]. Jensen et al.

applied the interpolation method of the second-order statistical model in a steady-state random process and achieved good results in edge preservation [112]. Leizza Rodrigues proposed a local adaptive non-linear interpolation method. For a point to be interpolated, the local standard deviation is calculated, and the result is compared with the preset threshold value, determining the method to complete the interpolation calculation of the point [113].

Zhang et al. improved the spatial resolution of images by fusion of auxiliary high-resolution optical images [114]. Li et al. have done a lot of research on super-resolution methods based on sequential images [107]. Hao et al. proposed a general spatial domain interpolation analog iterative method based on convex set projection, which improves the frequency domain of digital image spatial resolution [115]. Su et al. studied a hybrid super-resolution method based on projection onto convex sets and maximum a posteriori (MAP), achieving good results [116, 117].

The ultimate purpose of the super-resolution algorithm and subpixel mapping method is to improve the resolution of the original coarse image, but they have both relation and difference. The connection lies in that the first step of the traditional subpixel mapping method can be realized by applying some suitable super-resolution algorithms, thus realizing the effective combination of super-resolution technology and subpixel mapping technology in the field of RS image processing. However, the difference between the two technologies is also obvious. The super-resolution algorithm is used to alleviate the image blur caused by the imaging environment and imaging condition, and to reconstruct a high-resolution image by inputting one or more images with continuous gray value, that is, the output is also an image with continuous gray value. However, the subpixel mapping method is mainly applied to the mixed pixels to weaken the influence of the mixed pixels on the spatial resolution. The high-resolution images obtained by the subpixel mapping method have the land cover labels, that is, the output is the image with discrete gray value.

1.3   PROBLEMS IN SUBPIXEL MAPPING

On the analysis of the current research status of subpixel mapping technology, it can be found that the soft-then-hard subpixel mapping method has become one of the main research types of subpixel mapping methods, which has been widely concerned and studied by more and more scholars at home and abroad. This book focuses on the introduction and research of subpixel mapping methods related to this type. However, the mainstream subpixel mapping methods still have the following imperfect problems to be solved:

1. When performing subpixel mapping processing for a single RS image, the subpixel sharpening step processes the low-resolution abundance images. These low-resolution abundance images are directly extracted from the original coarse RS image through spectral unmixing technology. However, due to the low resolution of the original RS image, there are many uncertainties in the distribution of each land cover class in the image. At the same time, the existing spectral unmixing technology also has certain limitations, so that the supervision information in the original image, especially

the spatial information and spectral information, cannot be fully extracted and used. In addition, it is dealing with the problem of subpixel mapping in a single RS image. At this time, the spatial correlation information is not considered comprehensive enough, which reduces the accuracy of subpixel mapping.

2. In the existing subpixel mapping methods based on multiple subpixel shifted images, the original coarse multiple subpixel shifted images are obtained by first unmixing and then subpixel sharpening, which means the fine multiple subpixel shifted images cannot fully inherit the spatial-spectral information of the original multiple subpixel shifted images. At the same time, the subpixel sharpening method based on a single scale makes the scale information of the fine multiple subpixel shifted images simple, that is, the information of the fine multiple subpixel shifted images is not rich.

3. How to improve the spatial resolution of original RS images and supplement spatial-spectral information more effectively by using auxiliary data, especially panchromatic images with higher spatial resolution or spectral images in the same area, is also one of the problems worth studying.

4. Due to the limitations of the existing spectral unmixing technology, a large number of spectral unmixing errors will be generated in the process of spectral unmixing, which will be carried into the next subpixel mapping process, affecting the final subpixel mapping results.

5. In practical subpixel mapping applications, traditional methods only use the spatial information of RS images to obtain the final mapping results, but the spectral information of RS images often cannot be fully considered, affecting the final mapping accuracy.

1.4 MAIN RESEARCH CONTENTS AND CHAPTER ARRANGEMENT

This book focuses on the research of subpixel mapping technology. Aiming at the five problems of the technology mentioned, the authors propose new subpixel mapping methods from five chapters to solve the related problems.

The book is divided into seven chapters, the main research content and structure arrangement:

Chapter 1: Introduction. Firstly, the research background and significance of subpixel mapping are briefly introduced. Secondly, the development status of main types of subpixel mapping and related technologies at home and abroad is introduced. Thirdly, some problems existing in the current subpixel mapping technology are explained. Finally, the main research content and methods of this book are introduced, and the structure is arranged.

Chapter 2: Basic principles of subpixel mapping. Firstly, the basic principles of the spectral unmixing method and mainstream subpixel mapping technology are introduced, the realization process of subpixel mapping technology is explained in detail, and several typical subpixel sharpening methods and class allocation methods are introduced in detail. Finally, the evaluation methods and experimental models used in this book for the experimental results of subpixel mapping are introduced.

Chapter 3: Subpixel mapping based on single RS image. Firstly, the existing subpixel mapping methods based on a single RS image and the existing problems are introduced. Next, subpixel mapping based on spatial-spectral interpolation, subpixel mapping based on I-HNN, subpixel mapping based on extended random walker, and subpixel mapping based on spatial-spectral correlation (SSC) for spectral imagery are to solve the existing subpixel mapping method based on a single RS image and existing problems. Experiments show that the three methods can make full use of the spatial-spectral information of the original image or make the spatial correlation information more comprehensive.

Chapter 4: Subpixel mapping based on multiple subpixel shifted images. Firstly, the authors introduce in detail how the multiple subpixel shifted images are used as auxiliary data to improve the subpixel mapping accuracy and then introduce the basic principles and existing problems of the existing subpixel mapping based on multiple subpixel shifted images. Finally, using multiple subpixel shifted images with spatial-spectral information in subpixel mapping, subpixel mapping based on the spatial attraction model with multi-scale subpixel shifted images, utilizing parallel networks to produce subpixel shifted images with multi-scale spatial-spectral information for subpixel mapping, and spatiotemporal super-resolution mapping by considering the point spread function effect are proposed. Compared with the existing subpixel mapping methods based on multiple subpixel shifted images, it is found that the multiple subpixel shifted images in the proposed method contain more supervision information, and the subpixel mapping results are more accurate.

Chapter 5: Subpixel mapping of RS image based on pansharpening technology, soft-then-hard subpixel mapping based on the pansharpening technique for RS imaging, improving super-resolution mapping based on the spatial attraction model by utilizing the pansharpening technique, subpixel land cover mapping based on dual processing paths for hyperspectral image, and subpixel mapping based on multi-source RS fusion data for land cover classes are proposed. Experimental results show that the fusion technology can improve the resolution of the original image, supplementing more spatial-spectral information. The final subpixel mapping results are improved by fusing the panchromatic image or spectral image of higher spatial resolution with the original coarse RS image.

Chapter 6: Subpixel mapping of RS image based on reconstruction then classification. The authors first introduce the basic principles of the super-resolution method and fully supervised information classification method, and then propose subpixel mapping based on MAP super-resolution recovery and utilizing the pansharpening technique to produce a subpixel resolution thematic map from a coarse RS image. Experiments show that under certain conditions, these methods are not only a more effective use of the supervision of the original image information better than the soft-then-hard subpixel mapping method but at the same time also effectively avoid the mixed spectral solution steps.

Chapter 7: Application of subpixel mapping technology in an RS image. Firstly, the achievements and problems of subpixel mapping technology in practical application are introduced. Then, improving super-resolution flood-inundation mapping for a multispectral RS image by supplying more spectral information, subpixel mapping for urban building by using spatial-spectral information from spaceborne multispectral RS images, and multispectral image super-resolution burned-area mapping based on space-temperature information are proposed. The experimental results show that the proposed methods can fully use the spatial-spectral information of multispectral images and obtain more accurate mapping results in practical application.

REFERENCES

[1] Chen S P, Tong Q X, Guo H D. Mechanism of Remote Sensing Information[M]. Beijing: Science Press, 1998:18–20.

[2] Zhang B, Gao L R. Hyperspectral Image Classification and Target Detection[M]. Beijing: Science Press, 2011.

[3] Fangjie W. Research on Band Selection for Hyperspectral Imagery[D]. Harbin: Engineering University, 2013.

[4] Schowengerdt R A. Remote Sensing: Models and Methods for Image Processing[M]. San Diego, CA: Academic, 1997.

[5] Wang Q. Research on Sub-Pixel Mapping and Its RelatedTechniques for Remote Sensing Imagery[D]. Master Dissertation, Harbin Engineering University, 2012.

[6] Heylen R, Scheunders P. Multidimensional pixel purity index for convex hull estimation and endmember extraction[J]. IEEE Transactions on Geoscience and Remote Sensing, 2013, 51(7): 4059–4069.

[7] Plaza A, Martinez P, Perez R, Plaza J. A quantitative and comparative analysis of endmember extraction algorithms from hyperspectral data[J]. IEEE Transactions on Geoscience and Remote Sensing, 2004, 42(3): 650–663.

[8] Heinz D C, Chang C I. Fully constrained least squares linear spectral mixture analysis method for material quantification in hyperspectral imagery[J]. IEEE Transactions on Geoscience and Remote Sensing, 2001, 39(3): 529–545.

[9] Keshava N, Mustard J F. Spectral unmixing[J]. IEEE Signal Processing Magazine, 2002, 19: 44–57.

[10] Settle J J, Drake N A. Linear mixing and estimation of ground cover proportions[J]. International Journal of Remote Sensing, 1993, 14(6): 1159–1177.

[11] Wang L, Jia X. Integration of soft and hard classification using extended support vector machine[J]. IEEE Geoscience and Remote Sensing Letters, 2009, 6(3): 543–547.

[12] Schowengerdt R A. On the estimation of spatial-spectral mixing with classifier likelihood functions[J]. Pattern Recognition Letters, 1996, 17(13): 1379–1387.

[13] Carpenter G M, Gopal S, Macomber S, Martens S, Woodcock C E. A neural network method for mixture estimation for vegetation mapping[J]. Remote Sensing of Environment, 1999, 70: 138–152.

[14] Bastin L. Comparison of fuzzy c-means classification, linear mixture modeling and MLC probabilities as tools for unmixing coarse pixels[J]. International Journal of Remote Sensing, 1997, 18: 3629–3648.

[15] Atkinson P M. Mapping subpixel boundaries from remo tely sensed images. Innovations in GIS[M]. 1997, 4: 166–180.
[16] Feng L, Shengjun W, Fei X, et al. Sub-pixel mapping of remotely sensed imagery: A review[J]. Journal of Image and Graphics, 2011, 16(8):1335–1345.
[17] Tatem A J, Lewis H G, Atkinson P M, Nixon M S. Increasing the spatial resolution of agricultural land cover maps using a Hopfield neural network[J]. International Journal of Geographical Information Science, 2003, 17(7): 647–672.
[18] Thornton M W, Atkinson P M, Holland D A. Subpixel mapping of rural land cover objects from fine spatial resolution satellite sensor imagery using super-resolution pixel-swapping[J]. International Journal of Remote Sensing, 2006, 27(3): 473–491.
[19] Zhang H, Shi J, Liu S. Sub-pixel lakes mapping in Tibetan Plateau[J]. Advances in Water Science, 2006, 17(3): 376–382.
[20] Foody G M, Muslim A M, Atkinson P M. Super-resolution mapping of the waterline from remotely sensed data[J]. International Journal of Remote Sensing, 2005, 26(24): 5381–5392.
[21] Muslim A M, Foody G M, Atkinson P M. Shoreline mapping from coarse-spatial resolution remote sensing imagery of Seberang Takir, Malaysia[J]. Journal of Coastal Research, 2007, 23(6): 1399–1408.
[22] Saura S, Castro S. Scaling functions for landscape pattern metrics derived from remotely sensed data: Are their subpixel estimates really accurate?[J]. ISPRS Journal of Photogrammetry and Remote Sensing, 2007, 62(3): 201–216.
[23] Li X, Du Y, Ling F, Wu S, Feng Q. Using a subpixel mapping model to improve the accuracy of landscape pattern indices[J]. Ecological Indicators, 2011, 11(5): 1160–1170.
[24] Ling F, Li W, Du Y, Li X. Land cover change mapping at the subpixel scale with different spatial-resolution remotely sensed imagery[J]. IEEE Geoscience and Remote Sensing Letters, 2010, 8(1): 182–186.
[25] Foody G M, Doan H T X. Variability in soft classification prediction and its implications for subpixel scale change detection and super-resolution mapping[J]. Photogrammetric Engineering and Remote Sensing, 2007, 73(8): 923–933.
[26] Wang Q, Shi W, Wang L. Allocating classes for soft-then-hard subpixel mapping algorithms in units of class[J]. IEEE Transactions on Geoscience and Remote Sensing, 2014, 5(5): 2940–2959.
[27] Wang Q, Wang L, Liu D. Particle swarm optimization-based subpixel mapping for remote-sensing imagery[J]. International Journal of Remote Sensing, 2012, 33(20): 6480–6496.
[28] Wang Z. Research on Sub-Pixel Mapping for Remote Sensing Image[D]. Master Dissertation, Harbin Engineering University, 2014.
[29] Atkinson P M. Subpixel target mapping from soft-classified, remotely sensed imagery[J]. Photogrammetric Engineering and Remote Sensing, 2005, 71(7): 839–846.
[30] Shen Z, Qi J, Wang K. Modification of pixel-swapping algorithm with initialization from a subpixel/pixel spatial attraction model[J]. Photogrammetric Engineering and Remote Sensing, 2009, 75(5): 557–867.
[31] He D, Shi Q, Liu X, Zhong Y, Zhang X. Deep subpixel mapping based on semantic information modulated network for urban land use mapping [J]. IEEE Transactions on Geoscience and Remote Sensing, 2021, 59(12): 10628–10646.
[32] Makido Y. Land-Cover Mapping at Subpixel Scales[D]. Ph.D. Dissertation, Michigan State University, 2006.
[33] Makido Y, Shortridge A. Weighting function alternatives for a subpixel allocation model[J]. Photogrammetric Engineering and Remote Sensing, 2007, 73(11): 1233–1240.

[34] Makido Y, Shortridge A, Messina J P. Assessing alternatives for modeling the spatial distribution of multiple land-cover classes at subpixel scales[J]. Photogrammetric Engineering and Remote Sensing, 2007, 73(8): 935–943.

[35] Hu J, Ge Y, Chen Y, Li D. Super-resolution land cover mapping based on multiscale spatial regularization[J]. IEEE Journal of Selected Topics in Applied Earth Observations and Remote Sensing, 2015, 8(5): 2031–2039.

[36] Villa A, Chanussot J, Benediktsson J A, et al. Spectral unmixing for the classification of hyperspectral images at a finer spatial resolution[J]. IEEE Journal of Selected Topics in Signal Processing, 2011, 5(3): 521–533.

[37] Atkinson P M. Super-resolution land cover classification using the two-point histogram[C]. GeoENV IV: Geostitistics for Environmental Applications, 2004: 15–28.

[38] Mertens K C, Verbeke L P C, Ducheyne E I, Wulf R De. Using genetic algorithms in subpixel mapping[J]. International Journal of Remote Sensing, 2003, 24(21): 4241–4247.

[39] Wang Q, Wang L, Liu D. Integration of spatial attractions between and within pixels for subpixel mapping[J]. Journal of Systems Engineering and Electronics, 2012, 23(2): 293–303.

[40] Muslim A M, Foody G M, Atkinson P M. Shoreline mapping from coarse-spatial resolution remote sensing imagery of Seberang Takir, Malaysia[J]. Journal of Coastal Research, 2007, 23(6): 1399–1408.

[41] Atkinson P M. Super-resolution mapping using the two-point histogram and multi-source imagery[C]. GeoENV VI: Geostatistics for Environmental Applications, 2008: 307–321.

[42] Lin H, Bo Y, Wang J, Jia X. Landscape structure based superresolution mapping from remotely sensed imagery[C]. Geoscience and Remote Sensing Symposium, Vancouver, BC, July 24–29, 2011.

[43] Tatem A J. Super-resolution land cover mapping from remotely sensed imagery using a Hopfield neural network[D]. Ph.D. thesis, University of Southampton, UK, 2001.

[44] Tatem A J, Lewis H G, Atkinson P M, Nixon M S. Super-resolution target identification from remotely sensed images using a Hopfield neural network[J]. IEEE Transactions on Geoscience and Remote Sensing, 2001, 39(4): 781–796.

[45] Tatem A J, Lewis H G, Atkinson P M, Nixon M S. Land cover mapping at the subpixel scale using a Hopfield neural network[J]. International Journal of Applied Earth Observation and Geoinformation, 2001, 3(2): 184–190.

[46] Tatem A J, Lewis H G, Atkinson P M, Nixon M S. Super-resolution land cover pattern prediction using a Hopfield neural network[J]. Remote Sensing of Environment, 2002, 79(1): 1–14.

[47] Tatem A J, Lewis H G, Atkinson P M, Nixon M S. Super-resolution mapping of multiple scale land cover features using a Hopfield neural network[C]. Proceedings of the International Geoscience and Remote Sensing Symposium, IEEE, Sydney, 2001.

[48] Tatem A J, Lewis H G, Atkinson P M, Nixon M S. Increasing the spatial resolution of agricultural land cover maps using a Hopfield neural network[J]. International Journal of Geographical Information Science, 2003, 17(7): 647–672.

[49] Tatem A J, Lewis H G, Atkinson P M, Nixon M S. Super-resolution mapping of urban scenes from IKONOS imagery using a Hopfield neural network[C]. Proceedings of the International Geoscience and Remote Sensing Symposium, IEEE, Sydney, 2001.

[50] Muad A M, Foody G M. Impact of land cover patch size on the accuracy of patch area representation in HNN-based super resolution mapping[J]. IEEE Journal of Selected Topics in Applied Earth Observations and Remote Sensing, 2012, 5(5): 1418–1427.

[51] Mertens K C, Basets B D, Verbeke L P C, Wulf R De. A subpixel mapping algorithm based on subpixel/pixel spatial attraction models[J]. International Journal of Remote Sensing, 2006, 27(15): 3293–3310.

[52] Ling F, Li X, Du Y, Xiao F. subpixel mapping of remotely sensed imagery with hybrid intra- and inter-pixel dependence[J]. International Journal of Remote Sensing, 2013, 34(1): 341–357.

[53] Chen Y, Ge Y, Wang Q, Jiang Y. A subpixel mapping algorithm combining pixel-level and subpixel-level spatial dependences with binary integer programming[J]. Remote Sensing Letters, 2014, 5(10): 902–911.

[54] Mertens K C, Basets B D, Verbeke L P C, Wulf R De. Subpixel mapping with neural networks: Real-world spatial configurations learned from artificial shapes[C]. Proceedings of Fourth International Symposium on Remote Sensing of Urban Areas, 2003, pp. 117–121.

[55] Mertens K C, Verbeke L P C, Westra T, Wulf R De. Subpixel mapping and subpixel sharpening using neural network predicted wavelet coefficients[J]. Remote Sensing of Environment, 2004, 91(2): 225–236.

[56] Zhang L, Wu K, Zhong Y, et al. A new subpixel mapping algorithm based on a BP neural network with an observation model[J]. Neurocomputing, 2008, 71(10): 2046–2054.

[57] Gu Y, Zhang Y, Zhang J. Integration of spatial-spectral information for resolution enhancement in hyperspectral images[J]. IEEE Transactions on Geoscience and Remote Sensing, 2008, 46(5): 1347–1358.

[58] Zhang L, Li P. A sub-pixel mapping algorithm based on BP neural network with spatial autocorrelation function for remote sensing imagery[J]. Acta Geo-daetica et Canographwa Sinica, 2001, 40(3): 307–311.

[59] Jin H, Mountrakis G, Li P. A super-resolution mapping method using local indicator variograms[J]. International Journal of Remote Sensing, 2012, 33(24): 7747–7773.

[60] Verhoeye J, Wulf R De. Land-cover mapping at subpixel scales using linear optimization techniques[J]. Remote Sensing of Environment, 2002, 79(1): 96–104.

[61] Wang Q, Atkinson P M, Shi W. Indicator cokriging-based subpixel mapping without prior spatial structure information[J]. IEEE Transactions on Geoscience and Remote Sensing, 2015, 53(1): 309–323.

[62] Wang Q, Shi W, Atkinson P M. Subpixel mapping of remote sensing images based on radial basis function interpolation[J]. ISPRS Journal of Photogrammetry and Remote Sensing, 2014, 92(1): 1–15.

[63] Chen Y, Ge Y, Song D. Superresolution land-cover mapping based on high-accuracy surface modeling[J]. IEEE Geoscience and Remote Sensing Letters, 2015, 12(12): 2516–2520.

[64] Ling F, Foody G M, Ge Y, Li X, Du Y. An iterative interpolation deconvolution algorithm for superresolution land cover mapping[J]. IEEE Transactions on Geoscience and Remote Sensing, 2016, 54(12): 7210–7222.

[65] Wang L, Wang Z, Dou Z, Wang Y. Edge-directed interpolation-based subpixel mapping[J]. Remote Sensing Letters, 2013, 12(4): 1195–1203.

[66] Ling F, Du Y, Li X, Li W, Xiao F, Zhang Y. Interpolation-based super-resolution land cover mapping[J]. Remote Sensing Letters, 2013, 4(7): 629–638.

[67] Boucher A, Kyriakidis P C. Super-resolution land cover mapping with indicator geostatistics[J]. Remote Sensing of Environment, 2006, 104(3): 264–282.

[68] Boucher A, Kyriakidis P C, Cronkite-Ratcliff C. Geostatistical solutions for super-resolution land cover mapping[J]. IEEE Transactions on Geoscience and Remote Sensing, 2008, 46(1): 272–283.

[69] Boucher A. subpixel mapping of coarse satellite remote sensing images with stochastic simulations from training images[J]. Mathematical Geosciences, 2009, 41(3): 265–290.

[70] Jin H, Mountrakis G, Li P. A super-resolution mapping method using local indicator variograms[J]. International Journal of Remote Sensing, 2012, 33(24): 7747–7773.

[71] Wang Q, Shi W, Zhang H. Class allocation for soft-then-hard subpixel mapping algorithms with adaptive visiting order of classes[J]. IEEE Geoscience and Remote Sensing Letters, 2014, 11(9): 1494–1498.

[72] Chen Y, Ge Y, Heuvelink G B M, Hu J, Jiang Y. Hybrid constraints of pure and mixed pixels for soft-then-hard super-resolution mapping with multiple shifted images[J]. IEEE Journal of Selected Topics in Applied Earth Observations and Remote Sensing, 2015, 8(5): 2040–2052.

[73] Ge Y, Chen Y, Stein A, Li S, Hu J. Enhanced subpixel mapping with spatial distribution patterns of geographical objects[J]. IEEE Transactions on Geoscience and Remote Sensing, 2016, 54(4): 2356–2370.

[74] Foody G M, Muslim A M, Atkinson P M. Super-resolution mapping of the waterline from remotely sensed data[J]. International Journal of Remote Sensing, 2005, 26(24): 5381–5392.

[75] Su Y F, Foody G M, Muad A M, Cheng K S. Combining pixel swapping and contouring methods to enhance super-resolution mapping[J]. IEEE Journal of Selected Topics in Applied Earth Observations and Remote Sensing, 2012, 5(5): 1428–1437.

[76] Ge Y, Li S, Lakhan V C. Development and testing of a subpixel mapping algorithm[J]. IEEE Transactions on Geoscience and Remote Sensing, 2009, 47(7): 2155–2164.

[77] Kasetkasem T, Arora M K, Varshney P K. Super-resolution land-cover mapping using a Markov random field based approach[J]. Remote Sensing of Environment. 2005, 96(3/4): 302–314.

[78] Tolpekin V A, Hamm N A S. Fuzzy super resolution mapping based on Markov random fields[C]. Proceedings of International Geoscience and Remote Sensing Symposium, 2008, pp. 875–878.

[79] Tolpekin V A, Stein A. Effects of land cover class spectral separability and parameter estimation in super resolution mapping of an ASTER image[C]. ACRS 2008: Proceedings of the 29th Asian Conference on Remote Sensing, Colombo, Sri Lanka, 2008.

[80] Tolpekin V A, Stein A. Quantification of the effects of land-cover-class spectral separability on the accuracy of Markov-random-field based superresolution mapping[J]. IEEE Transactions on Geoscience and Remote Sensing, 2009, 47(9): 283–3297.

[81] Ardila Lopez J P, Tolpekin V A, Bijker W, Stein A. Markov-random-field-based super-resolution mapping for identification of urban trees in VHR images[J]. ISPRS Journal of Photogrammetry and Remote Sensing, 2011, 66(6): 762–775.

[82] Li X, Ling F, Du Y, et al. Building extraction at the sub-pixel scale from remotely sensed images based on anisotropic Markov random field[J]. Journal of Image and Graphics, 2012, 17(8):1042–1048.

[83] Li X, Du Y, Ling F. Spatially adaptive smoothing parameter selection for Markov random field based subpixel mapping of remotely sensed images[J]. International Journal of Remote Sensing, 2012, 33(24): 7886–7901.

[84] Wang L, Wang Q. Subpixel mapping using Markov random field with multiple spectral constraints from subpixel shifted remote sensing images[J]. IEEE Geoscience and Remote Sensing Letters, 2013, 10(3): 598–602.

[85] Atkinson P M. Downscaling in remote sensing[J]. International Journal of Applied Earth Observation and Geoinformation, 2013, 22(20): 106–114.

[86] Foody G M. Sharpening fuzzy classification output to refine the representation of subpixel land cover distribution[J]. International Journal of Remote Sensing, 1998, 19(13): 2593–2599.

[87] Aplin P, Atkinson P M. Subpixel land cover mapping for per-field classification[J]. International Journal of Remote Sensing, 2001, 22(14): 2853–2858.

[88] Nguyen M Q, Atkinson P M, Lewis H G. Superresolution mapping using a Hopfield neural network with LIDAR data[J]. IEEE Geoscience and Remote Sensing Letters, 2005, 3(2): 366–370.

[89] Ling F, Xiao F, Du Y, Xue H, Ren X. Waterline mapping at the subpixel scale from remote sensing imagery with high-resolution digital elevation models[J]. International Journal of Remote Sensing, 2008, 29(6): 1809–1815.

[90] Ling F, Li X, Xiao F, Fang S, Du Y. Object-based subpixel mapping of buildings incorporating the prior shape information from remotely sensed imagery[J]. International Journal of Applied Earth Observation and Geoinformation, 2012, 18(1): 283–292.

[91] Thornton M W, Atkinson P M, Holland D A. A linearised pixel swapping method for mapping rural linear land cover features from fine spatial resolution remotely sensed imagery[J]. Computers and Geosciences, 2007, 33(10): 1261–1272.

[92] Boucher A, Kyriakidis P C. Integrating fine scale information in super-resolution land-cover mapping[J]. Photogrammetric Engineering and Remote Sensing, 2007, 73(8): 913–921.

[93] Nguyen M Q, Atkinson P M, Lewis H G. Super-resolution mapping using Hopfield neural network with panchromatic image[J]. International Journal of Remote Sensing, 2011, 32(21): 6149–6176.

[94] Nguyen M Q, Atkinson P M, Lewis H G. Superresolution mapping using a Hopfield neural network with fused images[J]. IEEE Transactions on Geoscience and Remote Sensing, 2006, 44(3): 736–749.

[95] Ling F, Li X, Du Y, Xiao F. Super-resolution land cover mapping with spatial-temporal dependence by integrating a former fine resolution map[J]. IEEE Journal of Selected Topics in Applied Earth Observations and Remote Sensing, 2014, 7(5): 1816–1825.

[96] Li X, Du Y, Ling F. Super-resolution mapping of forests with bitemporal different spatial resolution images based on the spatial-temporal Markov random field[J]. IEEE Journal of Selected Topics in Applied Earth Observations and Remote Sensing, 2014, 7(1): 29–39.

[97] Ling F, Du Y, Xiao F, Xue H, Wu S. Super-resolution land-cover mapping using multiple subpixel shifted remotely sensed images[J]. International Journal of Remote Sensing, 2010, 31(19): 5023–5040.

[98] Xu X, Zhong Y, Zhang L, Zhang H. Subpixel mapping based on a MAP model with multiple shifted hyperspectral imagery[J]. IEEE Journal of Selected Topics in Applied Earth Observations and Remote Sensing, 2013, 6(2): 580–593.

[99] Wang Q, Shi W, Wang L. Indicator cokriging-based subpixel land cover mapping with shifted images[J]. IEEE Journal of Selected Topics in Applied Earth Observations and Remote Sensing, 2014, 7(1): 327–339.

[100] Shi W Z, Zhao Y L, Wang Q M. Sub-pixel mapping based on BP neural network with multiple shifted remote sensing images[J]. Journal of Infrared & Millimeter Waves, 2014, 33(5): 527.

[101] Wang Q, Shi W, Atkinson P M. Spatiotemporal subpixel mapping of time-series images[J]. IEEE Transactions on Geoscience and Remote Sensing, 2016, 54(9): 5397–5411.

[102] Xu X, Zhong Y, Zhang L. A subpixel mapping based on an attraction model for multiple shifted remotely sensed images[J]. Neurocomputing, 2014, 134(9): 79–91.

[103] Wang Q, Shi W. Utilizing multiple subpixel shifted images in subpixel mapping with image interpolation[J]. IEEE Geoscience and Remote Sensing Letters, 2014, 11(4): 798–802.

[104] Ling F, Foody G M. Super-resolution land cover mapping by deep learning [J]. Remote Sensing Letters, 2019, 10(6): 598–606.

[105] Jeng S, Tsai W. Improving quality of unwarped omni-images with irregularly-distributed unfilled pixels by a new edge-preserving interpolation technique [J]. Pattern Recognition Letters, 2007, 28(15): 1926–1936.

[106] Wald L, Ranchin T, Mangolini M. Fusion of satellite images of different spatial resolutions: Assessing the quality of resulting images[J]. Photogrammetric Engineering and Remote Sensing, 1997, 63(6): 691–699.

[107] Li J, Wei X, Ma Z. The recognition techniques of single moving target from two frames of sequence images[J]. Acta Electronica Sinica, 2005, 14(2): 229–234.

[108] Li J. A fast super-resolution algorithm based on ideal sample[J]. Journal of Harbin Institute of Technology, 2005(6): 822–825.

[109] L Pu, Jin W, Liu Y, Su B, Zhang N. Super-resolution interpolation algorithm based on mixed bicubic MPMAP algorithm[J]. Transaction of Beijing Institute of Technology, 2007, 27(2): 161–165.

[110] Xie S H. Study on image restoration method based on prior information and regularization technique[J]. Chinese Journal of Quantum Electronics, 2007, 24(4):429–433.

[111] Wolberg G. Digital Image Warping[J]. IEEE Computer Society Press, Los Alamitos, CA, 1992, 3(1): 81–95.

[112] Schultz R, Stevenson R. Extraction of high-resolution frames from video sequence[J]. IEEE Transactions on Image Processing, 1996, 2(4): 66–71.

[113] Chen T, DeFigueiredo R. Two-dimensional interpolation by generalized spline filters based on partial differential equation image models[J]. IEEE Transactions on Acoustics Speech and Signal Processing, 1985, 23(3): 631–642.

[114] Karayiannis B, Venetsanopolous A N. Image interpolation based on variational principles[J]. Signal Processing, 1991, 8(3): 259–288.

[115] Zhang J. Hyperspectral Image Classification Based on Information Fusion[D]. Master Dissertation, Harbin Institute of Technology, 2002.

[116] Hu M, Tan J, Zhao Q. Aptive rational image interpolation based on local gradient features[J]. Journal of Information and Computational Science, 2007(4): 59–67.

[117] Liu J. Research and Realization of Multisensor Image Fusion[D]. Master Dissertation, Harbin Institute of Technology, 2006.

2 Basic Principles of Subpixel Mapping

2.1 INTRODUCTION

Subpixel mapping is an important part of mixed pixels processing, which is a follow-up step of spectral unmixing. The mainstream subpixel mapping mainly includes two steps: subpixel sharpening and class allocation. This chapter starts from the basic principles of subpixel mapping. Firstly, the spectral unmixing methods that are used to obtain the low-resolution abundance images are summarized. Then the commonly used methods of subpixel sharpening and class allocation are introduced. Finally, the main evaluation methods and experimental models used in this paper are described.

2.2 SPECTRAL UNMIXING METHOD

Because there could be many land cover classes in each mixed pixel, it is not accurate to classify each mixed pixel as a class by traditional hard classification. The spectral unmixing method can also be called a soft classification method [1]. As one of the main technologies for processing mixed pixels, it determines the proportion occupied by each class in the mixed pixels according to the spectral unmixing model. Tong et al. divided the spectral unmixing model into the linear spectral unmixing model (LSMM) and non-linear spectral unmixing model (NSMM) [2] based on the interaction between different land cover classes in the mixed pixel. As a preliminary process of subpixel mapping, the effect of the low-resolution abundance images produced by spectral unmixing method has a very important influence on the final subpixel mapping results. The following is a brief introduction to the two spectral unmixing models.

2.2.1 LINEAR SPECTRAL UNMIXING MODEL

The linear spectral unmixing model considers that there is no interaction between different land cover classes in the mixed pixel. Due to its simple physical significance, this model is the mainstream spectral mixing model in spectral unmixing methods at present. The linear mixing model assumes that there is a linear combination relationship between the land cover class (spectral endmember) in the mixed pixel and the proportion of this class in the mixed pixel [3], which can be expressed by formula (2.1):

$$\mathbf{y} = \mathbf{E}\mathbf{x} + \mathbf{n} \qquad (2.1)$$

where $\mathbf{y}$ represents the spectral vector of the unmixed pixel of size $M \times 1$; M is the number of bands of the remote sensing image; $\mathbf{E}$ is the $M \times N$ matrix formed by

DOI: 10.1201/9781003279082-2

the spectral endmembers; N is the number of spectral endmembers in the mixed pixel; $\mathbf{x}$ is the proportion $N \times 1$ vector of each land cover class in the mixed pixel, namely, the abundance value; and $\mathbf{n}$ is the random noise. Since the number of bands in the remote sensing image is usually larger than the number of land cover classes, the number of unknowns is smaller than the number of formulas for formula (2.1). Therefore, the linear unmixing model needs to use the least squares method to seek the optimal estimation in the case of the minimum random noise and needs to attach full constraints (normalized and non-negative constraints) to meet the actual physical significance. The normalization constraint is that the sum of the abundance values of each class is 1. The non-negative constraint is that the abundance values of each class are greater than 0 and less than 1.

2.2.2 NON-LINEAR SPECTRAL UNMIXING MODEL

In the non-linear spectral unmixing model, there are multiple scattering interactions between various land cover classes, namely, the spectral endmember in the mixed pixel has a non-linear combination relationship with the abundance value of this class [4]. The most used non-linear spectral unmixing is the sum of quadratic polynomial and random noise $\mathbf{n}$ formed by spectral endmember matrix $\mathbf{E}$ and abundance value vector $\mathbf{x}$, as shown in formula (2.2):

$$\mathbf{y} = f(\mathbf{E}, \mathbf{x}) + \mathbf{n} \qquad (2.2)$$

where $f(\bullet)$ represents the quadratic polynomial of a non-linear function.

The linear unmixing model and the non-linear unmixing model often exist at the same time, and especially for large-scale land cover classes, the non-linear unmixing model can be completely regarded as the linear mixing model, so the linear mixing model occupies a dominant position [5]. In addition, due to the complex physical significance of the non-linear unmixing model, there are still many limitations and imperfections in the non-linear unmixing model at the present stage. Therefore, the low-resolution abundance images are obtained by the spectral unmixing method based on the linear unmixing model.

2.3 THEORETICAL BASIS OF SPATIAL CORRELATION

Spatial correlation is the theoretical basis of most subpixel mapping models, namely, when compared with the distant subpixel, the relatively close subpixel is more likely to belong to the same class. Figure 2.1 shows a simple example of what spatial correlation means.

Assume that the original image contains only two classes (class 1 and class 2) represented in gray and white, respectively. Figure 2.1(a) is a low-resolution abundance image obtained by spectral unmixing technology. The image contains a mixed pixel, and each pixel is marked with the proportion of class 1 in the pixel. If the ratio scale $S = 2$, each mixed pixel will be divided into four subpixels, and the number of subpixels of class 1 in each pixel can be calculated according to the proportion information of class 1 in each mixed pixel shown in Figure 2.1(a). For example, if the ratio scale

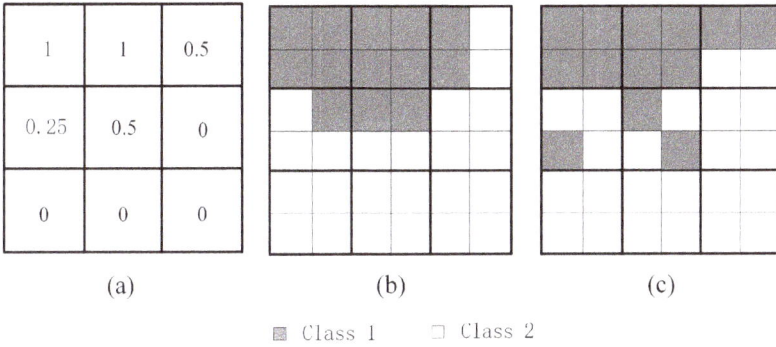

FIGURE 2.1   Spatial correlation principle: (a) abundance value of class 1, (b) distribution probability 1, and (c) distribution probability 2.

is 0.25, the number of corresponding subpixels is $2 \times 2 \times 0.25 = 1$. Figure 2.1(b)–(c) gives the two different spatial distribution states. According to the theoretical basis of spatial correlation, subpixels belonging to the same class should be closer to each other. Compared with Figure 2.1(c), the distribution in Figure 2.1(b) better satisfies the spatial correlation criterion and presents better spatial correlation, so the distribution of subpixels in Figure 2.1(b) may be closer to the real situation.

2.4   PROCESSING FLOW OF SUBPIXEL MAPPING

The basic principle diagram of the mainstream subpixel mapping technology is shown in Figure 2.2. Firstly, the low-resolution abundance image of each class is obtained from the original coarse remote sensing image through spectral unmixing. Then, a suitable subpixel sharpening method was used to transform the low-resolution abundance images into the high-resolution abundance images with subpixel abundance values (probability membership information) of each class. Finally, according to the information provided by the high-resolution abundance images and the limitation of the fixed number of subpixels in each class, the hard attribute value (class label) is assigned to each subpixel by the class allocation method, obtaining the subpixel mapping results. It can be seen that the subpixel sharpening method and class allocation method have great influence on the subpixel mapping results. Several common subpixel sharpening methods and class allocation methods are introduced next.

2.4.1   Subpixel Sharpening Method

Existing subpixel sharpening methods mainly include the Hopfield neural network, backpropagation (BP) neural network, spatial attraction model [50], and some appropriate super-resolution methods such as kriging and indicator cokriging (ICK). In particular, when using simple super-resolution methods such as bilinear interpolation and bicubic interpolation, subpixel mapping results can be obtained

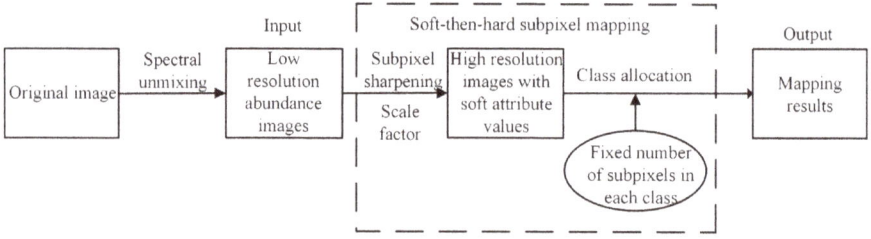

FIGURE 2.2   Subpixel mapping flowchart.

quickly. First, assume that the scale is S, then each mixed pixel will be divided into $S \times S$ subpixels. P_J ($J = 1, 2, \ldots, M$) (M is the number of pixels in the original coarse image) represents a mixed pixel; p_j ($j = 1, 2, \ldots, MS^2$) (MS^2 is the number of subpixels) represents a subpixel. The abundance value $H_k(p_j)$ represents the subpixel belonging to the kth class ($k = 1, 2, \ldots, K$) (K is the number of feature classes); $H_k(p_j)$ is also the value of the first subpixel sharpening output. In addition, we define $x_k(p_j)$:

$$x_k(p_j) = \begin{cases} 1, & \text{if subpixel } p_j \text{ belongs to class } k \\ 0, & \text{elsewise} \end{cases} \tag{2.3}$$

This section mainly introduces the basic principles of three common methods including the BP neural network, kriging interpolation, and ICK interpolation. The Hopfield neural network, spatial attraction model, and interpolation algorithm are introduced in the following chapters.

2.4.1.1 Backpropagation Neural Network

Due to its simple structure, good stability, and easy hardware implementation, the BP neural network has been widely applied in various engineering fields [6]. Mertens first introduced the BP neural network into subpixel mapping. The BP neural network consists of input layer, hidden layer, and output layer. In this model, land cover distribution information needs to be extracted from the high-resolution land cover classification map as training samples. Then the BP neural network is matched with the training samples. The input of the training samples is a vector composed of the proportion value occupied by a land cover class in all mixed pixels in a local window, and the output is a vector composed of the probability value of each subpixel belonging to this class. In the training process, the connection weights between hidden layer, input layer, and output layer are obtained by iterative feedback step by step. The trained BP neural network can process the low-resolution abundance images, and the final result is the abundance value of each class.

Subpixel mapping based on the BP neural network can get high precision results. However, when the training data are insufficient and the neural network training is insufficient, both the mapping accuracy and the detailed information will be greatly affected, so this method has many limitations.

2.4.1.2   Kriging Interpolation

Verhoeye et al. first introduced kriging interpolation into subpixel mapping, which assumed that the subpixel abundance value of each class was a weighted linear combination of the observed values, as shown in formula (2.4):

$$H_k(p_j) = \sum_{n=1}^{N} \beta_n L_k(P_n) \qquad (2.4)$$

where $L_k(P_n)$ represents a continuous variable indicating that pixels belong to the kth class, and β_n is the weight parameter. $L_k(P_n)$ can be described by the abundance value of the pixel P_n belonging to the kth class, and β_n is obtained through the kriging interpolation system [7]. Semi-variance functions in kriging interpolation systems can be obtained from the low-resolution abundance images without any prior structural information.

2.4.1.3   Indicator Kriging Interpolation

In the indicator kriging interpolation method, first all the abundance values belonging to the kth class form a vector F_k with size of $M \times 1$, for which ω_k is the average value of all elements in the vector F_k. Suppose there is already B fine pixels with the known class label, then these B indicator values will form a vector of j_k with size of $B \times 1$. The subpixel abundance value $H_k(p_j)$ can be estimated by the following formula:

$$H_k(p_j) = \eta_k(p_j)^T F_k + \lambda_k(p_j)^T j_k + \omega_k[1 - sum(\eta_k(p_j)^T) - sum(\lambda_k(p_j)^T)] \qquad (2.5)$$

where the size of vector $\eta_k(p_j)$ is $M \times 1$, and the size of vector $\lambda_k(p_j)$ is $B \times 1$, which are both ICK weight parameters belonging to the kth class; $sum(\bullet)$ represents the sum of all the elements of the vector. The sum of $\eta_k(p_j)$ and $\lambda_k(p_j)$ weight parameters need to be obtained by the ICK system. In the ICK system, the semi-variance function needs to be obtained from the known fine-resolution image, namely, the model needs prior information.

Sometimes, some subpixel abundance values $H_k(p_j)$, which are directly obtained through the subpixel sharpening method, may be less than 0 or do not meet the normalization conditions. Therefore, further processing is required to uniformly set these subpixel abundance values to 0. At the same time, the following formula is used to standardize the abundance values of these subpixel elements. Finally, the size of each subpixel abundance $\tilde{H}_k(p_j)$ value belongs to [0,1], and the sum of all subpixel abundance values in the same class is 1:

$$\tilde{H}_k(p_j) = \frac{H'_k(p_j)}{\sum_{k=1}^{K} H'_k(p_j)} \qquad (2.6)$$

Although the previous three subpixel sharpening methods can produce good mapping results, these subpixel sharpening methods have complex physical meanings and are burdened by iterative operation and parameter selection, and some methods also require prior information.

2.4.2 CLASS ALLOCATION METHOD

Since the ultimate goal of subpixel mapping is to generate hard classification images at the subpixel scale, the subpixel abundance value $H_k(p_j)$ ($j = 1, 2, \ldots, MS^2$; $k = 1, 2, \ldots, K$) obtained by the subpixel sharpening method is used to assign hard attribute values (class labels) to subpixels under the condition that the number of subpixels of each class is fixed, producing the subpixel mapping results. This step is called class allocation.

This chapter introduces three class allocation methods, namely, the linear optimization technique (LOT), units of subpixels (UOS), and highest soft attribute values first (HAVF). For the convenience of description in this chapter, it is assumed that all subpixels in the original coarse image with M pixels are divided into M groups, namely, p_j ($j = 1, 2, \ldots, MS^2$) are re-labeled as p_j^J ($j = 1, 2, \ldots, S^2$; $J = 1, 2, \ldots, M$), representing the subpixel within the pixel. The second-class indicator value $x_k(p_j)$ ($j = 1, 2, \ldots, S^2$; $k = 1, 2, \ldots, K$) will be rewritten as $x_k\left(p_j^J\right)$ ($j = 1, 2, \ldots, S^2$; $J = 1, 2, \ldots, M$; $k = 1, 2, \ldots, K$), and a subpixel mapping image R should have K gray values generated, namely,

$$R\left(p_j^J\right) = \sum_{k=1}^{K} k x_k\left(p_j^J\right) \tag{2.7}$$

According to the conclusion obtained by formula (2.3) and the principle that each subpixel in the subpixel mapping image can only have one class label, it can be known that $R\left(p_j^J\right) = 1, 2, \ldots, K$.

2.4.2.1 Linear Optimization Technique

The LOT model is firstly applied in subpixel mapping based on kriging. For each mixed pixel, class allocation is carried out under the condition that the mathematical model maximizes the sum of the abundance values of all subpixels according to formula (2.8):

$$\max t_J = \sum_{i=1}^{S^2} \sum_{k=1}^{K} x_k\left(p_j^J\right) H_k\left(p_j^J\right)$$

$$s.t. \sum_{k=1}^{K} x_k\left(p_j^J\right) = 1, j = 1,2,\ldots,S^2 \tag{2.8}$$

$$\sum_{j=1}^{S^2} x_k\left(p_j^J\right) = L_k\left(P_J\right)S^2, k = 1,2, \ldots ,K$$

where $L_k(P_J)$ is the abundance value of pixel P_J belonging to the kth class, which should satisfy $L_k(P_J) = 1$; $H_k\left(P_j^J\right)$ represents the subpixel abundance value obtained by the subpixel sharpening method.

The two constraints of formula (2.8) can be rewritten as two corresponding expressions, formulas (2.9) and (2.10):

$$X1_K = 1_{S^2} \tag{2.9}$$

$$\mathbf{X}^T \mathbf{1}_{S^2} = S^2 \mathbf{L} \tag{2.10}$$

where $\mathbf{X}$ is a matrix of size $S^2 \times K$; $\mathbf{L} = [L_1(P_J), L_2(P_J), \ldots, L_K(P_J)]^T$; and $\mathbf{1}_K$ and $\mathbf{1}_{S^2}$ represent the identity matrix of size $K \times 1$ and $S^2 \times 1$, respectively.

$$\mathbf{X} = \begin{bmatrix} x_1\left(p_1^J\right) & x_2\left(p_1^J\right) & \cdots & x_K\left(p_1^J\right) \\ x_1\left(p_2^J\right) & x_2\left(p_2^J\right) & \cdots & x_K\left(p_2^J\right) \\ \cdots & \cdots & \cdots & \cdots \\ x_1\left(p_{S^2}^J\right) & x_2\left(p_{S^2}^J\right) & \cdots & x_K\left(p_{S^2}^J\right) \end{bmatrix} \tag{2.11}$$

Formula (2.9) indicates that each subpixel can only be marked by a specific class label. Formula (2.10) indicates that the number of subpixels belonging to each class is fixed. In short, the LOT model makes the sum of the abundance values of all subpixels reach the maximum according to formula (2.8) under the restriction of formulas (2.9) and (2.10) and distributes all subpixels as a whole. Using the LOT model, formula (2.8) will produce the optimal subpixel mapping results.

2.4.2.2   Units of Subpixels

The UOS model considers a subpixel as an independent unit in order to allocate, its mathematical model is also formula (2.8), complying with formula (2.9) and formula (2.10) restrictions. However, its allocation principle is different from that of the LOT model. The LOT model makes class allocation of all subpixels as a whole, while the UOS model takes subpixels as an independent unit and compares them according to the abundance values of each subpixel, obtaining an ideal allocation order for class allocation.

The UOS model first determines the number of subpixels of each class according to formula (2.10). Under the limitation that the number of subpixels of each class is fixed, each subpixel is allocated a uniquely designated class label according to the subpixel abundance value. Then, an access path is defined that determines the order in which subpixels are accessed, assuming that the subpixel p_j^J is allocated in this path, and its K abundance values $H_1\left(P_j^J\right)$, $H_2\left(P_j^J\right)$, $\ldots$, $H_K\left(P_j^J\right)$ will be arranged in descending order according to the size. If the abundance value $H_{k_0}\left(P_j^J\right)$ belonging to the k_0th class is the largest, and the number of subpixels belonging to the k_0th class does not reach a fixed value $H_{k_0}\left(P_j^J\right)=1$, the rest will be $H_{k \neq k_0}\left(P_j^J\right)=0$. On the contrary, if the number of subpixels belonging to the k_0th class has reached a fixed value, the abundance $H_{k_0}\left(P_j^J\right)$ value belonging to the k_0th class will no longer be considered, and the subpixel p_j^J will be allocated to the remaining classes that do not reach the fixed value and have the highest abundance value. Finally, all subpixel class allocations are completed in this way. The UOS model is a single-path method without iterative process, so the calculation is small, and the calculation speed is fast.

2.4.2.3   Highest Soft Attribute Values First

Similar to the UOS model, another sequentially allocated model, HAVF, is also achieved by directly comparing the magnitude of the subpixel abundance values. However, in the process of each comparison, not only the K abundance values of each subpixel in the mixed pixel but also the abundance values of all subpixels in the mixed pixel are compared. The corresponding subpixel will be selected at the same time that the highest abundance value is found. If the number of subpixels in this class does not reach a fixed value, the corresponding subpixels will be marked as this class. Otherwise, if the fixed value of the number of subpixels in this class has been reached, all the S^2 abundance values belonging to this class will be set to be less than 0 and will be directly excluded in the following comparison process. All the subpixels complete allocation classes in this way. Unlike the random order of access subpixels in the UOS model, HAVF compares the abundance values of all KS^2 subpixels, so it has a fixed priority access order for the subpixels in the mixed pixel, and it avoids the influence of randomness on the subpixel mapping results.

It can be seen from the previous discussion that the three classical class allocation methods have their own advantages and disadvantages, so it is necessary to choose the class allocation method according to the actual situation. In addition to the three classical class allocation methods presented earlier, many remote sensing scholars have recently proposed new class allocation methods such as units of class, hybrid constraints of pure and mixed pixels, and subpixel mapping (SPM). In a word, the class allocation method has a very important influence on the final SPM results.

2.5 EVALUATION METHOD OF SUBPIXEL MAPPING ACCURACY

The performance evaluation of the SPM method can be considered from two aspects: operation speed and mapping accuracy. It is necessary to use the same performance evaluation method to evaluate each SPM method, reflecting the advantages and disadvantages of each SPM method fairly and reasonably. Currently, the number of error mapping pixels (EMPs) is the most intuitive method to evaluate the SPM accuracy of the two land cover classes problem [8]. EMPs can be calculated by subtracting the error image obtained from the reference image and the SPM result. The smaller EMP value means better performance of the SPM method. The root mean square error is also an important indicator of the SPM accuracy of the two land cover classes problem [9], which can be obtained from formula (2.12):

$$RMSE = \sqrt{\frac{\sum_{i=1}^{MS^2}(y_i - x_i)^2}{MS^2}} \qquad (2.12)$$

where y_i is the gray value of the subpixel i in the real reference image, and x_i is the gray value of the subpixel i in the subpixel mapping result. The smaller root mean square error value means that the SPM result is closer to the real reference image.

The SPM method is more widely used in a remote sensing image with many kinds of land cover classes. When evaluating each class of mapping accuracy, it is usually expressed by the proportion OA_i of the number of subpixels correctly mapped in the total number of subpixels belonging to this class. The overall accuracy evaluation can be expressed by the percentage of correctly classified (PCC, also named as OA) and kappa coefficient (Kappa) [10].

For the confusion matrix, the number of correctly mapped subpixels and incorrectly mapped subpixels are put into a matrix, assuming that the matrix **N** is a confusion matrix, whose definition is shown in formula (2.13):

$$\mathbf{N} = \begin{bmatrix} n_{11} & n_{12} & \cdots & n_{1K} \\ n_{21} & n_{22} & \cdots & n_{2K} \\ \cdots & \cdots & \cdots & \cdots \\ n_{K1} & n_{K2} & \cdots & n_{KK} \end{bmatrix} \tag{2.13}$$

where n_{ij} $(i, j = 1, 2 \ldots K)$ represents the number of subpixels belonging to class i in the reference image mapped as subpixels belonging to class j in the subpixel mapping result. From the confusion matrix, it can be seen that the elements on the main diagonal represent the number of correctly mapped subpixels. Then the mapping accuracy evaluation of each class can be expressed by formula (2.14):

$$\mathrm{OA}_i = \frac{n_{ii}}{\sum\limits_{j=1}^{K} n_{ij}} \tag{2.14}$$

where OA_i represents the percentage of correct subpixels in the ith class. The overall mapping accuracy PCC (%) [6] can also be obtained according to the confusion matrix, as shown in formula (2.15):

$$\mathrm{PCC} = \frac{\sum\limits_{i=1}^{K} n_{ii}}{\sum\limits_{i=1}^{K}\sum\limits_{j=1}^{K} n_{ij}} \tag{2.15}$$

In addition to these two evaluation methods, as a multivariate statistical method for mapping accuracy, the Kappa coefficient is also one of the commonly used performance evaluation indicators. The calculation formula of the Kappa coefficient is shown in Equation (2.16):

$$\mathrm{Kappa} = \frac{MS^2 \cdot \sum\limits_{i=1}^{K} n_{ii} - \sum\left(n_{i+} \cdot n_{+i}\right)}{\left(MS^2\right)^2 - \sum\left(n_{i+} \cdot n_{+i}\right)} \tag{2.16}$$

where n_{i+} and n_{+i} are the sum of the ith row and the ith column, respectively, and MS^2 represents the total number of subpixels. OA_i (%), PCC (%), and Kappa coefficients are higher, indicating the higher precision and the better performance of SPM results. This chapter mainly studies the SPM methods of various land cover classes, so OA_i (%), PCC (%), and Kappa coefficient are mainly selected as the accuracy evaluation indexes in the experiment.

After determining the SPM accuracy evaluation index, it is necessary to design the SPM experiment. If the design is carried out according to the real experiment in Figure 2.3, it is necessary to perform spectral unmixing of the original coarse remote

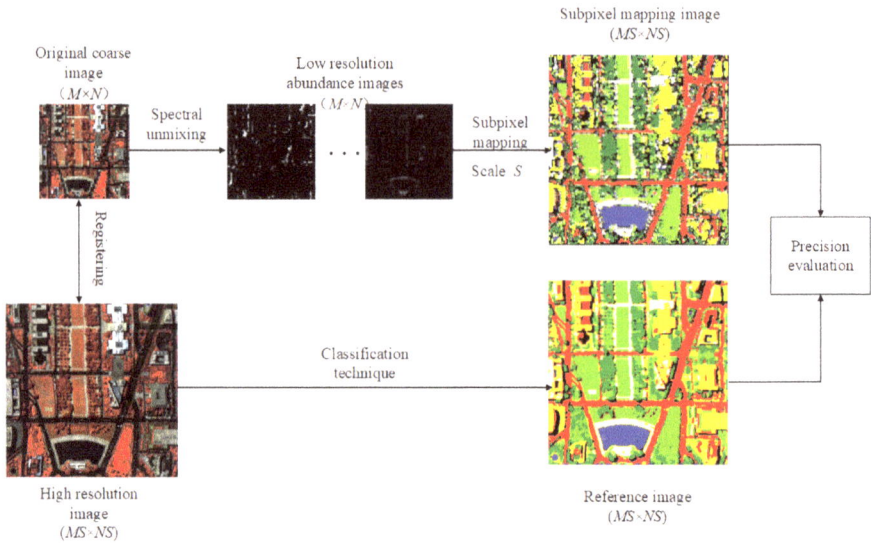

FIGURE 2.3 Flowchart of real experiment of subpixel mapping.

sensing image, then SPM, then a high-resolution remote sensing image is required to obtain the reference image through classification technology [11–13], and finally the accuracy of the two images is compared.

However, the real experiment is bound to bring many errors; the image registration error between the high-resolution image and the coarse image will especially have great influence on the final SPM results. In order to reduce the influence of errors on the experiment, the experimental design method adopted in this chapter is based on the two most commonly used simulation experiment methods for SPM, which are the experimental method based on downsampling of the reference image and the experimental method based on downsampling of the original high-resolution image, respectively.

Figure 2.4 presents the flowchart of the experimental method based on downsampling of the reference image. The reference image can be obtained by classifying a high-resolution remote sensing image; the simulated low-resolution abundance images can be directly produced by downsampling the reference image. This experimental method can effectively avoid the error brought by the spectral unmixing. It is convenient for different SPM methods to be compared directly, but this simulation method is too idealized and completely ignores the steps of spectral unmixing.

To fully simulate the whole process of mixed pixels processing, considering the steps of spectral unmixing and effectively avoiding the registration error between images, an experimental method based on downsampling of an original high-resolution image is proposed. As shown in Figure 2.5, in this simulation experimental method, first, the original high-resolution remote sensing image is downsampled to produce the simulated coarse remote sensing image, and the low-resolution abundance images are obtained by mixing the simulated coarse remote sensing image.

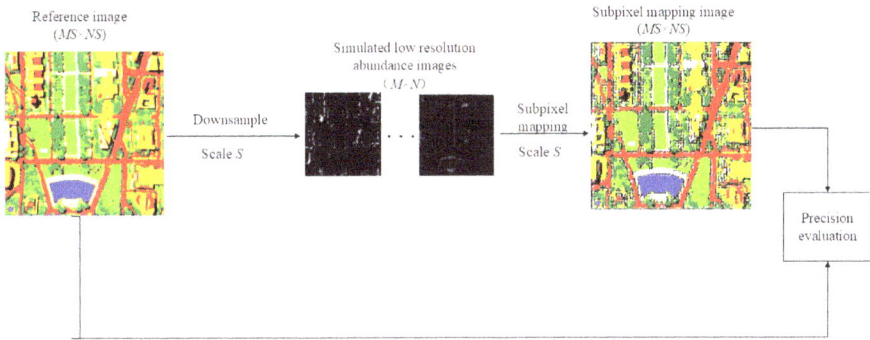

FIGURE 2.4 Experimental method based on downsampling of the reference image.

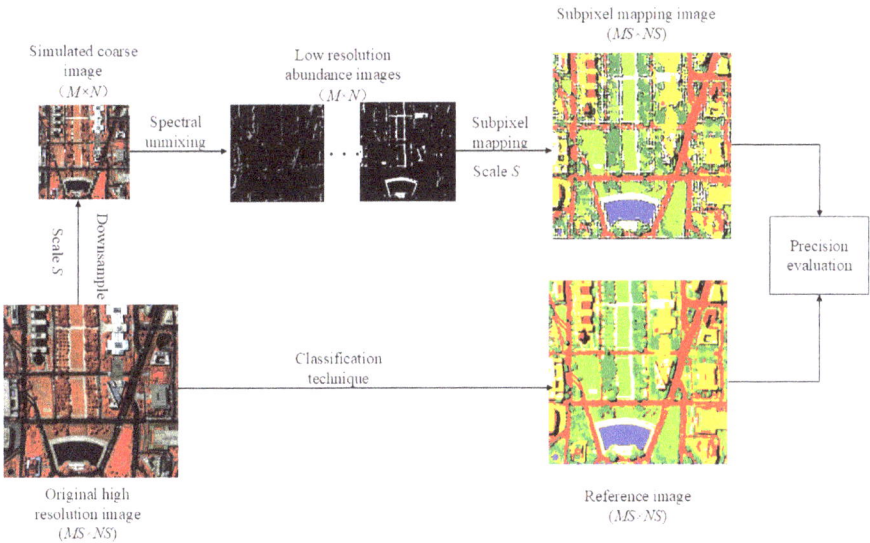

FIGURE 2.5 Experimental method based on downsampling of original high-resolution image.

This experimental method effectively avoids the problem between image registration and is closer to reality.

2.6 SUMMARY

This chapter first introduces the previous step of SPM, namely, spectral unmixing. The theoretical basis and basic principle of mainstream SPM is then introduced. Several typical subpixel sharpening methods and class allocation methods are also introduced. Finally, the common evaluation methods and the experimental model used in this book are introduced. These introductions lay a theoretical foundation for several SPM methods proposed in the following chapters.

REFERENCES

[1]   Ling F, Wu S, Xiao F, et al. Sub-pixel mapping of remotely sensed imagery: A review[J]. Journal of Image and Graphics, 2011, 16(8): 1335–1345.

[2]   Wang L G. Research on Hybrid Pixel Processing Technology of Hyperspectral Image [D]. Master Dissertation, Harbin Institute of Technology, 2005.

[3]   Tong Q, Zhang B, Zheng L, Hyperspectral Remote Sensing: Principle, Technology and Application[M]. Beijing: Higher Education Press, 2006.

[4]   Xu X. Research on the Theory and Method of Subpixel Location of Remote Sensing Image Considering the Spatial Characteristics of Ground Objects[D]. Wuhan: Wuhan University, 2012.

[5]   Hapke B. Bidirectional reflectance spctroscopy: 1. Theory[J]. Journal of Geophysical Research: Solid Earth (1978–2012), 1981, 86(B4): 3039–3054.

[6]   Wang L, Zhao C. Hyperspectral Image Processing Technology[M]. Beijing: Engineering Press, 2013.

[7]   Karathanssi V, Sykas D, Topouzelis K N. Development of a network-based method for unmixing of hyperspectral data[J]. IEEE Transactions on Geoscience and Remote Sensing, 2012, 50(3): 839–849.

[8]   Atkinson P M. Issues of uncertainty in super-resolution mapping and their implications for the design of an inter-comparison study[J]. International Journal of Remote Sensing, 2009, 30(20): 5293–5308.

[9]   Ling F, Fu B. Super-resolution mapping of urban buildings with remotely sensed imagery based on prior shape information[C]. Urban Remote Sensing Event, 2009 Joint. IEEE, 2009: 1–5.

[10] Cohen J. A coefficient of agreement for nominal scales[J]. Educational and Psychological Measurement, 1960, 20: 37–46.

[11] Hossain M A, Jia X, Benediktsson J A. One-class oriented feature selection and classification of heterogeneous remote sensing images[J]. IEEE Journal of Selected Topics in Applied Earth Observations and Remote Sensing, 2016, 9(4): 1606–1612.

[12] Hao S, Wang L, Bruzzone L, Wang Q. Spatial-dictionary for collaborative representation classification of hyperspectral images[J]. Multimedia Tools and Applications, 2016, 75(15): 9241–9254.

[13] Villa A, Chanussot J, Benediktsson J A, Jutten C, Dambreville R. Unsupervised methods for the classification of hyperspectral images with low spatial resolution[J]. Pattern Recognition, 2013, 46(6): 1556–1568.

3 Subpixel Mapping Based on Single Remote Sensing Image

3.1 INTRODUCTION

The existing subpixel mapping based on a single remote sensing image is largely dependent on the high-resolution abundance images from subpixel sharpening. However, due to the low resolution of the original remote sensing image and the limitations of the existing spectral unmixing technology, the spatial-spectral information of the original remote sensing image is not fully utilized, which affects the final subpixel mapping results. In addition, when dealing with the subpixel mapping of a single remote sensing image, the spatial correlation information is not considered comprehensively and accurately, which will also reduce the accuracy of subpixel mapping.

In view of these problems, this chapter proposes four new subpixel mapping methods based on a single remote sensing image. Subpixel mapping based on spatial-spectral interpolation (SPM-SS) and subpixel mapping based on Hopfield neural network with more prior information (I-HNN) solve the problem that the existing subpixel mapping methods based on interpolation and Hopfield neural network cannot make full use of the spatial-spectral information of the original image, improving the mapping accuracy of their respective types of subpixels.

Subpixel mapping based on extended random walker (SPMERW) and subpixel mapping based on spatial-spectral correlation (SSC) for spectral imagery generate more accurate spatial correlation, including spatial information between and within objects and spectral information, improving the final subpixel mapping results. The experimental results of the four methods show that when compared with the existing methods based on a single remote sensing image, the proposed method can obtain more accurate subpixel mapping results.

3.2 SUBPIXEL MAPPING BASED ON SPATIAL-SPECTRAL INTERPOLATION

Simple interpolation algorithms, such as bilinear interpolation and bicubic interpolation, have the advantages of no parameters, no iteration, fast operation, and no need for any prior structure information. When these simple interpolation algorithms are used as subpixel sharpening methods, the subpixel mapping results will be quickly obtained, so the subpixel mapping algorithm based on interpolation has been widely used in practical engineering applications.

DOI: 10.1201/9781003279082-3

3.2.1 Interpolation Problem

Because bilinear interpolation (BI) and bicubic interpolation (BIC) have the advantages of being non-parametric, being non-iterative, and allowing for fast calculation, in the proposed subpixel mapping algorithm based on spatial-spectral interpolation, the two interpolation algorithms are selected as interpolation algorithms of the proposed model. Here is a brief introduction to the two interpolation algorithms.

3.2.1.1 Bilinear Interpolation Algorithm

In the BI algorithm, the gray values of pixels interpolated will be linearly interpolated in horizontal and vertical directions [1, 2]. As shown in Figure 3.1, it is assumed that a certain point $f(i + u, j + v)$ is an inserted point in the high-resolution image, where i, j are all non-negative integers, and u, v are floating point numbers in the interval [0,1). The gray values of the four adjacent points f_1, f_2, f_3, f_4 are $f_1(i, j)$; $f_2(i, j + 1)$; $f_3(I + 1, j)$; $f_4(I + 1, j + 1)$, respectively. As shown in Figure 3.1, firstly, the gray values of points r_1 and r_2 are obtained from the four neighboring points $f_1, f_2, f_3,$ and f_4 in the x direction. Then, the gray values of points r_1 and r_2 determine the gray values of f in the y direction, which $f(i + u, j + v)$ can be obtained by formula (3.1):

$$f(i + u, j + v) = (1 - u) \times (1 - v) \times f(i, j) + (1 - u) \times v \times f(i, j+1)$$
$$+ u \times (1 - v) \times f(i + 1, j) + u \times v \times f(i + 1, j + 1) \qquad (3.1)$$

It can be seen from the formula that the BI algorithm has a simple structure, and the interpolation results are not affected by the interpolation order, so it can get the interpolation result quickly without any prior information and iteration. However, the

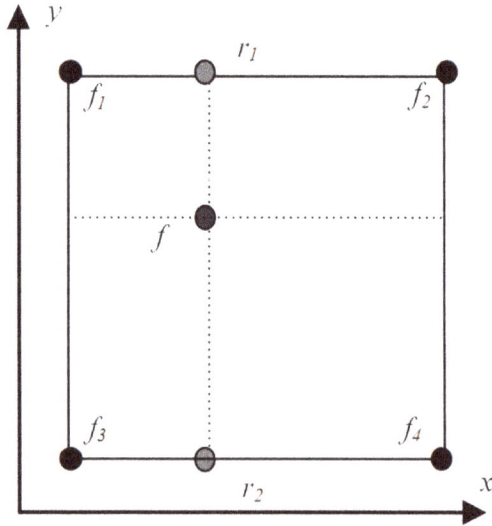

FIGURE 3.1   Diagram of bilinear interpolation.

interpolation algorithm has the characteristics of a low-pass filter, which damages the high-frequency component of the image, so the contour of the resulting image generated by BI will be blurred to a certain extent.

3.2.1.2   Bicubic Interpolation Algorithm

The BIC algorithm is more complex than the BI algorithm, so it can get a smoother image edge. It also considers the influence of the gray values of four direct adjacent points and the influence of the change rate of gray values between each adjacent point [3]. The BIC algorithm uses cubic polynomials $S(w)$ (w represents independent variables) to represent the relations among 16 known pixel points in the 4 × 4 field centered on the insertion point, and it estimates the gray value of the insertion point $f(i + u, j + v)$ by this relation. The mathematical expression of $S(w)$ is

$$S(w) = \begin{cases} 1 - 2|w|^2 + |w|^3 & 0 \le |w| < 1 \\ 4 - 8|w| + 5|w|^2 - |w|^3 & 1 \le |w| < 2 \\ 0 & |w| \ge 2 \end{cases} \tag{3.2}$$

The BIC expression is

$$f(i + u, j + v) = \mathbf{A} \times \mathbf{B} \times \mathbf{C} \tag{3.3}$$

where

$$\mathbf{A} = \begin{bmatrix} S(1+u) & S(u) & S(1-u) & S(2-u) \end{bmatrix} \tag{3.4}$$

$$\mathbf{B} = \begin{bmatrix} f(i-1,j-2) & f(i,j-2) & f(i+1,j-2) & f(i+2,j-2) \\ f(i-1,j-1) & f(i,j-1) & f(i+1,j-1) & f(i+2,j-1) \\ f(i-1,j) & f(i,j) & f(i+1,j) & f(i+2,j) \\ f(i-1,j+1) & f(i,j+1) & f(i+1,j+1) & f(i+2,j-1) \end{bmatrix} \tag{3.5}$$

$$\mathbf{C} = \begin{bmatrix} S(1+v) & S(v) & S(1-v) & S(2-v) \end{bmatrix}^T \tag{3.6}$$

It can be seen from these expressions that the BIC algorithm considers the correlation between pixels in a larger field around the points to be inserted, so the gray value of the points inserted is closer to the real value. Therefore, the image from the BIC is better than that of the BI algorithm.

3.2.2   Existing Subpixel Mapping Based on Interpolation

It is assumed that an original coarse remote sensing image g^1 can generate K low-resolution abundance images L_k^1 ($k = 1, 2, \ldots, K$; K represents the number of land cover classes); S is the ratio scale and then each pixel will be divided into $S \times S$ sub-pixels. $L_k^1(P_J)$ represents the abundance value of pixel P_J ($J = 1, 2, \ldots, M$, M is the number of pixels in the low-resolution abundance image) belonging to the kth class; $H_k^{spa}(p_j)$ represents the abundance value of the subpixel p_j ($j = 1, 2, \ldots, MS^2$, MS^2 is the total number of subpixels) belonging to the kth class.

The flowchart of the existing subpixel mapping based on the interpolation is shown in Figure 3.2. The original coarse remote sensing image g^1 is first unmixed to obtain the low-resolution abundance images L_k^1, which are as input data. The low-resolution abundance images are then upsampled through the interpolation algorithm, such as BI and BIC, to produce the high-resolution abundance images H_k^{spa} with the subpixel abundance value $H_k^{spa}(p_j)$. Finally, under the restriction of the previous formula (2.10) and formula (2.11), hard attribute values (class labels) are allocated to each subpixel by the appropriate class allocation method according to the subpixel abundance value, yielding the subpixel mapping results.

3.2.3 PROCESSING FLOW OF THE PROPOSED METHOD

By analyzing the process of the existing subpixel mapping based on interpolation, we can find that the high-resolution abundance images are obtained by spectral unmixing and then the interpolation. However, due to the uncertainty of the distribution of each land class in the original coarse image and the limitation of the existing spectral unmixing technology, the supervision information, especially the spectral information, in the original coarse remote sensing image cannot be fully utilized, affecting the accuracy of the subpixel mapping result. To make full use of spatial-spectral information in the original coarse image and improve the result of subpixel mapping, a new subpixel mapping method based on spatial-spectral interpolation (SPM-SS) is proposed in this chapter.

The model is mainly composed of two object terms, namely, spatial term and spectral term. The spatial term can be represented by the existing subpixel abundance values $H_k^{spa}(p_j)$ obtained by spectral unmixing then interpolation. The spectral term is obtained in the following way. First, the original coarse remote sensing image g^1 is transformed into high-resolution remote sensing image g^h by an interpolation algorithm. Then, the spectral unmixing method is used to unmix the high-resolution remote sensing image g^h to obtain other high-resolution abundance images H_k^{spe} ($k = 1, 2, \ldots, K$) with the subpixel abundance values $H_k^{spe}(p_j)$ ($k = 1, 2, \ldots, K$; $j = 1, 2, \ldots, MS^2$), which are as the spectral term. Next, the two object terms are

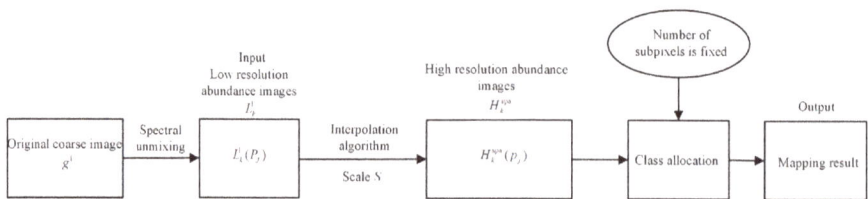

FIGURE 3.2 Flowchart of the existing subpixel mapping based on interpolation.

integrated to obtain the fine subpixel abundance values $H_k(p_j)$ with more spatial-spectral information of the original image by an appropriate parameter w, as shown in formula (3.7):

$$H_k\left(p_j\right)= wH_k^{\text{spe}}\left(p_j\right)+\left(1-w\right)H_k^{\text{spa}}\left(p_j\right) \tag{3.7}$$

Finally, according to the information from the subpixel abundance values $H_k(p_j)$, the class allocation method based on the linear optimization technique model is used to assign hard attribute values to each subpixel at the subpixel scale to obtain the final subpixel mapping result. In the SPM-SS method, the BI algorithm is selected as the interpolation algorithm of the two object terms at the same time, which is denoted as SPM-SS-BI, or the BIC algorithm is selected as the interpolation algorithm of the two object terms, which is denoted as SPM-SS-BIC.

SPM-SS can be implemented in four steps, as shown in Figure 3.3:

Step 1. K low-resolution abundance images L_k^1 are obtained from the original coarse remote sensing image g^1 through spectral unmixing. At the same time, the original image g^1 is interpolated to obtain the high-resolution image g^h.

Step 2. The spatial term $H_k^{\text{spa}}\left(p_j\right)$ is obtained by upsampling the low-resolution abundance value image L_k^1 by the interpolation method, while the spectral term $H_k^{\text{spe}}\left(p_j\right)$ is obtained by directly unmixing the high-resolution image g^h.

Step 3. The spatial term $H_k^{\text{spa}}\left(p_j\right)$ and spectral term $H_k^{\text{spe}}\left(p_j\right)$ are integrated to generate a target term $H_k(p_j)$ with more spatial-spectral information through formula (3.7).

Step 4. Under the limitation of the fixed number of subpixels in each class, according to the target term $H_k(p_j)$, the class labels are allocated to sub-pixels by the class allocation based on the linear optimization technique to complete the subpixel mapping.

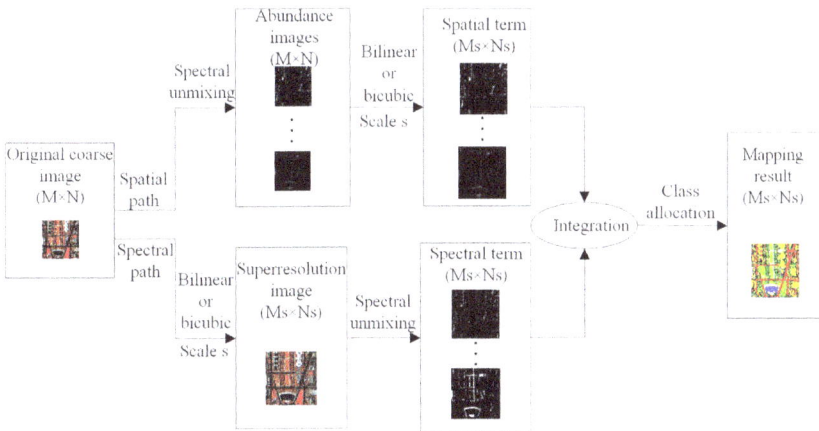

FIGURE 3.3 Flowchart of subpixel mapping based on mixed interpolation.

It can be seen from Figure 3.3 that the main contribution of spatial term $H_k^{\text{spa}}\left(p_j\right)$ is that the low-resolution abundance images are transformed into the high-resolution abundance images with rich spatial information through interpolation. But the spectral information of the original remote sensing image is very difficult to fully extract only through the single processing step. The spectral term $H_k^{\text{spe}}\left(p_j\right)$ plays a role by supplementing spectral information and makes the original low-resolution remote sensing image into a high-resolution image by interpolation. Due to improvement of the resolution of the original image, the spectral information can be extracted more easily. Finally, the two target terms with different information and subpixel abundance values are integrated by appropriate parameters to obtain the subpixel abundance value $H_k(p_j)$, leading to a more ideal subpixel mapping result.

3.2.4 EXPERIMENTAL CONTENT AND RESULT ANALYSIS

To be closer to the real experimental situation, this work selected two different real hyperspectral remote sensing image data for the experiment. In this paper, the experimental method based on downsampling of an original high-resolution image is selected to verify the performance of SPM-SS, namely, the original high-resolution hyperspectral remote sensing image was downsampled by the $S \times S$ mean filter to obtain the simulated low-resolution remote sensing image. Due to its simple physical significance and good performance, the N-FINDR algorithm [4] is selected as an endmember extraction method for both processing processes. Considering that the spectral unmixing algorithm [5] based on the support vector machine (SVM) has good spectral unmixing effect and good regression, this algorithm is selected as the spectral unmixing method in this experiment. To obtain a fairer comparison, only considering the influence of the newly added target term, namely, the spectral term $H_k^{\text{spe}}\left(p_j\right)$, on the mapping results. Therefore, in the SPM-SS method, the BI algorithm (SPM-SS-BI) is selected as the interpolation algorithm of two object terms at the same time, or the BIC algorithm (SPM-SS-BIC) is selected as the interpolation algorithm of two object terms at the same time. The existing subpixel mapping method based on BI or the subpixel mapping method based on BIC are used for comparison. OA_i, PCC (%), and Kappa coefficients were selected as the accuracy evaluation indexes in the experiment.

3.2.4.1 Experiment 1

In the first experiment, the experimental data set is selected from a Washington, DC, hyperspectral remote sensing image data set that was obtained by a HYDICE (Hyperspectral Digital Imagery Collection Experiment) imager. The area with a size of 240 × 240 is selected as the research data of this experiment, as shown in Figure 3.4(a). Seven classes can be observed from the area, including background, water, road, tree, grass, roof, and path. Figure 3.4(b) shows the reference image obtained by classification method. In order to simulate the real environment, the original remote sensing image is downsampled with scale $S = 2$ to obtain the simulated low-resolution image, as shown in Figure 3.4(c). It can be observed from Figure 3.4(c) that the spatial distribution of each class becomes blurred due to the

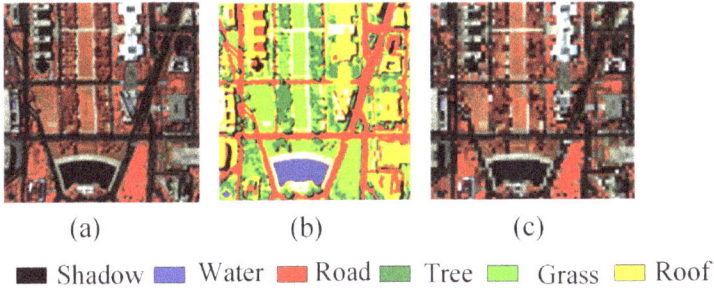

■ Shadow ■ Water ■ Road ■ Tree □ Grass □ Roof

FIGURE 3.4 Washington, DC, data set: (a) false color composite image (RGB band: 65, 52, 36); (b) reference image; and (c) low-resolution image ($S = 2$).

0   1

FIGURE 3.5 Abundance images: from left to right—background, water, road, trees, grass, roof, and path.

decrease of image resolution, so many land cover classes are difficult to clearly distinguish. In this case, subpixel mapping technology is needed to provide accurate distribution information of land cover classes. The parameter $w = 0.5$ is selected in both SPM-SS-BI and SPM-SS-BIC.

The abundance images of seven land cover classes obtained by using the spectral unmixing method based on SVM are shown in Figure 3.5. Although the proportion information of each class in the mixed pixels can be obtained through spectral unmixing, it is difficult to obtain the accurate spatial distribution information of each class from these low-resolution abundance images. Therefore, as a follow-up step of spectral unmixing, subpixel mapping is used to obtain accurate spatial distribution information of each class and obtain the better spatial resolution mapping result.

The mapping results of the four subpixel mapping methods are shown in Figure 3.6. The visual comparison results show that the mapping results of SPM-SS-BI and SPM-SS-BIC proposed in this chapter are significantly better than the existing mapping results of BI and BIC. The existing subpixel mapping based on interpolation does not make full use of the supervisory information of the original image, especially the spectral information, resulting in many obvious broken holes on the road and many rough burrs on the water boundary. Under the spectral term $H_k^{spe}\left(p_j\right)$, the supervision information of the original image is further fully utilized, so the previously discussed mislocution phenomenon can be alleviated in the image

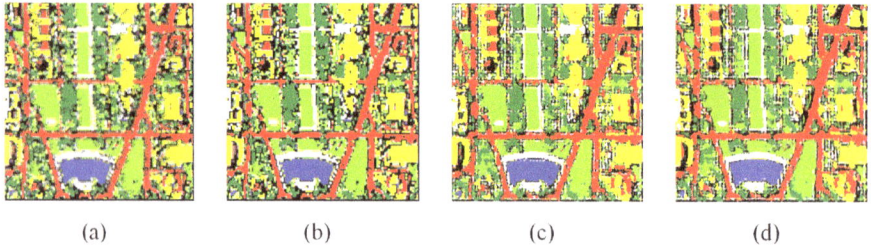

(a)   (b)   (c)   (d)

FIGURE 3.6   Subpixel mapping results: (a) BI, (b) BIC, (c) SPM-SS-BI, and (d) SPM-SS-BIC.

TABLE 3.1

Mapping Accuracy (%) of Four Methods in Experiment 1 ($S = 2$)

Land Cover Class	BI	BIC	SPM-SS-BI	SPM-SS-BIC
Background	73.44	75.03	81.62	85.59
Water	85.56	88.97	94.45	94.73
Road	70.55	72.74	76.29	78.35
Tree	72.45	75.45	76.61	77.89
Lawn	74.70	78.60	82.74	84.46
Roof	70.67	72.98	77.18	78.33
Path	73.88	75.58	79.08	82.89
PCC	76.82	77.47	80.72	82.23

generated by SPM-SS. Because the BIC algorithm is more efficient than the BI algorithm, SPM-SS-BIC produces the higher continuity of each class and the smoother class boundary than SPM-SS-BI.

In addition to visual comparison, the performance of each subpixel mapping method can be proved by comparing the accuracy of the results produced by each subpixel mapping method. the mapping accuracy of each class OA_i, and the overall mapping accuracy PCC (%) of the four subpixel mapping methods as shown in Table 3.1. By checking the accuracy values in Table 3.1, it can be found that the classes of the two proposed methods SPM-SS-BI and SPM-SS-BIC, namely, the mapping accuracy of background, water, road, tree, grass, roof, and path increases about 10.5%, 5.8%, 5.6%, 2.4%, 5.9%, 5.3%, and 7.3%, respectively, compared with the mapping accuracy generated by existing BI and BIC. Due to the higher efficiency of BIC, the overall mapping accuracy of SPM-SS-BIC is about 1.5% higher than that of SPM-SS-BI, and the mapping accuracy is the highest among the four subpixel mapping methods.

The scale S is one of the important factors affecting the mapping results. Four subpixel mapping methods are also tested at the other two scales $S = 4$ and $S = 6$, and their PCC (%) and Kappa coefficients at different scale S are recorded, as shown in Figure 3.7. It can be noticed that the PCC (%) and Kappa coefficients of the four subpixel mapping methods gradually decrease with the increase of scale S. This is

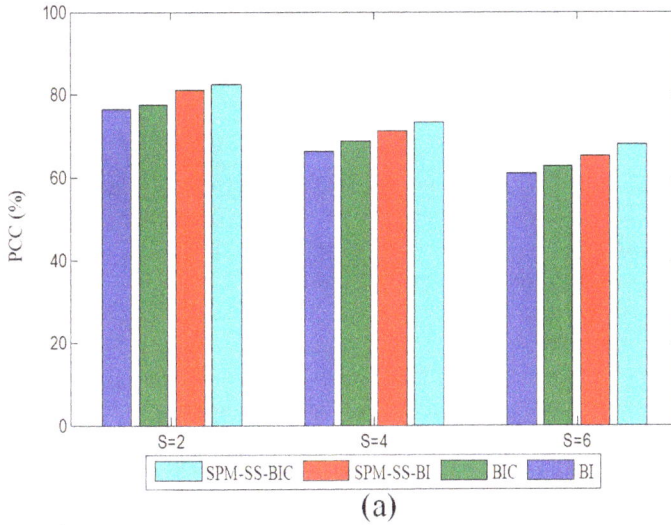

(a)

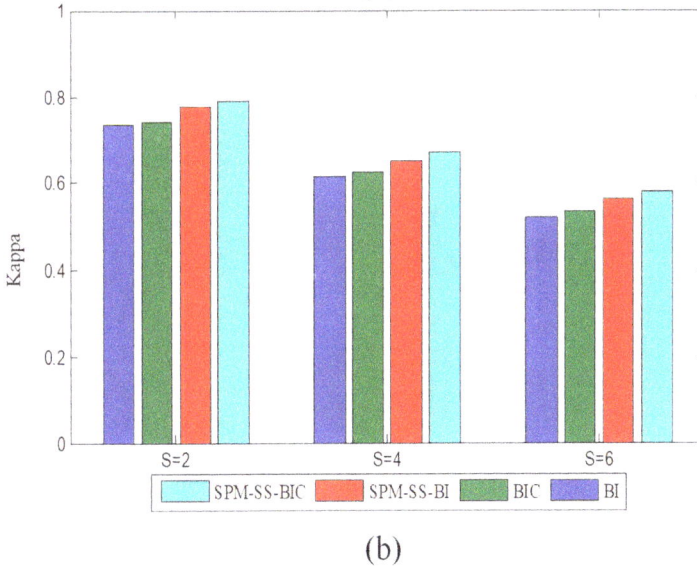

(b)

FIGURE 3.7   (a), (b) The relationship between PCC (%) and Kappa of four methods and the scale S.

because when the scale S increases, the simulation of the original coarse image resolution will be lower and the distribution of land cover class is more complex so that it is more difficult to restore the internal distribution structure of mixed pixels by the subpixel mapping method. As the scale S increases, the decreasing trend of mapping accuracy will be more obvious. Therefore, it is of great significance to select an appropriate scale for subpixel mapping. However, similar to the conclusion obtained in Table 3.1, the proposed subpixel mapping method SPM-SS can still generate higher PCC (%) and Kappa coefficients at different scales, which further proves that the performance of the proposed subpixel mapping method is better than that of the existing interpolation-based subpixel mapping method.

3.2.4.2 Experiment 2

The data set in Experiment 2 is collected using the Reflective Optics System Imaging Spectrometer (ROSIS) series of spectrometers in two densely populated residential areas in the southern Italian city of Pavia. The area with a size of 400×400 pixels is selected as the research area, as shown in Figure 3.8(a). Figure 3.8(b) is the reference image obtained by the classification method. Six classes of background, water, road, tree, grass, and roof can be observed from the reference image. The original remote sensing image is downsampled to simulate the low-resolution image. Figure 3.8(c) shows the simulated low-resolution image obtained from the original remote sensing image through the scale $S = 2$. In this experiment, the parameter w is selected as $w = 0.7$ in SPM-SS-BI and $w = 0.6$ in SPM-SS-BIC. The low-resolution abundance images of six land cover classes are shown in Figure 3.9, and subpixel mapping utilizes these abundance images to obtain the high spatial resolution mapping image.

As shown in Figure 3.10(a)–(b), there are many linear artifacts in the mapping results in the existing subpixel mapping methods based on interpolation. These phenomena are significantly improved by the proposed SPM-SS method. In particular,

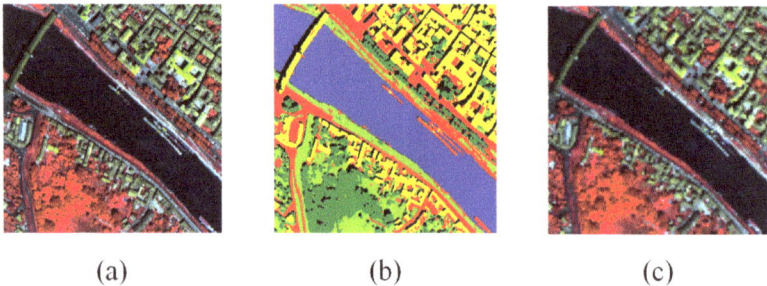

(a)   (b)   (c)

FIGURE 3.8   Pavia City Centre data set: (a) false color composite image (RGB band: 102, 56, 31); (b) reference image; and (c) low-resolution image ($S = 2$).

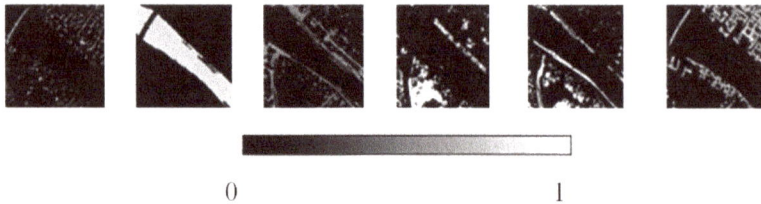

0   1

FIGURE 3.9   Abundance images: from left to right, background, water, road, trees, grass, roof, and path.

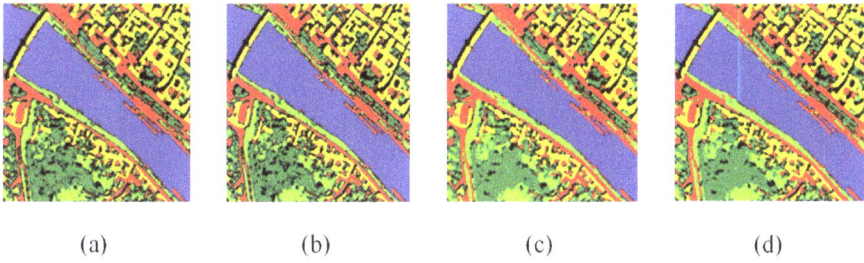

| (a) | (b) | (c) | (d) |

FIGURE 3.10   Subpixel mapping results: (a) result of BI, (b) result of BIC. (c) result of SPM-SS-BI, and (d) result of SPM-SS-BIC.

TABLE 3.2
Mapping Accuracy (%) of the Four Methods in Experiment 2 ($S = 2$)

Land Cover Class	BI	BIC	SPM-SS-BI	SPM-SS-BIC
Background	77.59	82.36	84.97	85.23
Water	95.84	96.29	97.66	98.22
Road	71.69	74.51	80.76	81.88
Tree	74.28	75.63	80.90	81.79
Lawn	69.23	71.40	78.26	79.68
Roof	79.75	82.07	84.39	85.10
PCC	80.45	82.40	85.37	86.24

the subpixel mapping results obtained by SPM-SS-BIC are closer to the reference image than those obtained by the other three subpixel mapping methods.

Table 3.2 shows the mapping accuracy of each class OA_i and the overall mapping accuracy PCC (%) of the four subpixel mapping methods. Similar to the conclusion of experiment 1, the mapping accuracy of SPM-SS method is significantly higher than that of the existing subpixel mapping methods based on interpolation, and the PCC value of SPM-SS-BI method is about 4.9% higher than that of the existing BI method. The PCC value of the SPM-SS-BIC method is 3.8% higher than that of the existing BIC method. In addition, Figure 3.11(a)–(b) represents the PCC (%) and Kappa coefficients of the four subpixel mapping methods at three proportional scales ($S = 2$, $S = 5$, and $S = 8$). Similar to the results of experiment 1, the values of PCC (%) and Kappa in SPM-SS method are higher, and the subpixel mapping accuracy is more accurate.

3.2.4.3   Analysis of the Weight Parameter w

The parameter w is used as a weight parameter to balance the influence of spatial term $H_k^{\mathrm{spa}}\left(p_j\right)$ and spectral term $H_k^{\mathrm{spe}}\left(p_j\right)$ on formula (3.8). If the weight parameter w is too small, the spectral information of the original image cannot be fully utilized. On the contrary, if the weight parameter w is too large, the spectral term $H_k^{\mathrm{spe}}\left(p_j\right)$ will play a dominant role, and the subpixel mapping result may not have rich spatial

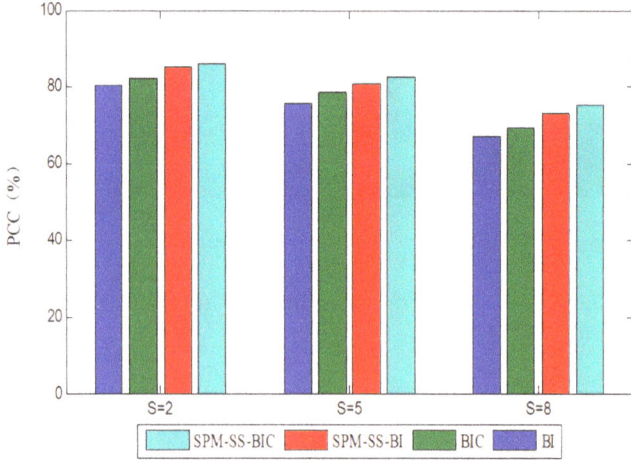

(a)

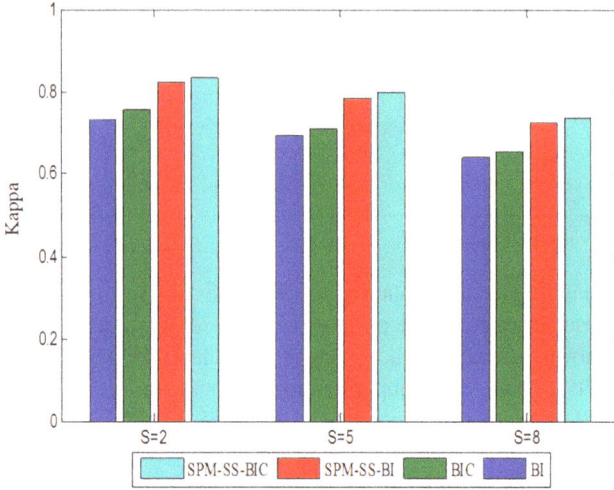

(b)

FIGURE 3.11   The relationship between PCC (%) and Kappa of four methods and the scale S.

information. Therefore, it is very important to select the appropriate weight param-eter w to obtain the ideal subpixel mapping result. To analyze the appropriate weight parameter w in SPM-SS-BIC, 10 groups of data with the weight parameter w increas-ing 0.1 each time in the range of [0, 0.9] are tested using the data sets of experiment 1 with $S = 2$ and experiment 2 with $S = 2$. The method of parameter selection in SPM-SS-BI is the same as that in SPM-SS-BIC.

Figure 3.12 shows the relationship between the different weight parameters w and PCC (%). It can be found that when $w = 0$, the spectral term has no effect, which is

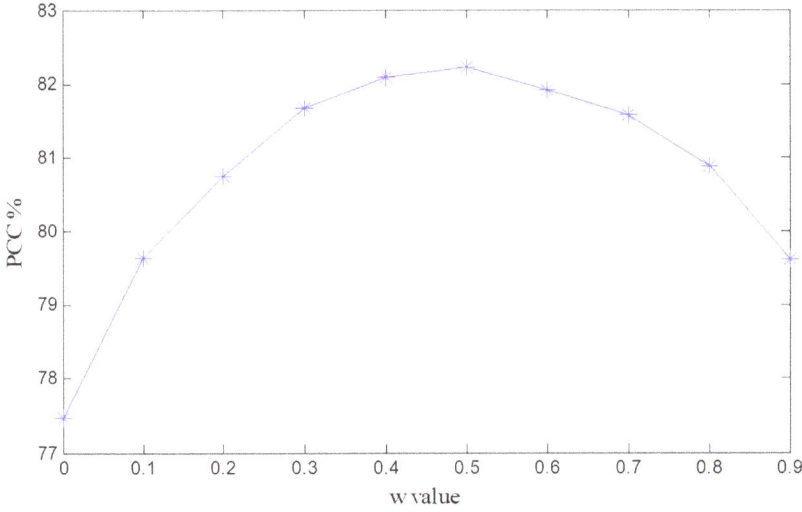

(a)

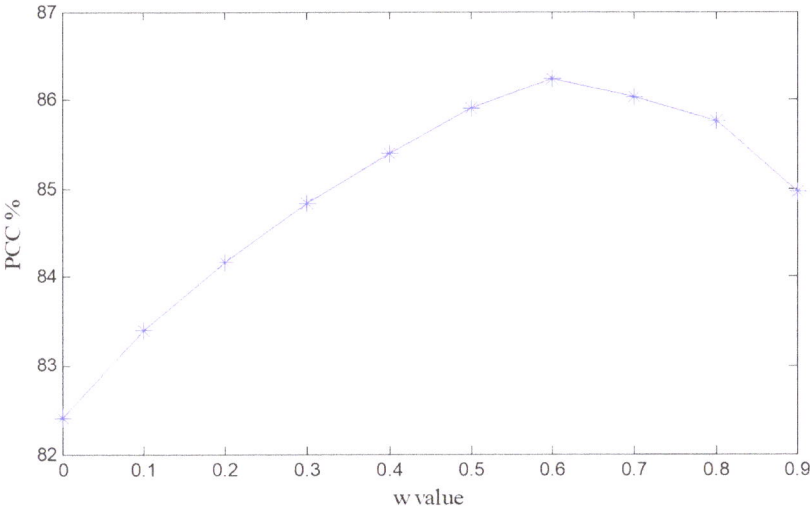

(b)

FIGURE 3.12   The relationship between PCC (%) and w in experiment 1 and experiment 2 ($S = 2$).

equivalent to the existing BIC method. As the w increases, PCC (%) also increases, because the spectral term $H_k^{spe}(p_j)$ supplements the rich spectral information of the original image. With the help of more space-spectral information, the subpixel mapping result will be improved. When $w = 0.5$ in experiment 1 and $w = 0.1$ in experiment 2, the highest PCC value can be obtained. However, when the value continues to rise, the spatial term $H_k^{spa}(p_j)$ will gradually lose its role, leading to the loss of some spatial information of the original image, thus affecting the accuracy of the final subpixel mapping result.

3.3 SUBPIXEL MAPPING BASED ON HOPFIELD NEURAL NETWORK WITH MORE SUPERVISION INFORMATION

Hopfield from the California Institute of Technology first introduced the concept of energy function to the network model in 1982 and created the Hopfield neural network [6]. The Hopfield neural network (HNN) model belongs to a feedback network, namely, each neural unit is connected to each other, the output and input are continuously fed back and iterated, and the final feedback result is output when it becomes stable. Tatem first applied this model to subpixel mapping and established an energy function in the subpixel mapping based on the HNN. Because HNN keeps the constraints as much as possible, it can reduce the soft classification error to a certain extent. Compared with other subpixel mapping methods, this is the advantage of HNN [7]. However, the current HNN method has two problems: 1) Due to the uncertainty of the original coarse image, such as the diversity of land cover classes and the limitation of the resolution of satellite sensors, it is difficult for HNN to collect all the prior information of the original image, which directly affects the subpixel mapping results. 2) Although the auxiliary data such as panchromatic images, fusion images, multi-shift remote sensing images, and light detection and ranging (LiDAR) data can be applied to HNN to improve the subpixel mapping results, these data are difficult to obtain, which limits their application. In order to solve these problems and obtain the more accurate results, a subpixel mapping method based on the HNN with more supervision information (I-HNN) is proposed.

3.3.1 TRADITIONAL SUBPIXEL MAPPING METHOD BASED ON HOPFIELD NEURAL NETWORK

Suppose (h, i, j) is a neuron located in row and column (h =1, 2, . . ., K; K is the number of land cover classes, which represents the subpixel (i, j) of the network layer corresponding to the feature category h) and v_{hij} is the predicted output value of the subpixel (i, j) belonging to the class h. When using HNN as an optimization tool, the v_{hij} value for each subpixel is updated. The network energy function can be defined as

$$E = -\sum_{h}\sum_{i}\sum_{j}\left(w_1 G1_{hij} + w_2 G2_{hij} + w_3 P_{hij} + w_4 M_{hij}\right) \tag{3.8}$$

where w_1, w_2, w_3, w_4 are four weight parameters; $G1$ and $G2$ are the value of spatial clustering; P is a proportional constraint; and M is a multi-class constraint.

$G1$ and $G2$ can be expressed as

$$\frac{dG1_{hij}}{dv_{hij}} = \frac{1}{2}\left[1 + \tanh\left(\frac{1}{8}\sum_{\substack{b=i-1\\b\neq i}}^{i+1}\sum_{\substack{c=j-1\\c\neq j}}^{j+1} v_{hbc} - 0.5\right)\lambda\right]\left(v_{hij} - 1\right) \tag{3.9}$$

$$\frac{dG2_{hij}}{dv_{hij}} = \frac{1}{2}\left[1 - \tanh\left(\frac{1}{8}\sum_{\substack{b=i-1 \\ b\neq i}}^{i+1}\sum_{\substack{c=j-1 \\ c\neq j}}^{j+1} v_{hbc} - 0.5\right)\lambda\right]v_{hij} \tag{3.10}$$

where λ is the number of iterations. In particular, the anisotropic spatial correlation model is used to improve the spatial correlation model in HNN [8].

The schematic diagram of HNN is shown in Figure 3.13. The low-resolution abundance images are obtained from the original coarse remote sensing image after spectral unmixing. Then high-resolution abundance images are obtained from HNN. Finally, the class allocation method is used to convert the high-resolution abundance image into subpixel mapping results. Here we chose the least squares linear mixture model (LSLMM) for spectral unmixing, because it has good unmixing and regression performance.

3.3.2 HOPFIELD NEURAL NETWORK WITH MORE PRIOR INFORMATION

As shown in Figure 3.13, the existing HNN is applied to the low-resolution abundance images obtained from the original coarse remote sensing image through spectral unmixing. Due to the diversity of land cover classes and the limited resolution of satellite sensors, the supervision information of the original image is not fully utilized, which affects the results of subpixel mapping based on HNN. The proposed I-HNN method adds a new processing path to HNN, namely, spectral unmixing after BIC. In the new processing path, BIC is used to improve the original coarse image. Then LSLMM is used to obtain a new prior item from the improved resolution remote sensing image. Due to the increase in resolution, the new prior item can extract more supervision information of the original image. Then the prior term is added to the network energy function, and the formula (3.8) can be re-expressed as

$$E = -\sum_{h}\sum_{i}\sum_{j}\left(\begin{array}{c} w_1 G1_{hij} + w_2 G2_{hij} \\ + w_3 P_{hij} + w_4 M_{hij} + w_5 H_{hij} \end{array}\right) \tag{3.11}$$

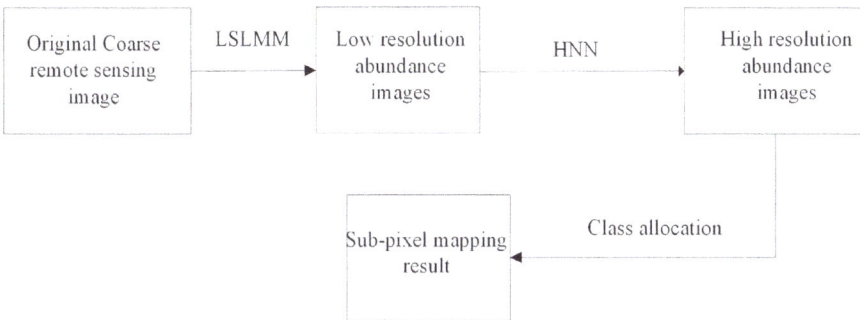

FIGURE 3.13 The existing Hopfield neural network.

$$\frac{dH_{hij}}{dv_{hij}} = O_{hij} \tag{3.12}$$

where O_{hij} is the proportional value of the subpixel (i, j) belonging to the h class in the improved resolution image. P is the proportional constraint for the class h, and the following formula is given by

$$\frac{dP_{hij}}{dv_{hij}} = \frac{1}{2S^2} \sum_{b=x\bullet S+1}^{x\bullet S+S} \sum_{c=y\bullet S+1}^{y\bullet S+S} \left[1 + \tanh\left(v_{hbc} - 0.5\right)\lambda\right] - F_{hxy} \tag{3.13}$$

where x and y are the coordinates of the mixed pixel containing the subpixel (i, j); F_{hxy} is the proportional value of the mixed pixel (x, y) belonging to the h class from spectral unmixing; S is the scale of proportions; and each mixed pixel is divided into $S \times S$ subpixels.

The multi-category M constraints are used to ensure that the sum of the proportional values of subpixel (i, j) belonging to each category should be equal to 1. Therefore, M can be calculated as

$$\frac{dM_{hij}}{dv_{hij}} = \left(\sum_{k=1}^{K} v_{kij}\right) - 1 \tag{3.14}$$

The schematic diagram of I-HNN is shown in Figure 3.14. Comparing Figure 3.14 with Figure 3.13, because the resolution of the original image in I-HNN is improved by the prior term H_{hij}, some uncertainty in the original image is reduced. Comparing formula (3.11) with formula (3.8), it can be considered as H_{hij} provides auxiliary information to HNN, improving the subpixel mapping result.

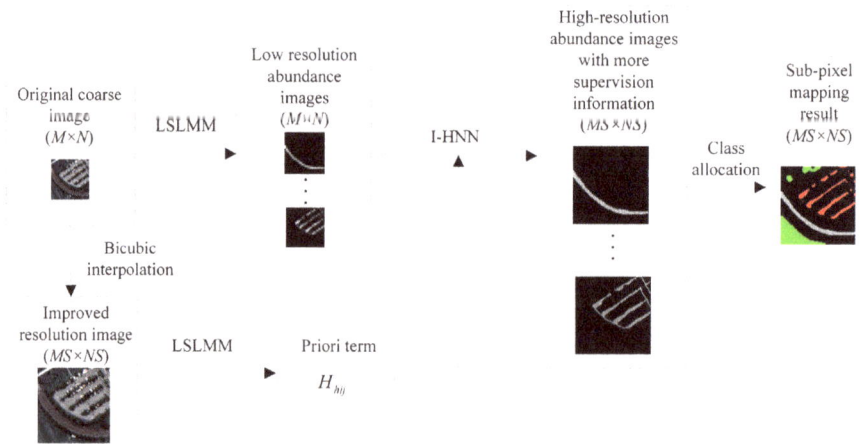

FIGURE 3.14   The Hopfield neural network with more prior information.

3.3.3 Experiment Content and Result Analysis

To verify the performance of I-HNN, two experiments were carried out by using real hyperspectral remote sensing images. To carry out quantitative evaluation, the original high-resolution image is downsampled through $S \times S$ mean filtering to obtain the simulated low-resolution remote sensing image. The principle of class allocation based on the highest abundance value (HAVF) is selected as the class allocation method. Since there is no rule for determining the optimal weight parameter set in HNN, the weight parameters in HNN and I-HNN are both set to 1. The number of iterations in both HNN and I-HNN is set to 100. The experiment considered four subpixel methods for comparison: the new-style subpixel mapping based on bicubic interpolation (NBIC) [9], the subpixel mapping of the hybrid spatial attraction model (HSAM) [10], the subpixel mapping based on HNN with anisotropic spatial correlation (HNNA) [8], and the proposed I-HNN. PCC and Kappa were used to quantitatively evaluate the accuracy of subpixel mapping results.

3.3.3.1 Experiment 1

In the first experiment, the data set was collected by the ROSIS sensor during the flight activities of the faculty of engineering of the University of Pavia. The test area in Figure 3.15(a) is 100 × 100. Figure 3.15(b) is the simulated low-resolution image by $S = 5$ downsampling, which is as the initial input image for I-HNN. The reference image is shown in Figure 3.15(c) and contains three land cover classes, namely, asphalt, grass, and brick.

The subpixel mapping results of the four methods are shown in Figure 3.16(a)–(d). Visual comparison shows that the proposed I-HNN provides better mapping results than the other subpixel mapping methods. As shown in Figure 3.16(a)–(b), due to the influence of spectral unmixing errors, there are obvious burrs at the junction of asphalt and bricks. Due to the stronger constraints of HNN, the roughness phenomenon is alleviated in Figure 3.16(c)–(d). It can be observed that since the proposed I-HNN provides more supervision information of the original image through a priori term, the mapping result of I-HNN in Figure 3.16(d) has a

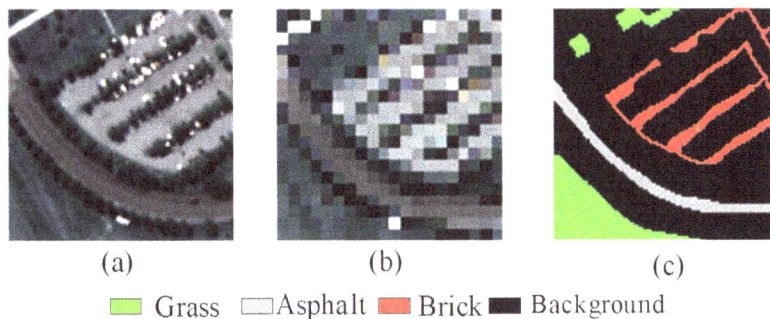

(a)   (b)   (c)

▢ Grass ▢ Asphalt ▮ Brick ▮ Background

FIGURE 3.15   Pavia University data set: (a) false color composite image (RGB bands: 19, 30, 44); (b) low-resolution image ($S = 5$); and (c) reference image.

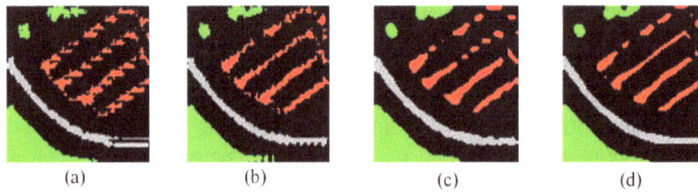

FIGURE 3.16   Subpixel mapping results: (a) NBIC, (b) HSAM, (c) HNNA, and (d) I-HNN.

TABLE 3.3
Mapping Accuracy of the Four Methods in Experiment 1 ($S = 5$)

Land Cover Class	NBIC	HSAM	HNNA	I-HNN
Grass (%)	95.65	95.92	97.46	98.10
Asphalt (%)	84.09	93.13	98.01	99.82
Tree (%)	61.70	73.97	76.03	81.54
Brick (%)	81.41	88.45	89.91	92.76
PCC (%)	75.10	83.84	85.68	89.51
Kappa	0.9565	0.9592	0.9746	0.9810

smoother boundary and is closer to the reference image than that of the HNNA in Figure 3.16(c).

The performance of the four methods in experiment 1 is quantitatively evaluated by the mapping accuracy of each class (%), PCC (%), and Kappa. Table 3.3 summarizes the accuracy of the four methods. It can be seen from Table 3.3 that when compared with HNNA, the accuracy of I-HNN for grass, asphalt, and brick has been improved by about 0.6%, 1.8%, and 5.6%, respectively. In addition, I-HNN produces the highest PCC and Kappa.

3.3.3.2   Experiment 2

In the second experiment, an airborne visible/infrared imaging spectrometer (AVIRIS) collected a hyperspectral remote sensing data set at the Indian Pine Proving Ground in June 1992. As shown in Figure 3.17(a), an area the size of 100 × 100 is selected as the test area. To evaluate the performance of I-HNN at different S scales, the original hyperspectral image in Figure 3.17(a) is downsampled with $S = 3$, 6, and 8 to generate the low-resolution remote sensing images. Figure 3.17(b) shows the low-resolution image produced at $S = 3$. The reference image including 16 land cover classes is shown in Figure 3.17(c).

The mapping results of the four subpixel mapping methods ($S = 3$) are shown in Figure 3.18(a)–(d). Since more supervision information is used in I-HNN, the subpixel mapping result in I-HNN is closer to the reference image than the other methods. Figure 3.19(a)–(b) shows the PCC (%) and Kappa in three scales of the four methods. Similar to the result of experiment 1, I-HNN still has the best PCC and Kappa.

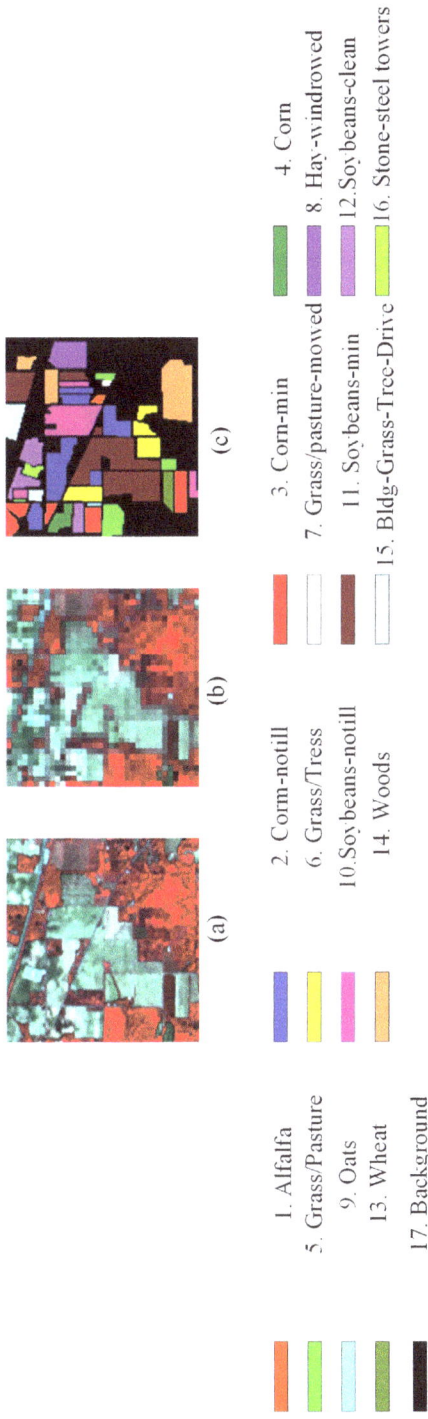

1. Alfalfa   2. Corm-notill   3. Corn-min   4. Corn
5. Grass/Pasture   6. Grass/Tress   7. Grass/pasture-mowed   8. Hay.-windrowed
9. Oats   10.Soy.beans-notill   11. Soy.beans-min   12.Soy.beans-clean
13. Wheat   14. Woods   15. Bldg.-Grass-Tree-Drive 16. Stone-steel towers
17. Background

FIGURE 3.17 Indian agriculture and forestry data set: (a) false color composite image (RGB band: 57, 27, 36); (b) low-resolution image ($S = 3$); and (c) reference image.

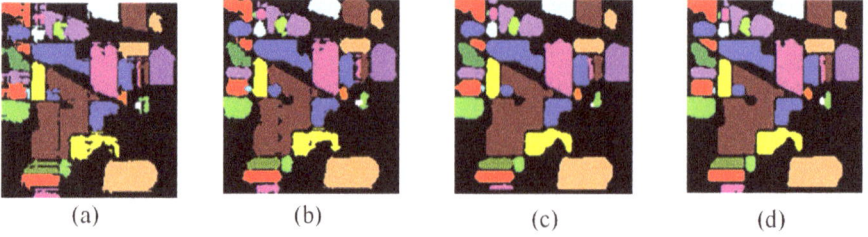

(a)   (b)   (c)   (d)

FIGURE 3.18   Subpixel mapping results: (a) NBIC, (b) HSAM, (c) HNNA, and (d) I-HNN.

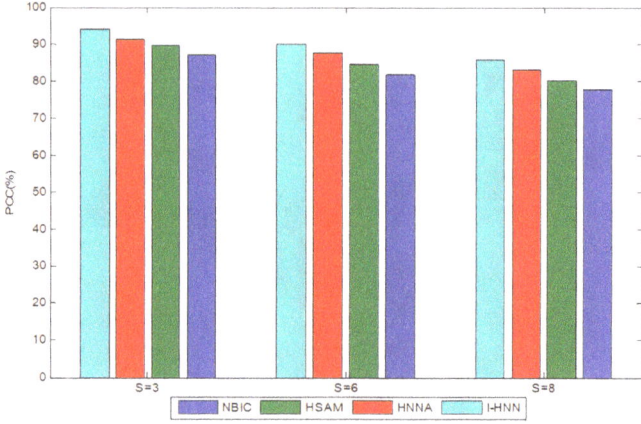

(a)

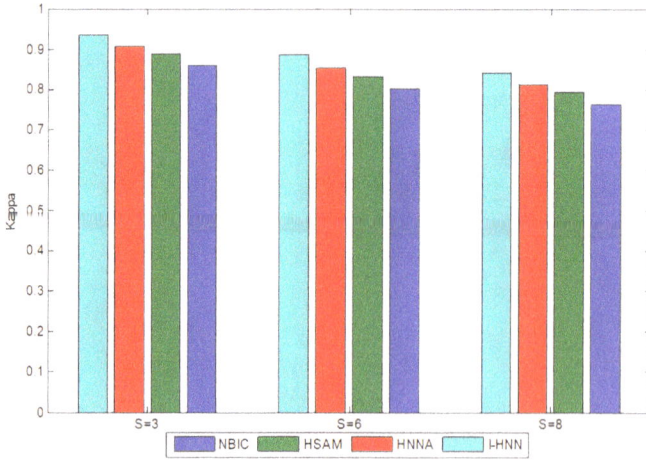

(b)

FIGURE 3.19   (a), (b) The relationship between PCC (%), Kappa, and scale S of the four methods.

3.4   SUBPIXEL MAPPING BASED ON EXTENDED RANDOM WALK

Traditional subpixel mapping methods are based on pixel spatial correlation or subpixel spatial correlation. In order to improve the final mapping results, Chen et al. proposed a subpixel mapping method based on object spatial correlation (SPM based on object spatial correlation, OSPM) [11]. However, the object in the OSPM method is generated by segmenting the original coarse image. The coarse resolution of the original image will reduce the accuracy of the segmentation result and affect the final mapping result. In addition, the object spatial correlation in OSPM only considers the spatial information between objects, and it does not fully consider the spatial information in each object area.

To solve these problems, subpixel mapping based on extended random walker (SPMERW) is proposed. The schematic diagram of SPMERW is shown in Figure 3.20. First, the BIC is used to improve the resolution of the original image. Second, the abundance value of the subpixel (the proportion value belonging to each class) is obtained by unmixing the upsampled image. The object is generated by segmenting the first main component of the upsampled image [12]. Third, the subpixel abundance values of each class are merged to generate the abundance values of the object. The extended random walk (ERW) algorithm is used to calculate the abundance values of the object to generate the spatial correlation of the object. At this time, the object spatial correlation includes spatial information between and within objects. Finally, according to the spatial correlation of the object, the class allocation method is used to obtain the final subpixel mapping result.

Compared with OSPM, the proposed SPMERW has the following advantages. First, because the BIC improves the resolution of the segmented image, the segmentation result of SPMERW is more accurate than that of OSPM. A more accurate segmentation result means that the abundance value of the obtained object is more accurate. Secondly, the object spatial correlation obtained by the ERW algorithm contains the spatial information among and within the objects. Therefore, the spatial correlation information considered in SPMERW is more comprehensive than that in OSPM, which improves the subpixel mapping results. The specific implementation process of the method is described in detail next.

3.4.1   Multi-Scale Segmentation Algorithm

In SPMERW, the original coarse image is firstly upsampled by BIC. The spectral unmixing is applied to the upsampled image to obtain the abundance value of each subpixel. At the same time, because the first principal component of the upsampled image that is extracted by principal component analysis (PCA) contains most of the spatial information, the multi-scale segmentation method is used in the first principal component to generate the objects. The segmentation scale parameter T is defined as the termination condition of merging to determine the size of the generated objects.

The segmentation method based on multi-scale adaptation is defined as

$$M = \alpha M^{\text{spectral}} + (1 - \alpha) M^{\text{shape}} \tag{3.15}$$

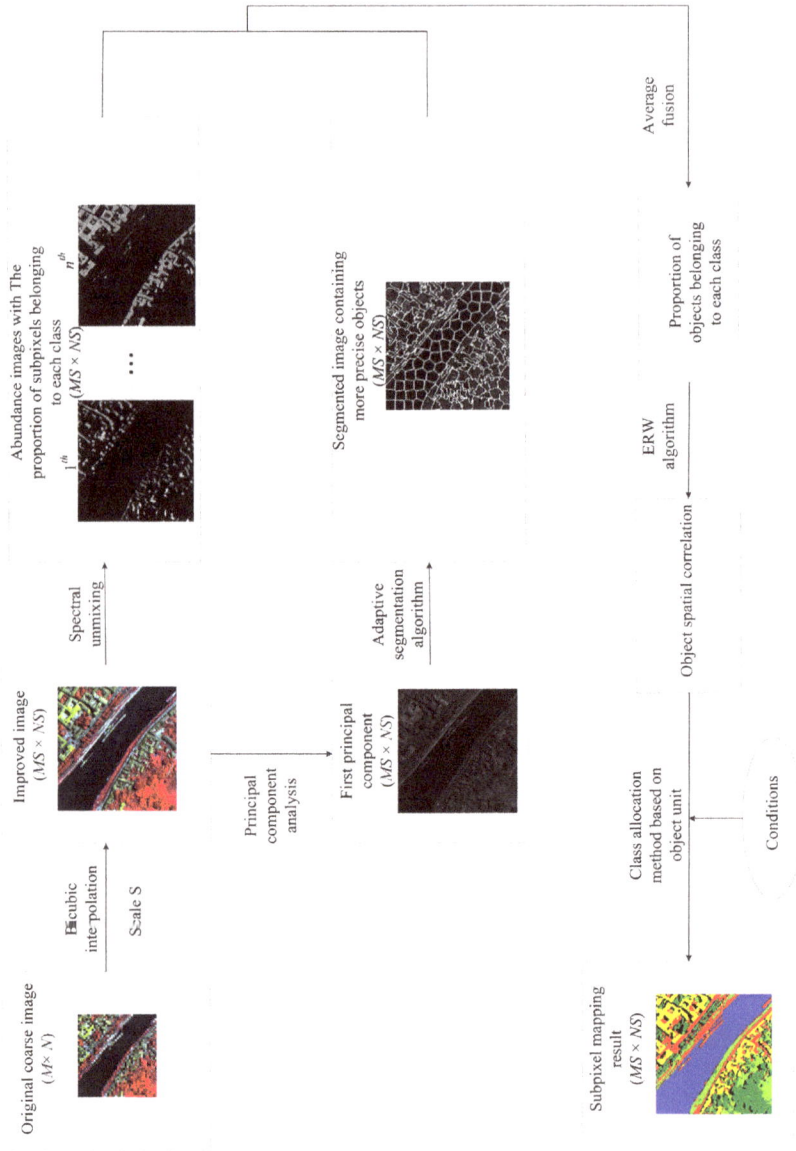

FIGURE 3.20 Subpixel mapping based on extended random walker.

where M is the regional heterogeneity, α is the weighting parameter that balances the spectral heterogeneity M^{spectral} and the shape heterogeneity M^{shape}, and M^{spectral} and M^{shape} are obtained by the following formulas, respectively:

$$M^{\text{spectral}} = \sum_{b=1}^{B} \alpha_b^{\text{spectral}} V_b \qquad (3.16)$$

$$M^{\text{shape}} = \alpha^{\text{shape}} \times l \big/ \sqrt{K^1} + \left(1 - \alpha^{\text{shape}}\right) \times l / R \qquad (3.17)$$

where V_b is the standard deviation of the spectral value of the b band in the segmented area; $\alpha_b^{\text{spectral}}$ is the spectral weight parameter of the b band; l is the actual boundary length in the segmented area; K^1 is the number of subpixels in the segmented area; R represents the length of the rectangular boundary in the segmented area; $l \big/ \sqrt{K^1}$ and l/R are used to calculate the smoothness and compactness of the segmented area, respectively; and α^{shape} is the balance weight parameter.

Finally, the two regions are merged with the least heterogeneity in adjacent object regions. When the heterogeneity of the merged region M is greater than the pre-defined segmentation scale parameter T, the merge process is terminated, and the objects are generated.

3.4.2 EXTENDED RANDOM WALK ALGORITHM

In order to more comprehensively consider the spatial correlation between the neighboring and interior of the objects, an improved extended random walking algorithm is used to calculate the abundance value of the objects, and then the spatial correlation of the objects is obtained. The first principal component of the remote sensing image improved by BIC is segmented to obtain I objects O_i ($i = 1, 2, \ldots, I$), where O_i contains K_i^1 subpixel. $U_n(O_i)$ is the proportion value of the O_i object belonging to the nth class, which is obtained by $H_n(p_j)$ denoting the average proportion values of all the subpixels p_j contained in this object belonging to the n class, as shown in formula (3.18):

$$U_n(O_i) = \frac{\sum_{j=1}^{K_i^1} H_n(p_j)}{K_i^1} \qquad (3.18)$$

$H_n(p_j)$ is obtained through unmixing the improved image.

Further, the formula (3.18) is normalized to obtain formula (3.19):

$$\hat{U}_n(O_i) = \frac{U_n(O_i)}{\sum_{n=1}^{N} U_n(O_i)} \qquad (3.19)$$

Then, the ERW algorithm is used to generate the spatial correlation of the object $E_n(p_j)$ whose subpixel p_a belongs to the nth class, as shown in formula (3.20):

$$E_n(p_j) = w E_n^{\text{among}}(\hat{U}_n) + (1-w) E_n^{\text{within}}(\hat{U}_n) \qquad (3.20)$$

where $E_n^{among}\left(\hat{\mathbf{U}}_n\right)$ considers the spatial information between the adjacent object regions; $E_n^{within}\left(\hat{\mathbf{U}}_n\right)$ represents the spatial information within each object region; $\hat{\mathbf{U}}_n = \left[\hat{U}_n\left(O_1\right), \hat{U}_n\left(O_2\right),...,\hat{U}_n\left(O_i\right)\right]$ is a column vector; and ω is a weight parameter. $E_n^{among}\left(\hat{\mathbf{U}}_n\right)$ can be calculated by the following formula:

$$E_n^{among}\left(\hat{\mathbf{U}}_n\right) = \hat{\mathbf{U}}_n^T \mathbf{L} \hat{\mathbf{U}}_n \qquad (3.21)$$

$\mathbf{L}$ is the Laplace matrix, and the specific expression is shown in formula (3.22):

$$\mathbf{L} = \begin{cases} \sum -z_{iq}, & \text{if } i = q \\ -z_{iq}, & \text{if } i \text{ and } q \text{ are adjacent objects} \\ 0, & \text{otherwise} \end{cases} \qquad (3.22)$$

where $z_{iq} = \exp\left(-\theta\left(\hat{y}_i - \hat{y}_q\right)^2\right)$ represents the difference in spectral value between the ith object O_i and the qth object O_q; θ is a free parameter and is set to 0.6. The spectral value $\hat{y}_i$ of the ith object O_i is given by the following formula:

$$\hat{y}_i = \sum_{a=1}^{K_i^l} y_a \Big/ K_i^l \qquad (3.23)$$

where y_a represents the spectral value of the ath subpixel in the object. $E_n^{within}\left(\hat{\mathbf{U}}_n\right)$ is defined as the following formula (3.24):

$$E_n^{within}\left(\hat{\mathbf{U}}_n\right) = \sum_{m=1, m \neq n}^{N} \hat{\mathbf{U}}_m^T \Lambda_m \hat{\mathbf{U}}_m + \left(\hat{\mathbf{U}}_n - 1\right)^T \Lambda_n \left(\hat{\mathbf{U}}_n - 1\right) \qquad (3.24)$$

where Λ_m is a diagonal matrix, where the abundance value of each object belongs to the mth class on the diagonal.

3.4.3 CLASS ALLOCATION METHOD BASED ON OBJECT UNIT

In order to obtain the final subpixel mapping result, a class allocation method based on object unit is adopted. A linear optimization model is used to minimize the spatial correlation of objects in each class. Class labels are assigned to subpixels within each object according to this model. The linear optimization model is given by the following formula:

$$E = \min \sum_{j=1}^{K_i^l} \sum_{n=1}^{N} x_n\left(p_j\right) \times E_n\left(p_j\right) \qquad (3.25)$$

$$x_n\left(p_j\right) = \begin{cases} 1, \text{if the subpixel } p_j \text{ belongs to } n\text{th class} \\ 0, \text{otherwise} \end{cases} \qquad (3.26)$$

In addition, when using the class allocation method, the following constraint function will be enforced:

$$\text{s.t.} \begin{cases} \sum_{n=1}^{N} x_n(p_j) = 1 \\ \sum_{a=1}^{K_i^l} x_n(p_j) = U_n(O_i) \times K_i^l \end{cases} \tag{3.27}$$

The first formula indicates that each subpixel belongs to only one class, and the second formula is defined as the number of subpixels in each object that should be proportional to the abundance value of the class.

This method uses a random walk process to obtain the final subpixel mapping result. First, a group of pure subpixels (subpixels with abundance value of 1) are labeled as nth label ($n = 1, 2, \ldots, N$). Secondly, for the marked subpixel, the ERW algorithm can calculate the probability that the unmarked subpixel starts to move and arrives at the marked subpixel for the first time. Based on the connection of the ERW network, the probability of ERW can be directly calculated by minimizing the linear optimization model. Finally, for unlabeled subpixels, the algorithm based on ERW assigns the label with the greatest probability. It is worth noting that when there are only pure subpixels in an object, in order to save calculation time, the labels of the same class are directly assigned to all subpixels in this object.

3.4.4 EXPERIMENTAL CONTENT AND RESULT ANALYSIS

Two remote sensing data sets are used to evaluate the performance of SPMERW. In order to carry out quantitative evaluation, the experimental process was designed to obtain the simulated coarse remote sensing images by downsampling the original fine remote sensing data set through $S \times S$ mean filtering. In the case of downsampling, the land cover classes at the subpixel level are known, which directly assess the impact of image registration errors on the technology. The least squares support vector machine (LSSVM) is selected as the spectral unmixing method. To avoid the influence of the parameters of the segmentation algorithm on the final result, only the parameters of the ERW algorithm are considered. Through a large number of experiments, the appropriate parameters of the two data sets in the segmentation algorithm are obtained. For the two data sets, the selection of $\alpha_b^{\text{spectral}}$ has little effect on the segmentation, and $\alpha_b^{\text{spectral}}$ are both set to 1. The α_b^{shape} of the two data sets are set to 0.4 and 0.5, respectively, and the α of two data sets are set to 0.6 and 0.5, respectively. The segmentation scale parameters T of the two data sets are set to 10 and 5, respectively. Four subpixel mapping methods are compared: the subpixel method based on dual-path bicubic interpolation (DPBIC) [9], the subpixel method based on the spatial correlation of pixels and subpixels (PSSD) [10], the OSPM method [11], and the proposed SPMERW method. The four SPM methods are evaluated by each class mapping accuracy, overall accuracy (PCC), and Kappa coefficient (Kappa). All experiments are tested using the MATLAB 2018a software package.

3.4.4.1 Test Data

A multispectral data set and a hyperspectral data set were tested. The tested multi-spectral data set was obtained from Landsat8 OLI in Rome, Italy. The multispectral data set has six bands (blue, green, red, near-infrared, short-wave infrared 1, short-wave infrared 2), 300 × 300 pixels, and 30 m spatial resolution. The RGB synthesis of the multispectral data set (red, green, and blue are short-wave infrared 2 bands, short-wave infrared 1 bands, and red bands, respectively) is shown in Figure 3.21(a). The tested hyperspectral data set was obtained at the Faculty of Engineering of the University of Pavia, with 320 × 320 pixels, 103 bands, and a spatial resolution of 1.3 m. The RGB composition of the hyperspectral data set (red, green, and blue bands 102, 56, and 31, respectively) is shown in Figure 3.22(a).

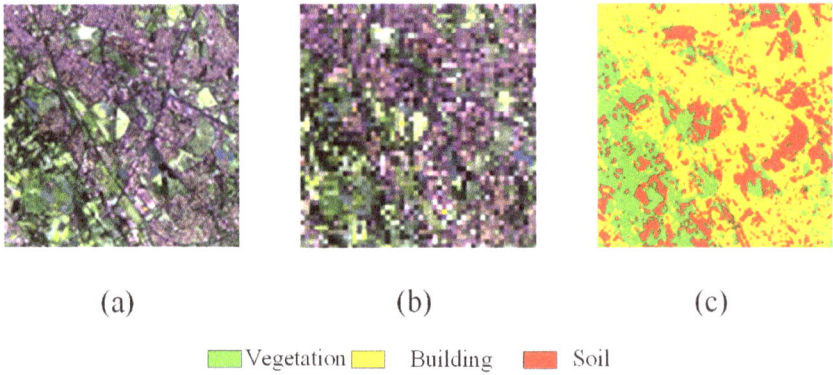

(a)   (b)   (c)

Vegetation   Building   Soil

FIGURE 3.21   Roman data set: (a) false color composite image (RGB band: short-wave infrared 2 band, short-wave infrared 1 band, red band); (b) low-resolution image ($S = 6$); and (c) reference image.

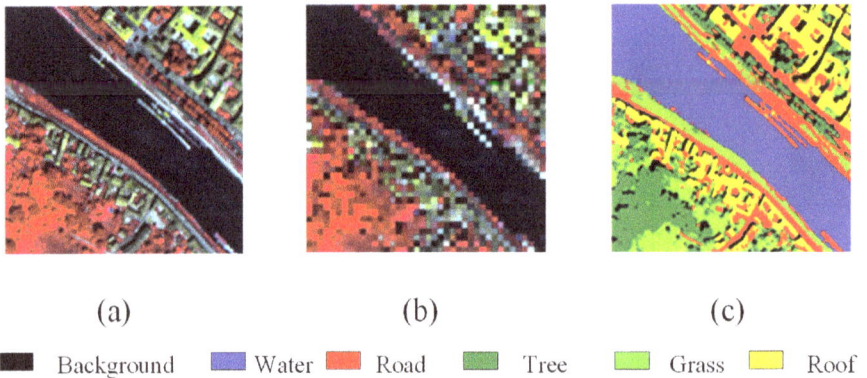

(a)   (b)   (c)

Background   Water   Road   Tree   Grass   Roof

FIGURE 3.22   Pavia city center data set: (a) false color composite image (RGB band: 102, 56, 31); (b) low-resolution image ($S = 8$); and (c) reference image.

As shown in Figure 3.21(b) and Figure 3.22(b), the tested multispectral data set and the tested hyperspectral data set are separately downsampled $S = 6$ and $S = 8$ to generate the simulated low-resolution images. The reference images of the two data sets are shown in Figure 3.21(c) and 3.22(c). The multispectral reference image includes three classes (vegetation, building, and soil), and the hyperspectral reference image includes six classes (shadow, water, road, trees, grass, and roof). The weight parameter ω of the multispectral data set is set to 0.6, and the weight parameter ω of the hyperspectral data set is set to 0.5.

3.4.4.2 Results and Analysis

The results of the four subpixel mapping methods for the multispectral data set and hyperspectral data set are shown in Figure 3.23 and Figure 3.24, respectively. As shown in Figure 3.21(b) and Figure 3.22(b), the simulated coarse images generated by the scale S are very coarse, which bring many difficulties to subpixel mapping. For example, there is a grid in Figure 3.23(b) because PSSD tends to treat most adjacent subpixels as the same class, forming the aggregated and homogenized patches. Because the proposed SPMERW method considers more comprehensive spatial information between and within objects, the SPMERW method is superior to the other three methods. The SPMERW method obtains a smoother boundary and a more continuous area, and the result is closer to the reference image. In addition, we use the mapping accuracy (%) of each class, PCC (%), and Kappa to evaluate the performance of the four subpixel mapping methods. Through the evaluation indicators

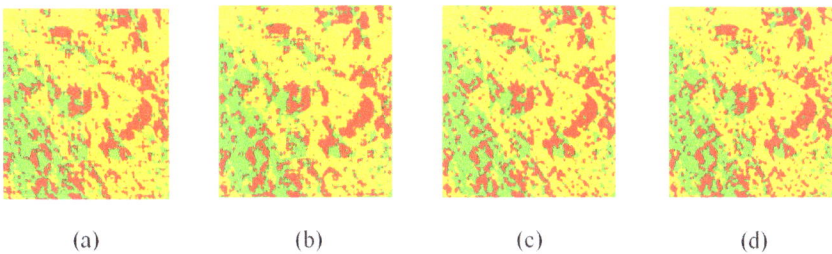

| (a) | (b) | (c) | (d) |

FIGURE 3.23 Subpixel mapping results: (a) DPBIC, (b) PSSD, (c) OSPM, and (d) SPMERW.

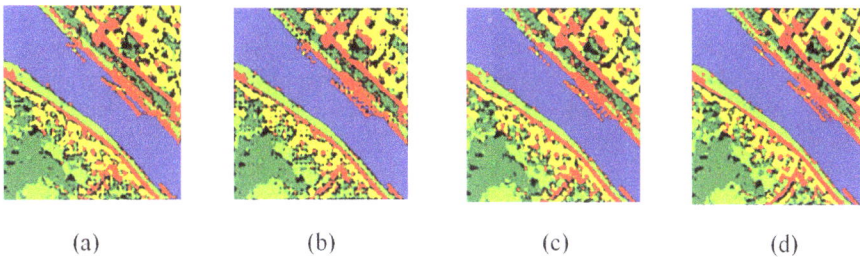

| (a) | (b) | (c) | (d) |

FIGURE 3.24 Subpixel mapping results: (a) DPBIC, (b) PSSD, (c) OSPM, and (d) SPMERW.

TABLE 3.4
Mapping Accuracy of the Four Methods in Experiment 1 (S = 6)

Land Cover Class	DPBIC	PSSD	OSPM	SPMERW
Vegetation (%)	66.99	69.28	71.20	73.39
Architecture (%)	76.06	74.27	78.26	80.72
Soil (%)	61.66	64.45	67.44	69.64
PCC (%)	70.10	71.45	73.73	76.05
Kappa	0.5250	0.5455	0.5830	0.6225

TABLE 3.5
Mapping Accuracy of the Four Methods in Experiment 2 (S = 8)

Land Cover Class	DPBIC	PSSD	OSPM	SPMERW
Background (%)	49.23	55.56	57.44	60.29
Water (%)	96.85	96.68	97.01	97.28
Road (%)	64.99	62.08	68.64	70.37
Tree (%)	75.11	75.96	78.70	80.25
Grass (%)	71.00	74.33	75.49	78.39
Roof (%)	76.32	78.87	80.59	83.06
PCC (%)	77.94	78.88	81.15	83.19
Kappa	0.7265	0.7387	0.7659	0.7977

in Table 3.4 and Table 3.5, the mapping accuracies of all classes in the proposed SPMERW are higher than the other three methods. The PCC (%) of SPMERW was 2.4% and 2.0% higher than OSPM. According to the definition of PCC (%), since the pixels of the two test areas are 300 × 300 and 320 × 320, the number of correct pixels in SPMERW is about 2160 and 2048 higher than the number of correct pixels in OSPM. Similarly, SPMERW get the highest Kappa.

3.4.4.3   Discussion

To analyze the impact of the scale on the performance of the proposed method, the multispectral data set was downsampled with $S = 3$, 6, and 10 to generate the simulated low-resolution images. Figure 3.25 shows the PCC (%) of the four subpixel mapping methods for all three scales. It is worth noting that as the S increases, the PCC (%) of the four methods decreases. This is because the higher S means the coarser image produced, the more challenging it is for subpixel mapping. But consistent with the results in Table 3.4 and Table 3.5, the proposed SPMERW produces a higher PCC (%) than the other three subpixel mapping methods.

In addition, the choice of parameters w is also discussed here. The hyperspectral data set ($S = 8$) tested 11 combinations in the range of [0, 1] with an interval of 0.1. The parameter w selection of the multispectral data set uses the same method. As shown in Figure 3.26, when $w = 0$, $E_n^{among}\left(\hat{U}_n\right)$ displays no effect, the mapping result

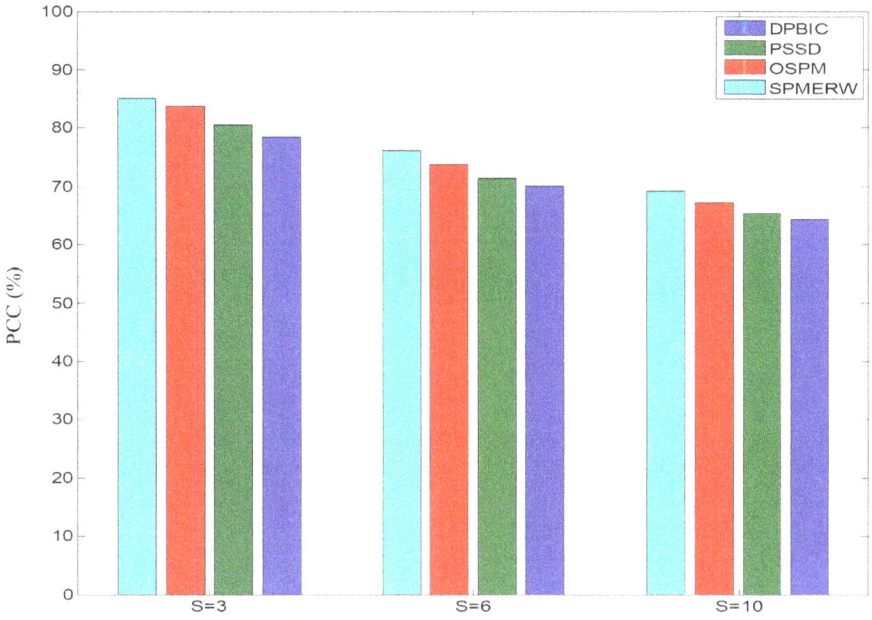

FIGURE 3.25   Relationship between PCC (%) and *S* scale of the four methods.

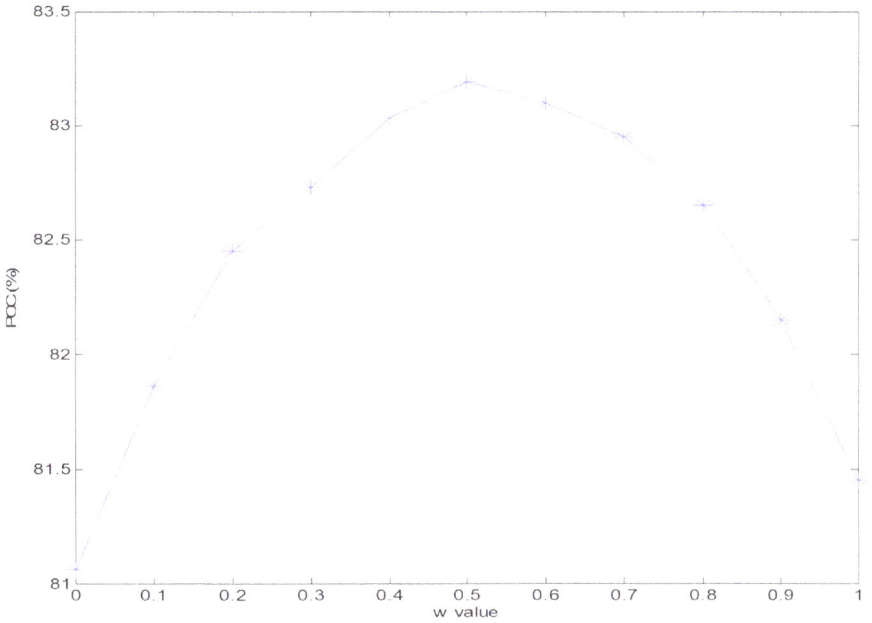

FIGURE 3.26   Relationship between PCC (%) and weight parameter λ of subpixel flood-inundation mapping for multispectral remote sensing image by supplying more spectral information method.

is not ideal. As w increases, the PCC (%) increases. This is because $E_n^{\text{among}}\left(\hat{\mathbf{U}}_n\right)$ provides the spatial information between objects. When $w = 0.5$, PCC (%) reaches the highest value. However, when w continues to increase, the contribution of $E_n^{\text{within}}\left(\hat{\mathbf{U}}_n\right)$ to formula (3.20) is small. The reduction of the spatial information from $E_n^{\text{within}}\left(\hat{\mathbf{U}}_n\right)$ will affect the final subpixel mapping result.

3.5 SUBPIXEL MAPPING BASED ON SPATIAL-SPECTRAL CORRELATION FOR SPECTRAL IMAGERY

The traditional subpixel mapping methods mostly rely on the spatial correlation based on linear distance, which ignores the influences of non-linear imaging conditions. In addition, spectral unmixing errors affect the accuracy of utilized spectral properties. To overcome the influence of linear and non-linear imaging conditions and utilize more accurate spectral properties, the subpixel mapping based on spatial-spectral correlation (SSC) is proposed in this work. Spatial correlation is obtained using the mixed spatial attraction model (MSAM) based on linear Euclidean distance. In addition, a spectral correlation that utilizes spectral properties based on the non-linear Kullback-Leibler distance (KLD) is proposed. Spatial and spectral correlations are combined to reduce the influences of linear and non-linear imaging conditions, which results in an improved mapping result. The utilized spectral properties are extracted directly by spectral imagery, thus avoiding the spectral unmixing errors. Experimental results on the three spectral images show that the proposed SSC yields better mapping results than state-of-the-art methods.

Suppose Y is the input spectral imagery with K pixels, N land cover classes, and B spectral bands. The abundance image F_n ($n = 1, 2, \ldots, N$) is obtained from Y by spectral unmixing. The abundance image F_n contains the proportion of the nth land cover class $F_n(P_A)$ for pixel P_A ($A = 1, 2, \ldots, K$). Assume S is the scale factor, then a pixel P_A into $S \times S$ subpixels p_a ($a = 1, 2, \ldots, KS^2$).

3.5.1 SPATIAL CORRELATION

Spatial correlation based on a linear distance is the theoretical foundation of most of the traditional subpixel mapping methods. It is assumed that two subpixels that belong to the same land cover class have a greater correlation than those that belong to different land cover classes. This assumption is illustrated in Figure 3.26 by a simple example of two land cover classes denoted as Class A and Class B. As shown in Figure 3.26(a), the abundance image includes nine mixed pixels marked with the proportions of Class A (red number) and Class B (blue number). When $S = 4$, a mixed pixel is segmented into 16 subpixels. For instance, 0.25 in the central pixel indicates that $4 \times 4 \times 0.25 = 4$ subpixels belong to Class A. Figure 3.26(b)–(c) show the two possible mapping results. Given a principle of spatial correlation, Figure 3.26(b) is considered as optimal.

The MSAM has been one of the main mapping methods, as it takes into account pixel-scale and subpixel-scale spatial correlations in the calculation of linear Euclidean distance. Thus, in the SSC, the MSAM is employed to obtain the spatial correlation C^{spa} and reduce the influences of linear imaging conditions. In the MSAM, the spatial correlation $C_n^{\text{spa}}(p_a)$ of the nth land cover class of subpixel p_a is calculated by

$$C_n^{\text{spa}}(p_a)=\delta \times C_n^{\text{pixel}}(p_a)+(1-\delta)\times C_n^{\text{subpixel}}(p_a) \qquad (3.28)$$

where δ denotes the weight parameter, and in this work, it is set to an empirical value of 0.4, while $C_n^{\text{pixel}}(p_a)$ and $C_n^{\text{subpixel}}(p_a)$ denote the pixel-scale and subpixel-scale spatial correlations of nth land cover class of p_a, and they are derived by (3.29) and (3.30), respectively:

$$C_n^{\text{pixel}}(p_a)=\frac{1}{K^{\text{I}}}\sum_{J=1}^{K^{\text{I}}}F_n(P_J)\times \exp(-d(p_a,P_J)^2/\varepsilon_1) \qquad (3.29)$$

$$C_n^{\text{subpixel}}(p_a)=\frac{1}{K^{\text{II}}}\sum_{j=1}^{K^{\text{II}}}x_{aj}\times \exp(-d(p_a,p_j)^2/\varepsilon_2) \qquad (3.30)$$

Further, $F_n(P_J)$ denotes the proportion of the nth land cover class of pixel P_J, which is a neighboring pixel of the central subpixel P_a. As given by (3.31), x_{aj} shows whether subpixel P_a and subpixel P_J belong to the same land cover class; and K^{I} and K^{II} denote the numbers of neighboring pixels and subpixels, respectively. Eight neighboring pixels or eight neighboring subpixels are considered here. In Figure 3.27, $d(p_a, P_J)$ represents linear Euclidean distance between central subpixel p_a and neighboring pixel P_J, while $d(p_a, p_j)$ denotes the distance between p_a and neighboring subpixel p_j; ε_1 and ε_2 are the exponential model parameters.

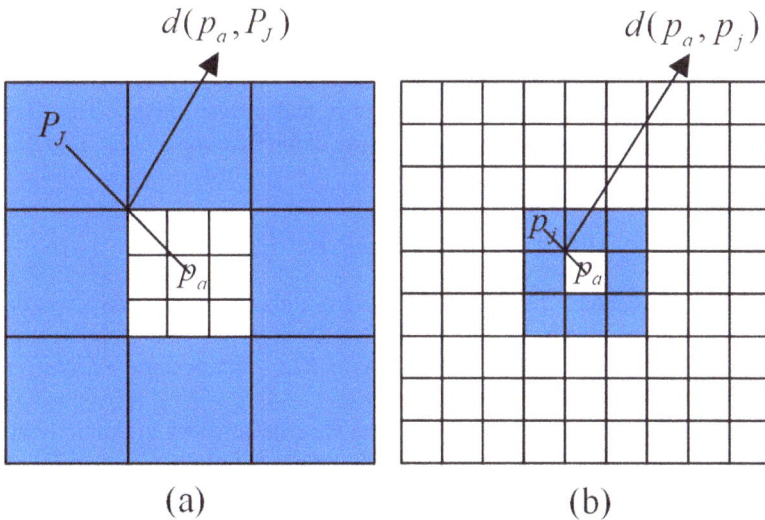

FIGURE 3.27 Euclidean distance between central subpixel p_a and (a) its eight neighboring pixels P_J and (b) its eight neighboring subpixels p_j.

$$x_{aj} = \begin{cases} 1, & p_a \text{ and } p_j \text{ are assigned to the same land cover class} \\ 0, & \text{otherwise} \end{cases} \tag{3.31}$$

Finally, the spatial correlation C^{spa} is maximized by

$$C^{\text{spa}} = \max \sum_{a=1}^{KS^2} \sum_{n=1}^{N} x_n(p_a) \times C_n^{\text{spa}}(p_a) \tag{3.32}$$

In (3.33), $x_n(p_a)$ is defined as

$$x_n(p_a) = \begin{cases} 1, & \text{if sub-pixel } p_a \text{ belongs to land cover class } n \\ 0, & \text{otherwise} \end{cases} \tag{3.33}$$

3.5.2 SPECTRAL CORRELATION

The Kullback-Leibler (KL) divergence has been widely used to measure the shape correlation between two probability density functions (PDFs). A smaller KL divergence corresponds to a higher correlation between two PDFs. Under the non-linear imaging conditions, pixels do not have a one-to-one linear relationship due to sensor differences, solar angle, and seasonal effects. When a large number of pixels are used to estimate a PDF, an advantage of the KL divergence is that it does not require a one-to-one linear relationship between the corresponding pixels and thus has good non-linear performance. Due to simple physical meaning and fast calculation, the original spectral imagery Y is first upsampled by BIC with a scale factor S, so there are KS^2 subpixels P_a ($a = 1, 2, \ldots, KS^2$) in the upsampled imagery. The KL divergence $KL(p_a; p_j)$ from p_a to its neighboring p_j is defined as

$$KL(p_a; p_j) = \int \log\left[\frac{f_{p_a}(x)}{f_{p_j}(x)}\right] \times f_{p_j}(x) dx \tag{3.34}$$

where $f_{p_a}(x)$ and $f_{p_j}(x)$ denote the PDFs of p_a and p_j, respectively. The KL divergence formula is not symmetrical due to the use of log function and ratio, so the KLD is used, and it is defined as

$$KLD(p_a; p_j) = KL(p_a; p_j) + KL(p_j; p_a) \tag{3.35}$$

where $KL(p_j; p_a)$ denotes the KL divergence from p_j to p_a, and $KLD(p_a; p_j)$ denotes the KLD between p_a and p_j. In this work, the KLD is introduced to obtain a spectral correlation C^{spe}. We focus on the calculation of $KL(p_a; p_j)$ because $KLD(p_a; p_j)$ can be easily derived from $KL(p_a; p_j)$.

The spectral properties obtained from the spectral imagery are directly utilized to establish PDF $f_{p_i}(x)$ of each subpixel p_i. Suppose $f_{p_i}(x)$ obeys a Gaussian distribution. The local part of the eight-neighborhood centered at p_i is $V_{pi} \cdot f_{p_i}(x)$ can be defined as

$$f_{p_i}(x) = \frac{1}{(2\pi)^{\frac{B}{2}} \sqrt{|\mathbf{M}|}} \times \exp\left(-(x-e)^T \mathbf{M}^{-1}(x-e)/2\right) \tag{3.36}$$

In (3.36), B denotes the number of spectral bands, and e and $\mathbf{M}$ are calculated from the spectral subpixel values of a local part V_{pi}; $e \varepsilon R^B$ is the mean vector, where $e_b = \mathrm{E}\left[V_{p_i}^{(b)} \right]$ ($b = 1, 2, \ldots, B$); $\mathbf{M}$ denotes the covariance matrix consisting of $[M_{b,c}]$, where $M_{b,c} = \mathrm{E}\left[\left(V_{p_i}^{(b)} - e^{(b)} \right)\left(V_{p_i}^{(c)} - e^{(c)} \right) \right]$, and is an invertible symmetric matrix with a positive determinant, that is, $|\mathbf{M}| > 0$.

By combining (3.34) and (3.36), $KL(p_a; p_j)$ can be derived as

$$KL\left(p_a; p_j \right) = \int \left[\frac{1}{2} \log \frac{|\mathbf{M}_{p_a}|}{|\mathbf{M}_{p_j}|} - \frac{1}{2}\left((x - e_{p_a})^T \mathbf{M}_{p_a}^{-1} (x - e_{p_a}) - (x - e_{p_j})^T \mathbf{M}_{p_j}^{-1} (x - e_{p_j}) \right) \right]$$

$$\times \frac{1}{(2\pi)^{\frac{B}{2}} \sqrt{|\mathbf{M}_{p_j}|}} \exp(-(x - e_{p_j})^T \mathbf{M}_{p_j}^{-1} (x - e_{p_j})/2) dx, \tag{3.37}$$

Formula (3.37) can be further simplified to

$$KL\left(p_a; p_j \right) = \frac{1}{2} \log \left| \mathbf{M}_{p_a}^{-1} \mathbf{M}_{p_j} \right| + \frac{1}{2} tr\left(\left(\mathbf{M}_{p_a}^{-1} \mathbf{M}_{p_j} \right)^{-1} \right) - \frac{B}{2}$$

$$+ \frac{1}{2} \left(e_{p_a} - e_{p_j} \right)^T \mathbf{M}_{p_j}^{-1} \left(e_{p_a} - e_{p_j} \right) \tag{3.38}$$

Further, $KL\left(p_j; p_a \right)$ can be derived using this process. Furthermore, the spectral correlation $C_n^{\mathrm{spe}}\left(p_a \right)$ of the nth land cover class of p_a is expressed as

$$C_n^{\mathrm{spe}}\left(p_a \right) = \sum_{j=1}^{K^{\mathrm{II}}} y_n\left(p_a; p_j \right) \times \frac{1}{KLD\left(p_a; p_j \right)} \tag{3.39}$$

where the number of neighboring subpixels K^{II} is set to eight; $y_n\left(p_a; p_j \right)$ shows whether p_a and p_j belong to the same land cover class based on an empirical threshold T [48], and it is defined as

$$y_n\left(p_a; p_j \right) = \begin{cases} 1, & \text{if } KLD\left(p_a; p_j \right) \leq T \\ 0, & \text{if } KLD\left(p_a; p_j \right) > T \end{cases} \tag{3.40}$$

where $y_n\left(p_a; p_j \right) = 1$ means that the PDFs of the spectral properties between p_a and p_j are similar; thus, the two subpixels are considered to belong to the same nth land cover class, otherwise, $y_n (p_a; p_j) = 0$.

Since a small KLD between two subpixels corresponds to a high spectral correlation, the spectral correlation C^{spe} can be defined as

$$C^{\mathrm{spe}} = \max \sum_{a=1}^{KS^2} \sum_{n=1}^{N} C_n^{\mathrm{spe}}\left(p_a \right) \tag{3.41}$$

3.5.3 SPATIAL-SPECTRAL CORRELATION IMPLEMENTATION

The proposed SCC is implemented as shown in Figure 3.28 and analyzed as follows.

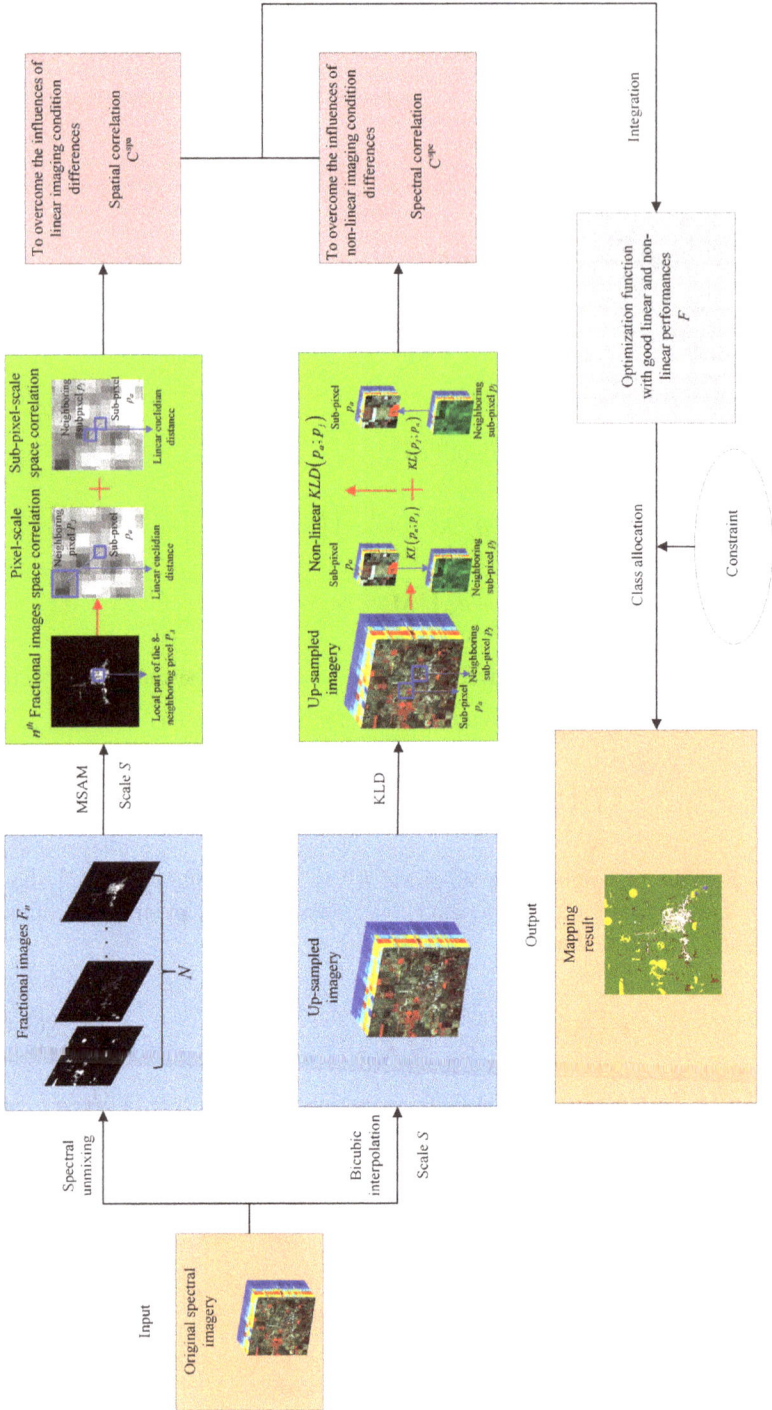

FIGURE 3.28　The spatial-spectral correlation implementation.

Step 1. In order to reduce the influences of linear imaging conditions, the spatial correlation C^{spa} is obtained using the MSAM. First, the abundance images F_n are generated from the original spectral imagery Y by spectral unmixing. Due to its simple structure and effectiveness, the LSSVM is used as a spectral unmixing method. The pixel-scale correlation $C_n^{\text{pixel}}\left(p_a\right)$ and subpixel-scale spatial correlation $C_n^{\text{subpixel}}\left(p_a\right)$ are then obtained from the abundance images using the MSAM. Finally, $C_n^{\text{pixel}}\left(p_a\right)$ and $C_n^{\text{subpixel}}\left(p_a\right)$ are combined to obtain C^{spe}.

Step 2. With the aim to utilize more accurate spectral properties and reduce the influences of non-linear imaging conditions, a spectral correlation C^{spe} is derived by the KLD. The original spectral imagery Y is first upsampled using the BIC with a scale factor S to obtain KS^2 subpixels p_a. Then, $KLD\left(p_a;p_j\right)$ is derived by calculating the spectral properties, which are obtained directly from the spectral imagery. Finally, the spectral correlation C^{spe} is obtained by formulas (3.39) to (3.41).

Step 3. Spatial correlation C^{spa} and spectral correlation C^{spe} are combined to produce an optimization function F, which has good linear and non-linear performances. The goal of the proposed SSC is to maximize F, which can be expressed as

$$\max F = \left(1-\theta\right)C^{\text{spa}} + \theta C^{\text{spe}} \tag{3.42}$$

where θ $\left(0 \le \theta < 1\right)$ is a weight parameter used to balance the effect between C^{spa} and C^{spe}.

Step 4. The class allocation method based on simulated annealing is used to assign land cover class labels to each subpixel and obtain final mapping results by the maximized function F. In addition, when the land cover class labels are assigned to each subpixel, the following function constraints should be met:

$$\text{s.t.} \begin{cases} \displaystyle\sum_{n=1}^{N} x_n\left(p_a\right) = 1 \\[2em] \displaystyle\sum_{a=1}^{K} x_n\left(p_a\right) = F_n\left(P_A\right)\times S^2 \end{cases} \tag{3.43}$$

where the first constraint means that a subpixel belongs only to one land cover class, and the second constraint ensures that the number of subpixels of the nth land cover class is fixed.

3.5.4 EXPERIMENTAL CONTENT AND RESULT ANALYSIS

3.5.4.1 Experimental Setup

The performance of the proposed SCC was qualitatively and quantitatively evaluated by the experiments using two simulated spectral images and one real spectral imagery. In the experiments, the coarse imagery is usually obtained by downsampling the fine spectral imagery. Therefore, the results are not accurate enough because the influences of non-linear imaging conditions are not taken into account. When the weather and illumination conditions change, a change in intensity is one of the most common non-linear influences, which leads to coarse spectral imagery. However, the intensity change influence cannot be accurately simulated by a linear mean filter. Therefore, the simulated experiments were conducted by simulating the intensity change; namely, the fine spectral imagery was first downsampled by a $S \times S$ linear mean filter to obtain the downsampled imagery, and S was set to 6. Then, the simulated coarse imagery was derived from the downsampled imagery by changing the intensity.

The simulated coarse imagery was achieved by utilizing the CIE1976 LAB space. The red (R), green (G), and blue (B) pixel values of the downsampled imagery were transformed into CIE1976 LAB space. The lightness L of the downsampled imagery was defined as

$$L = 116g\left(\frac{Z}{100}\right) - 16 \tag{3.44}$$

where Z denoted the transformation result from the RGB color space to the XYZ color space, and $g(t)$ was defined as

$$g(t) = \begin{cases} \sqrt[3]{t}, \text{ if } t > \lambda^3 \\ \dfrac{t}{3\lambda^2} + \dfrac{4}{29}, \text{ otherwise} \end{cases} \tag{3.45}$$

where $\lambda = \dfrac{6}{29}$. The imagery intensity could be changed by adjusting the value of L. To simulate the influence of intensity, L was reduced to 50%. Finally, the CIE1976 LAB space was transformed back to the RGB space, producing the simulated coarse spectral imagery.

The fine abundance images are obtained by the spectral unmixing method based on the LSSVM. The reference images of three experiments are obtained from the fine image by classification method based on LSSVM. Five subpixel mapping methods were compared: subpixel-scale spatial attraction model (SSAM) [10], spatial-spectral interpolation (SSI) [13], MSAM [14], object spatial dependence (OSD) [11], and the proposed SSC. The weight parameter θ in the three experiments was set to 0.4, 0.6, and 0.4, respectively. The performances of the five methods were evaluated based on the mapping accuracy of each land cover class, the overall accuracy (OA), and the Kappa coefficient (Kappa). MATLAB 2018a software was utilized to conduct the experiments on a Pentium dual-core processor (2.20 GHz).

3.5.4.2 Simulated Multispectral Imagery

In the first experiment, the five subpixel mapping methods were tested in simulated multispectral imagery. Fine multispectral imagery was acquired from Landsat-8 OLI over Ulysses, Kansas, in 2013. The imagery had a resolution of 360 × 360 pixels and included eight visible spectral bands at a spatial resolution of 30 m, and a panchromatic band at the spatial resolution of 15 m, as shown in Figure 3.29(a). As shown in Figure 3.29(b), the simulated coarse imagery was obtained by the method described in experimental setup, namely, by downsampling at $S = 6$ then intensity changing by setting $L = 50\%$. It should be noted that the mapping information of the land cover

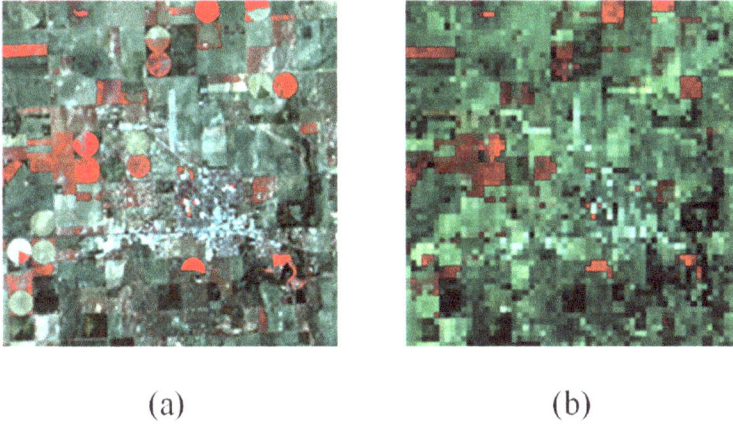

(a) (b)

FIGURE 3.29 Experiment 1: (a) false color imagery (bands 5, 4, and 3 for red, green, and blue) and (b) simulated coarse imagery ($S = 6$ and $L = 50\%$).

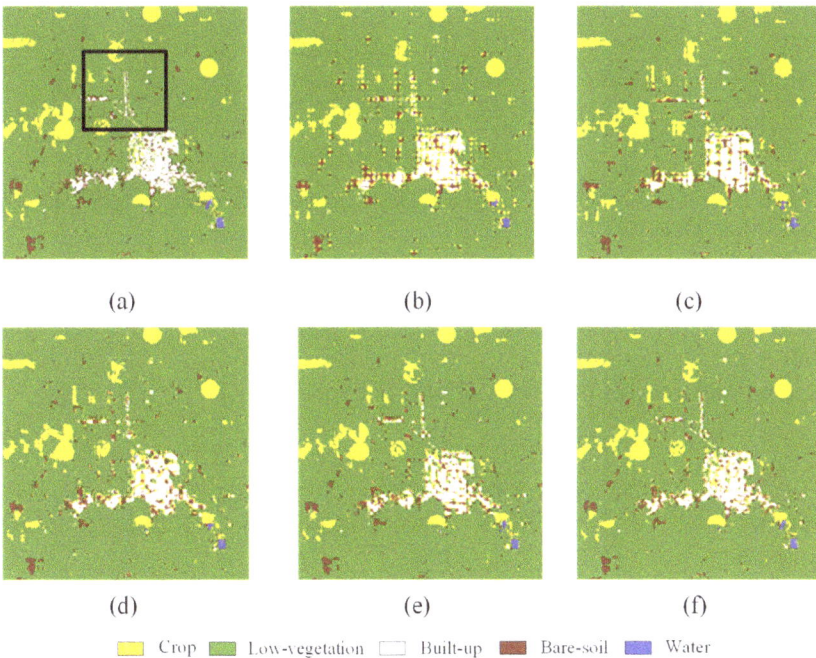

(a) (b) (c)

(d) (e) (f)

Crop ▢ Low-vegetation ▢ Built-up ▢ Bare-soil ▢ Water

FIGURE 3.30 Mapping results of experiment 1: (a) reference imagery, (b) subpixel-pixel spatial attraction model (SPSAM), (c) SSI, (d) MSAM, (e) OSD, and (f) SSC.

classes was difficult to obtain due to many mixed pixels in the simulated coarse imagery. Thus, we used the subpixel mapping methods to solve this issue.

As shown in Figure 3.30(a), the reference imagery was obtained by classifying the fine imagery, including five land cover classes (crop, low vegetation, built-up, bare soil, and water). The land cover class mapping results of the five subpixel mapping methods are shown in Figure 3.30(b)–(f). As presented in Figure 3.30(f), the result obtained by the proposed SSC was more similar to the reference imagery than those obtained by the other four subpixel mapping methods, which are illustrated in Figure 3.30(b)–(e). The salient region is marked with a black frame in Figure 3.30 and shown in Figure 3.31. As displayed in Figure 31(b)–(e), the subpixel mapping results of the four traditional methods were far away from ideal. There were many boundaries of protruding burrs and areas of disconnected shapes in the land cover classes, which was because the simulated coarse imagery was obtained by considering the influences of both linear and non-linear imaging conditions, which further posed larger challenges to the traditional subpixel mapping methods. In addition, the spectral properties could not be utilized accurately due to the errors of the spectral unmixing process in the four subpixel mapping methods. As shown in Figure 3.31(f), this phenomenon was alleviated in the proposed SSC because the influences of both linear and non-linear imaging conditions were reduced, and more accurate spectral

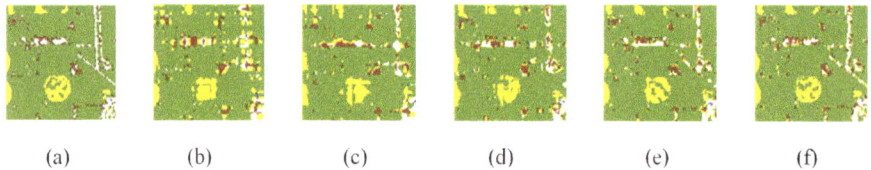

(a)   (b)   (c)   (d)   (e)   (f)

FIGURE 3.31   Salient region from Figure 3.30: (a) reference imagery, (b) SPSAM, (c) SSI, (d) MSAM, (e) OSD, and (f) SSC.

TABLE 3.6
Mapping Accuracy of the Five Methods in Experiment 1

Land Cover Class	SPSAM	SSI	MSAM	OSD	SSC
Crop (%)	75.97	73.32	79.09	81.47	90.40
Low vegetation (%)	95.20	95.49	95.87	96.50	97.57
Built-up (%)	48.05	50.25	53.68	55.68	67.14
Bare soil (%)	30.79	26.79	30.47	34.43	50.48
Water (%)	65.63	56.77	59.38	65.63	79.69
OA (%)	90.75	90.85	91.69	92.55	94.82
Kappa	0.5890	0.5807	0.6218	0.6563	0.7027

properties were utilized. Thus, in Figure 3.31(f), there are smoother boundaries and more continuous areas than in Figure 3.31 (b)–(e).

Table 3.6 shows the three evaluation indices, the mapping accuracy of each land cover class (%), OA (%), and Kappa. The results presented in Table 3.6 show that the proposed SSC performed better than the other four subpixel mapping methods.

For instance, compared with the mapping accuracy of each land cover class in the OSD, the accuracy of buildup, bare soil, and water in the SSC increased by about 11.5%, 16.1%, and 14.1%, respectively. Moreover, the SSC had the highest OA (%) of 84.82%, and Kappa of 0.7027.

3.5.4.3 Synthetic Hyperspectral Data

In Experiment 2, the hyperspectral imagery with more complex land cover class distributions was tested. This imagery was collected over the Kennedy Space Center (KSC), Florida, USA, in 1996 by NASA's AVIRIS instrument. As shown in Figure 3.32(a), the tested region had a resolution of 480 × 600 pixels and a spatial resolution of 18 m, and included 176 spectral bands. The simulated coarse imagery was generated by the method used in experiment 1, and it is shown in Figure 3.32(b). By simulating the influences of both linear and non-linear imaging conditions at the same time, the simulated imagery was coarser, posing greater challenges to the subpixel mapping methods.

The image presented in Figure 3.33(a), which was obtained from the fine imagery by classification, was considered as reference imagery. There were 13 land cover classes in the reference imagery. Figure 3.33 (b)–(f) show the subpixel mapping results of the SPSAM, the SSI, the MSAM, the OSD, and the proposed SSC, respectively. The SSC achieved a better land cover class mapping result than the other four traditional methods. Figure 3.34 shows the salient region of the subpixel mapping results of the five methods. As shown in Figure 3.34(b)–(e), the mapping result of the hardwood class was not continuous, and the graminoid marsh class was hardly mapped using the four traditional subpixel mapping methods. Because the proposed SCC considered the influences of imaging conditions more comprehensive through the spatial-spectral correlation, these phenomena were alleviated. The subpixel mapping result of the SCC is presented in Figure 3.34(f), where it can be seen that it was the closest to the reference imagery that is shown in Figure 3.34(a) among all the obtained results.

The mapping accuracies of each land cover class (%), OA (%), and Kappa of the five methods are given in Table 3.7, where it can be seen that the proposed SCC

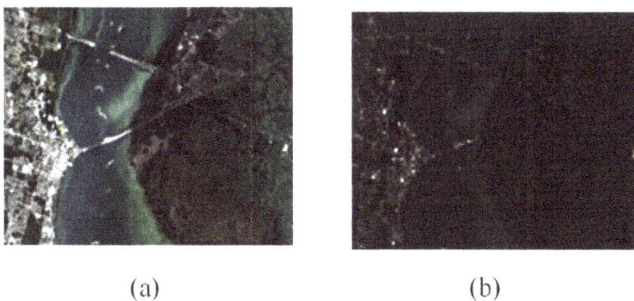

(a)   (b)

FIGURE 3.32   Experiment 2: (a) false color imagery (bands 28, 19, and 10 for red, green, and blue) and (b) simulated coarse imagery ($S = 6$, $L = 50\%$).

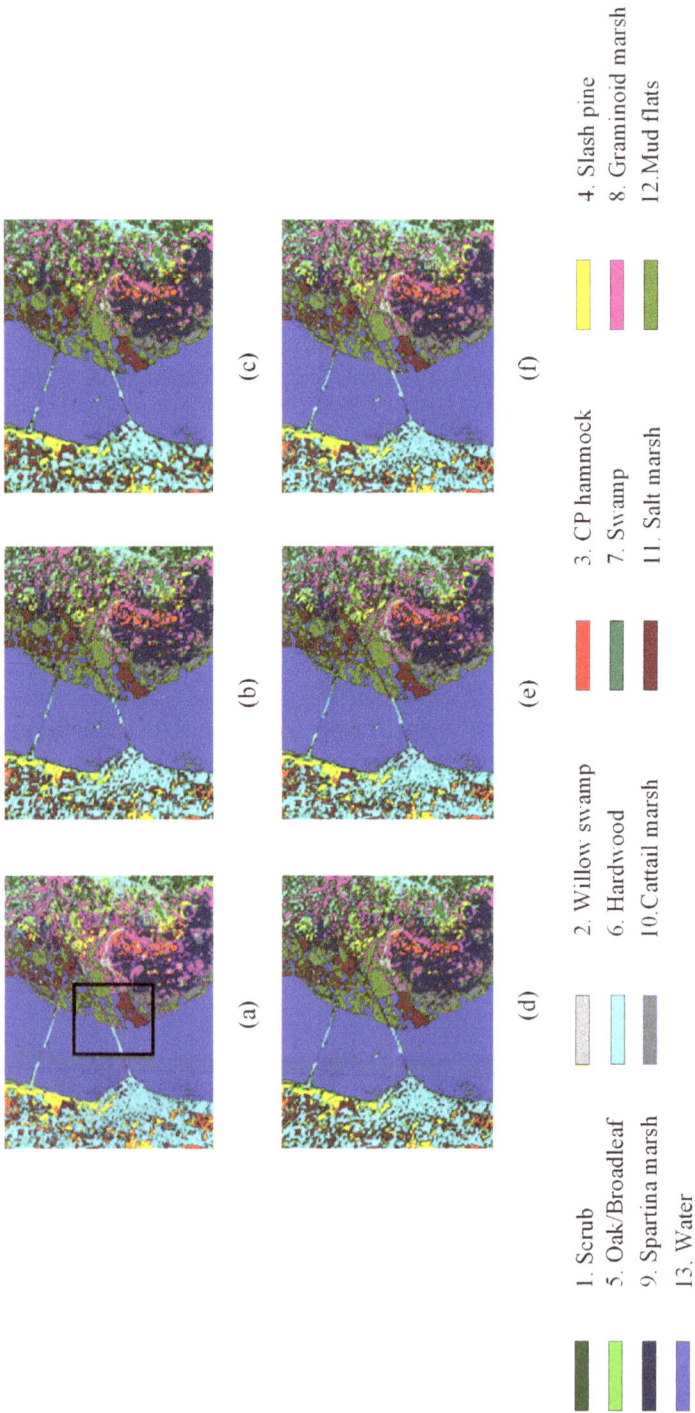

FIGURE 3.33 Mapping results of experiment 2: (a) reference imagery, (b) SPSAM, (c) SSI, (d) MSAM, (e) OSD, and (f) SSC.

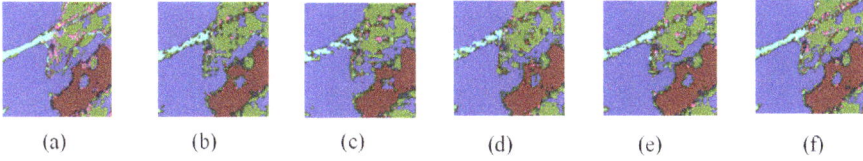

FIGURE 3.34   Salient region from Figure 3.33: (a) reference imagery, (b) SPSAM, (c) SSI, (d) MSAM, (e) OSD, and (f) SSC.

TABLE 3.7
Mapping Accuracy of the Five Methods in Experiment 2

Land Cover Class	SPSAM	SSI	MSAM	OSD	SSC
Scrub (%)	69.85	66.53	70.27	71.36	74.03
Willow swamp (%)	27.24	25.61	28.47	34.09	44.08
CP hammock (%)	23.91	24.37	25.68	30.01	40.50
Slash pine (%)	34.77	34.49	35.97	39.81	48.45
Oak/Broadleaf (%)	14.69	13.59	13.90	18.79	27.29
Hardwood (%)	52.25	54.01	56.05	57.80	64.54
Swamp (%)	55.01	48.90	53.50	58.44	64.67
Graminoid marsh (%)	33.15	32.35	35.80	39.69	48.06
Spartina marsh (%)	57.09	56.01	57.68	60.66	66.39
Cattail marsh (%)	37.16	36.36	37.87	41.58	48.24
Salt marsh (%)	50.92	48.85	52.08	55.16	62.73
Mud flats (%)	58.80	57.64	60.18	63.67	68.96
Water (%)	95.88	95.30	96.00	96.52	97.43
OA (%)	63.06	62.38	64.23	66.49	71.50
Kappa	0.5625	0.5546	0.5765	0.6035	0.6545

obtained the highest values of the three evaluation indices. Compared to the OSD, the OA (%) of the SSC increased by about 5%. According to the definition of OA (%), because there were 480×600 pixels in the tested imagery, an increase of about 5% means that the mapping result of the SSC produced 14,400 more correct mapping pixels, which represented a significant improvement.

3.5.4.4   Real Hyperspectral Imagery

In experiment 3, the real hyperspectral imagery was used to demonstrate the effectiveness of the proposed SSC. The imagery was obtained by the Hyperion imaging spectrometer over Rome, Italy, in 2002. As shown in Figure 3.35(a), the tested region had 240×160 pixels, the spatial resolution was 30 m, and there were 198 spectral bands. It should be noted that the real imagery was coarse due to both linear and nonlinear imaging influences, so the distributions of land cover classes were difficult to obtain. The proposed SSC was directly applied to the imagery to obtain the land cover class mapping information.

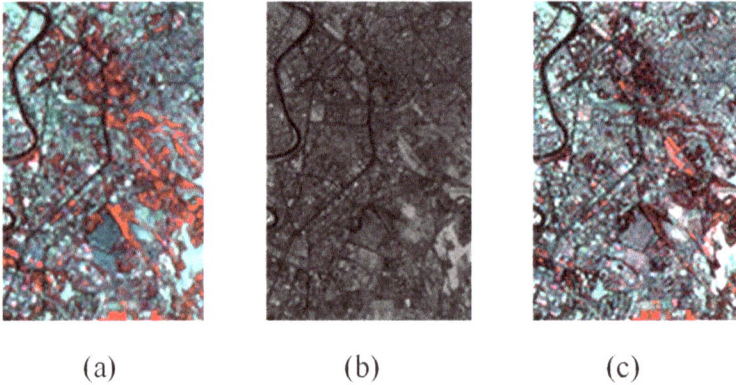

(a)   (b)   (c)

FIGURE 3.35 Experiment 3: (a) false color imagery (bands 150, 10, and 24 for red, green, and blue), (b) panchromatic image, and (c) pansharpened result.

As shown in Figure 3.35(b), a panchromatic imagery with a 15-m spatial resolution from Landsat-8OLI was obtained over the same area as the hyperspectral imagery. To obtain the reference imagery, fine pansharpening imagery with a 15-m spatial resolution was produced by fusing the real hyperspectral imagery and the panchromatic image. The fine pansharpening imagery is shown in Figure 3.35(c); the scale between the real hyperspectral imagery and the fine imagery was $S = 2$.

As shown in Figure 3.36(a), the reference imagery was obtained from the classification result of the pansharpening imagery. Figure 3.36(b)–(f) show the subpixel mapping results of the five methods. The salient region of the subpixel mapping results is shown in Figure 3.37. Similar to experiments 1 and 2, the proposed SSC achieved more continuous and smoother land cover class mapping results than the other four methods. Table 3.8 lists the three evaluation indices of the five subpixel mapping methods, where it can also be seen that the proposed SSC outperformed the other four subpixel mapping methods and obtained the highest values of the three evaluation indices.

3.5.4.5 Scale Factor

The simulated coarse imagery was obtained by downsampling the fine spectral imagery by a scale factor S. A larger S meant that while the simulated coarser imagery had been obtaining, larger influences of linear imaging conditions had occurred. Different values of the scale factor S were used to verify the performance of the proposed SSC under the influences of different degrees of linear imaging conditions. The five subpixel mapping methods were tested by repeating experiments 1 and 2 for three scale factor values (i.e., 3, 6, and 10). In order to ensure a comparable quantitative evaluation, the lightness L was kept at 50%.

The OA (%) values of the five methods at different scale factors are shown in Figure 3.38. Because a higher value of S indicated coarser input imagery, the OA (%) of the five methods decreased with S, but the proposed SSC still obtained the highest

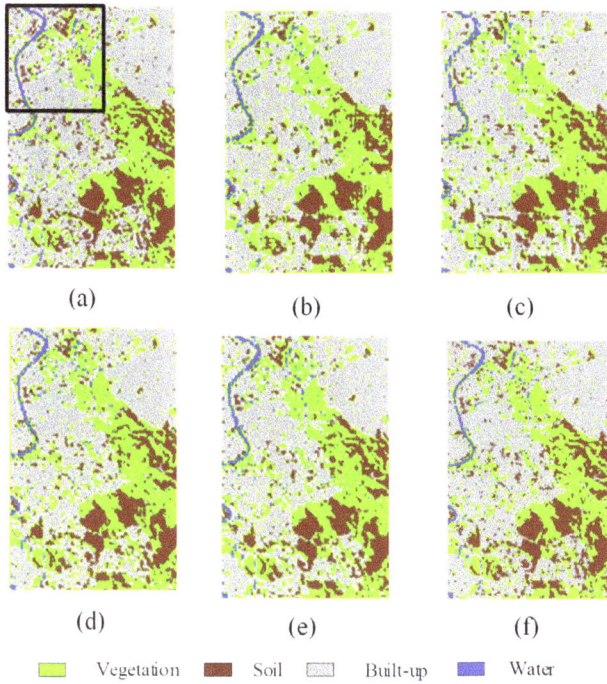

FIGURE 3.36 SRM results of experiment 3: (a) reference imagery, (b) SPSAM, (c) SSI, (d) MSAM, (e) OSD, and (f) SSC.

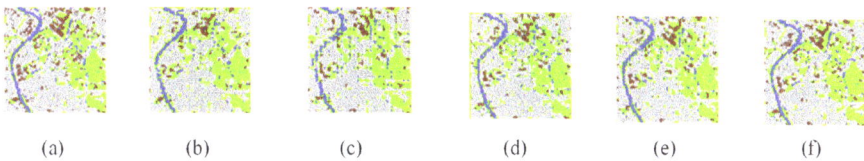

FIGURE 3.37 Salient region from Figure 3.36: (a) reference imagery, (b) SPSAM, (c) SSI, (d) MSAM, (e) OSD, and (f) SSC.

TABLE 3.8

Mapping Accuracy of the Five Methods in the Third Experiment

Land Cover Class	SPSAM	SSI	MSAM	OSD	SSC
Vegetation (%)	74.72	72.20	78.15	79.22	83.92
Soil (%)	61.83	61.19	67.47	67.56	72.57
Buildup (%)	85.14	83.65	86.03	87.19	92.21
Water (%)	53.81	42.26	50.78	54.26	75.34
OA (%)	77.36	75.52	79.71	80.70	84.97
Kappa	0.6327	0.6030	0.6714	0.6871	0.7274

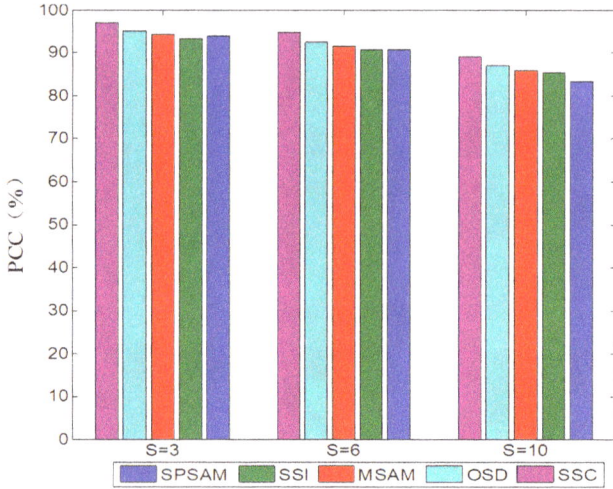

(a)

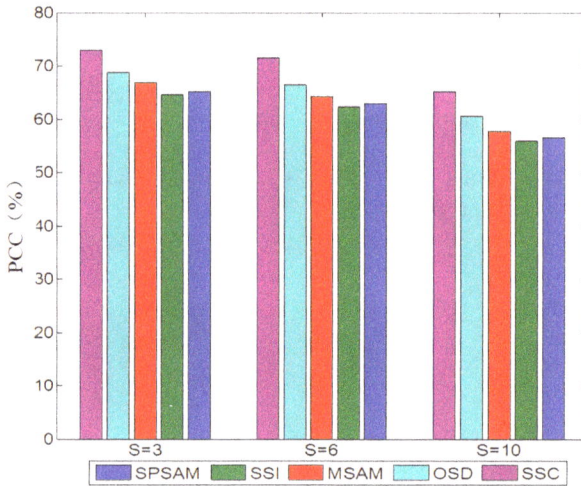

(b)

FIGURE 3.38   (a) OA (%) of the five subpixel mapping methods in relation to different scale factors s (a) in experiment 1 and (b) in experiment 2.

OA (%) value among all the methods in all three cases. In other words, the SSC had the best performance at the different-degree influence of linear imaging conditions among all the tested methods.

3.5.4.6   Lightness
We evaluated the performance of the proposed SSC by changing lightness L to determine the influence of different degrees of non-linear imaging condition on image

quality. When the value of L decreased, the lightness of the coarse imagery also decreased, which indicated an increased influence of non-linear imaging conditions. At the scale factor S of 6, the five subpixel mapping methods were tested by repeating experiments 1 and 2 at different lightness values, namely, 70%, 50%, and 30%.

The OA (%) values of the five subpixel mapping methods at the three lightness values of the two experiments are shown in Figure 3.39. Because the proposed SSC

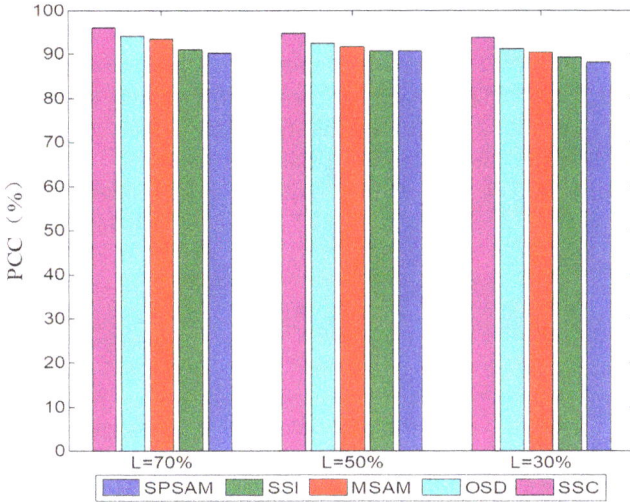

(a)

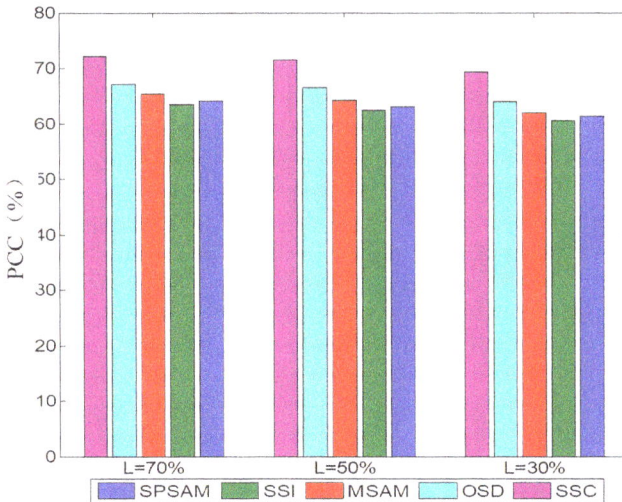

(b)

FIGURE 3.39   (a) OA (%) of the five subpixel mapping methods in relation to different lightness L (a) in experiment 1 and (b) in experiment 2.

could reduce the influence of non-linear imaging conditions using the spectral correlation, the experimental results show that the SSC obtained the highest OA (%) among all the methods at different values of L.

3.5.4.7　Weight Parameter

The weight parameter θ is introduced to balance the influence of spatial correlation and spectral correlation in the proposed SSC. Experiments 1 and 2 ($S = 6$ and $L = 50\%$) and experiment 3 were repeated to evaluate the OA (%) for ten combinations of θ in the range of [0, 0.9] at an interval of 0.1 in order to determine the most suitable value of θ and because the performance of spatial correlation and spectral correlation on improving the mapping results can be changed by changing the parameter value. Therefore, this study also demonstrates that spatial correlation and spectral correlation can improve the mapping results, respectively.

As shown in Figure 3.40, at $\theta = 0$, there was only spatial correlation in the SSC, so the influences of linear imaging conditions were reduced; thus, the mapping results can be obtained. As θ increased, the value of OA (%) also increased because the spectral correlation was added to the SCC, the influence of the non-linear imaging conditions was taken into account, and more accurate spectral properties were

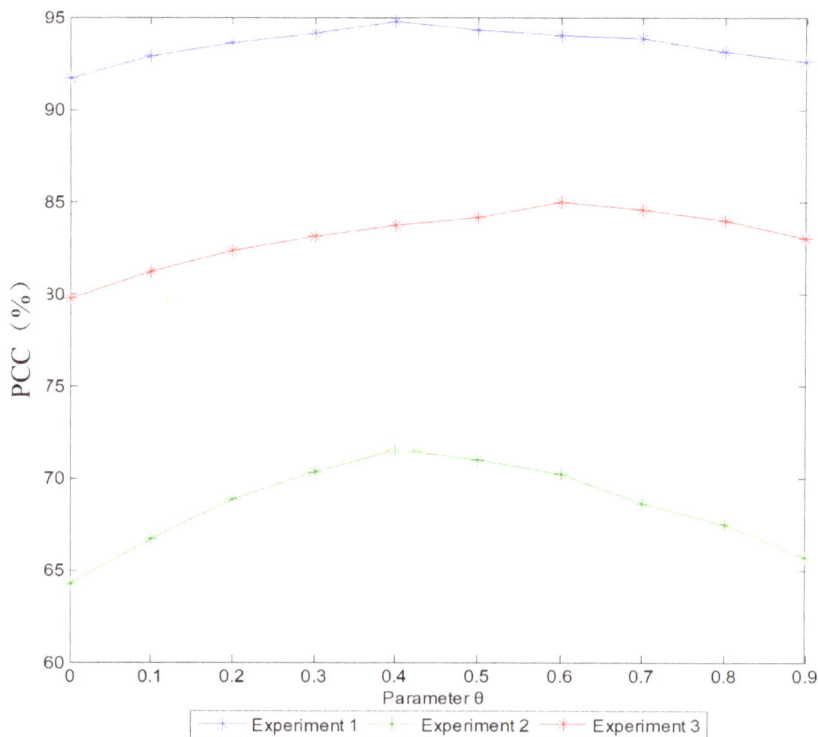

FIGURE 3.40　The OA (%) value of the SSC as a function of the relation weight parameter θ of the three experiments.

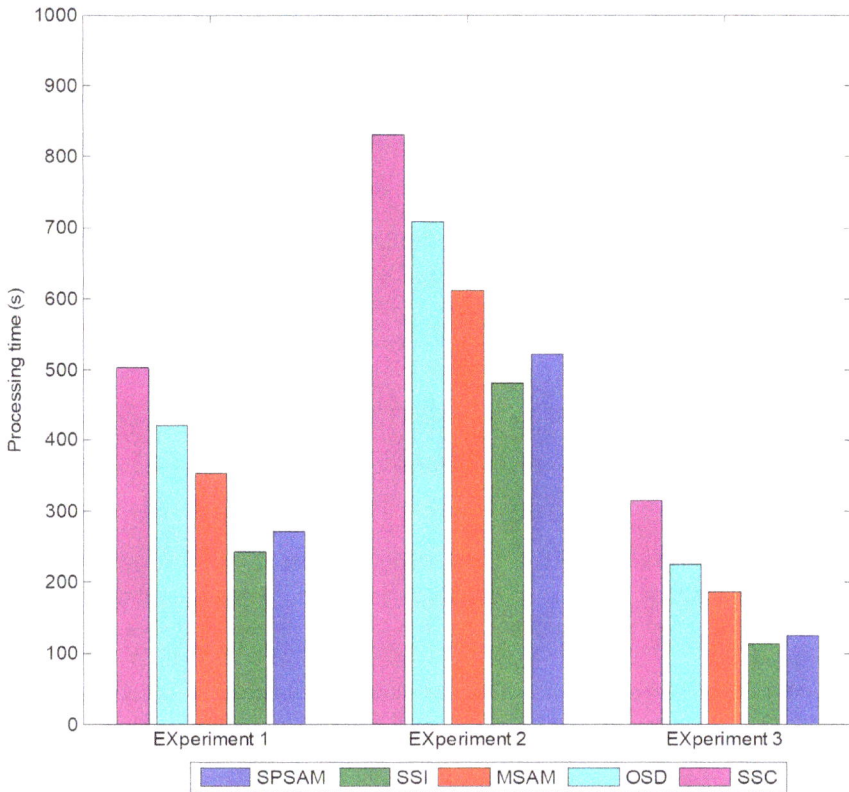

FIGURE 3.41   The processing time of the five subpixel mapping methods in the three experiments.

utilized. In Figure 3.40, it can also be seen that the most appropriate θ value of experiments 1, 2, and 3 were 0.4, 0.6, and 0.4, respectively; the most appropriate value was the one at which the OA (%) was the highest.

3.5.4.8   Processing Time

The processing time represents an important index to evaluate the performance of the subpixel mapping methods. The processing times of the five subpixel mapping methods in the three experiments are presented in Figure 3.41. Because the SSC involved more processing steps than the other four traditional methods, it required more time to obtain the final subpixel mapping result.

3.6   SUMMARY

The proposed SPM-SS method has the following characteristics: 1) It inherits the advantages of the existing interpolation-based subpixel mapping method, which has fast calculation speed and does not require any prior information and 2) compared with the existing method, by introducing the spectral term, the spatial-spectral information of the original image can be more fully utilized, and then more accurate mapping result can be obtained. The results of two sets of experiments show that the proposed SPM-SS method has significantly improved the accuracy of each land cover class. This fully shows that the proposed SPM-SS method can more effectively use the spatial-spectral information of the original image.

The proposed I-HNN adds a priori terms from a new type of processing path, making full use of the additional priori information of the original image. The experimental results show that the proposed I-HNN method has certain advantages compared with the existing HNN method. The results show that the proposed I-HNN can obtain better subpixel mapping results. In addition, the proposed I-HNN does not require any auxiliary data.

In the proposed SPMERW, BIC is first used to upsample the original coarse image. Then through spectral unmixing and segmentation of the upsampled image, the abundance values of the object are obtained. Third, ERW is used to calculate the abundance values of the object and generate the spatial correlation of the object, including the spatial information between the objects and the internal spatial information of the objects. Finally, according to the object spatial correlation, the class allocation method based on object unit is used to obtain the subpixel mapping result. Both visual and quantitative evaluations show that the proposed SPMERW can significantly improve the accuracy of subpixel mapping results. This fully shows that the proposed SPMERW method can obtain the more comprehensive spatial correlation information.

The SCC that combines spatial and spectral correlations is proposed to improve the accuracy of subpixel mapping results. Besides, to reduce the influences of linear imaging conditions, spatial correlation is conducted using the MSAM that is based on linear Euclidean distance, while spectral correlation is generated by the non-linear KLD. An objective function with good linear and non-linear performances is obtained by combining spatial and spectral correlations. Finally, a class allocation method based on the simulated annealing is utilized to obtain final subpixel mapping results. The proposed method was evaluated experimentally and compared to four state-of-the-art subpixel mapping methods. The results indicated that the proposed method achieved better subpixel mapping results than all the other methods, which was because it took into account the influences of both linear and non-linear imaging conditions. Thus, the proposed method can reduce the influences of imaging conditions on image quality and also improve the accuracy of the subpixel mapping results.

REFERENCES

[1] Kang X, Li S, Fang L, Li M, Benediktsson J A. Extended random walker-based classification of hyperspectral images[J]. IEEE Transactions on Geoscience and Remote Sensing, 2015, 53(1): 144–153.

[2] Deng C. Research of Image Interpolation Algorithm[D]. 2011. (Cai Deng. Research on Image Interpolation Algorithm [D]. Chongqing University).

[3] Wang L, Wei F, Liu D, Wang Q. Fast implementation of maximum simplex volume-based endmember extraction in original hyperspectral data space[J]. IEEE Journal of Selected Topics in Applied Earth Observations and Remote Sensing, 2013, 6(2): 516–521.

[4]   Wang L, Liu D, Wang Q. Spectral unmixing model based on least squares support vector machine with unmixing residue constraints[J]. IEEE Geoscience and Remote Sensing Letters, 2013, 10(6): 1592–1596.

[5]   Landgrebe D A. Signal Theory Methods in Multispectral Remote Sensing[M]. New Jersey: Wiley-Interscience, 2005.

[6]   Hopfield J J. Neural networks and physical systems with emergent collective computational abilities[J]. Proceedings of the National Academy of Sciences, 1982, 79(8): 2554–2558.

[7]   Wang Q, Shi W, Atkinson PM, Li Z. Land cover change detection at subpixel resolution with a hopfield neural network[J]. IEEE Journal of Selected Topics in Applied Earth Observations and Remote Sensing, 2014, 8(3): 1339–1352.

[8]   Li X, Du Y, Ling F, Feng Q, Fu B. Superresolution mapping of remotely sensed image based on hopfield neural network with anisotropic spatial dependence model[J]. IEEE Geoscience and Remote Sensing Letters, 2014, 11(7): 1265–1269.

[9]   Wang P, Wang L, Chanussot J. Soft-then-hard subpixel land cover mapping based on spatial-spectral interpolation[J]. IEEE Geoscience and Remote Sensing Letters, 2016, 13(12): 1851–1854.

[10]   Ling F, Li X, Du Y, Xiao F. Sub-pixel mapping of remotely sensed imagery with hybrid intra- and inter-pixel dependence[J]. International Journal of Remote Sensing, 2013, 34(1): 341–357.

[11]   Chen Y, Ge Y, Heuvelink G B M, An R, Chen Y. Object-based superresolution land-cover mapping from remotely sensed imagery[J]. IEEE Transactions on Geoscience and Remote Sensing, 2018, 56(1): 328–340.

[12]   Cui B, Xie X, Ma X, Ren G, Ma Y. Superpixel-based extended random walker for hyperspectral image classification[J]. IEEE Transactions on Geoscience and Remote Sensing, 2018, 59(6): 3233–3243.

[13]   Wang P, Wang L, Chanussot J. Soft-then-hard subpixel land cover mapping based on spatial-spectral interpolation[J]. IEEE Geoscience and Remote Sensing Letters, 2016, 13(12): 1851–1854.

[14]   Lu L, Hang Y, Di L, Huang D. A new spatial attraction model for improving subpixel land cover classification[J]. Remote Sensing, 2017, 9: 360.

4 Subpixel Mapping Based on Multi-Shift Remote Sensing Images

4.1 INTRODUCTION

The traditional subpixel mapping based on a single remote sensing image lacks sufficient prior information, and it is difficult to accurately obtain the spatial distribution information of all land cover classes in the mixed pixel. Multi-shift images (MSIs) from the same area at the subpixel scale can be used as auxiliary data to improve the accuracy of subpixel mapping. However, the existing fine MSIs have the defects of insufficient spatial-spectral information of the original image, single-scale information, incomplete information types, and inaccurate temporal information, which affect the final subpixel mapping result.

In response to these problems, a conversion subpixel mapping method based on MSIs with spatial-spectral information (using MSIs with spatial-spectral information in soft-then-hard subpixel mapping, MSI-SS) is proposed. The generated fine MSIs have the richer spatial-spectral information of the original image, which improves the accuracy of subpixel mapping.

Subpixel mapping based on the spatial attraction model with multi-scale subpixel shifted images (SAM-MSSI) is proposed, which is generated by the coarse-scale spatial attraction model and the fine-scale spatial attraction model. Fine MSIs with multi-scale information improve the final mapping results.

Furthermore, utilizing parallel networks to produce subpixel shifted images with multi-scale spatial-spectral information for subpixel mapping (SSI-MSSI) is proposed. The fine MSIs generated by the parallel path have multiple types of information, and the mapping result with higher mapping accuracy can be obtained.

Finally, spatiotemporal subpixel mapping by considering the point spread function effect (FCSTD) is proposed. Because the scales and properties of the two abundance images from original coarse spectral image and prior fine spectral image are the same regardless of the fine scale or coarse scale, the temporal dependence information is more accurate and richer. In addition, the point spread function effect is considered and introduced into spatiotemporal subpixel mapping.

4.2 THEORETICAL BASIS

4.2.1 MULTI-SHIFT IMAGES PROBLEM

Traditional low-resolution abundance images are obtained from a single original coarse remote sensing image through spectral unmixing technology. Since subpixel

DOI: 10.1201/9781003279082-4

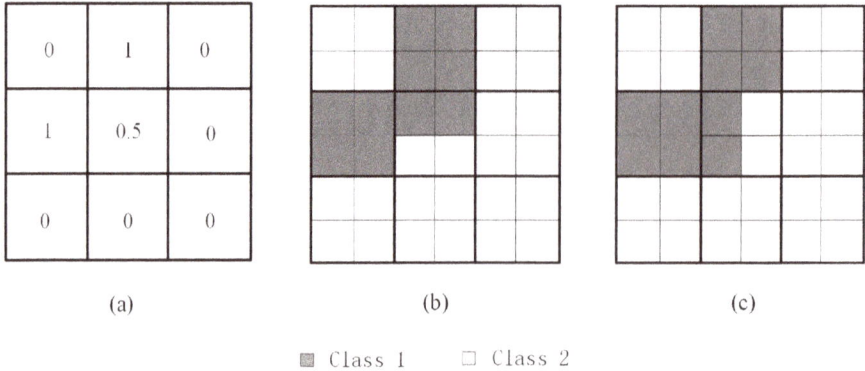

0	1	0
1	0.5	0
0	0	0

(a)   (b)   (c)

▨ Class 1   ☐ Class 2

FIGURE 4.1   Subpixel mapping based on a single remote sensing image: (a) class 1 abundance value, (b) distribution probability 1, and (c) distribution probability 2.

mapping is an ill-conditioned inverse problem, it is difficult to use only one low-resolution abundance image to accurately construct a high spatial resolution image. Figure 4.1 shows an example of subpixel mapping based on a single remote sensing image. Figure 4.1(a) is a low-resolution abundance image of class 1 obtained by spectral unmixing. According to the abundance value information and spatial correlation, the two subpixel mapping results are shown in Figure 4.1(b)–(c). The two results both satisfy the abundance value constraints and spatial correlation of class 1, so the subpixel mapping based on a single remote sensing image often cannot determine the uniqueness of the subpixel mapping results. In order to improve the accuracy of subpixel mapping results, it is often necessary to add auxiliary data to provide more prior information. Common auxiliary data include light detection and ranging (LiDAR) data, elevation model, fusion image, panchromatic image, prior shape information, and fine-scale information [1].

In addition to the previous auxiliary data, MSIs are widely used as a kind of auxiliary data in subpixel mapping [2]. In the subpixel mapping model, the MSIs are usually obtained by the same sensor. Therefore, the MSIs are not only easy to obtain but also can effectively avoid the problems of geometric correction and reflectance inversion between images with different spatial resolutions. At the same time, due to the slight orbital translation and the rotation of the earth, the obtained MSIs usually have subpixel-level translation. Therefore, the MSIs are different from each other, and the coverage of the same observation area will not overlap. The reciprocity of the MSIs could be used as auxiliary data to provide more prior information to help subpixel mapping to obtain higher mapping accuracy results. In addition, a large number of previous studies have shown that the more suitable MSIs mean higher accuracy of the mapping results. The following examples illustrate how the use of MSIs can improve the subpixel mapping results.

This book uses the most commonly used method of simulating MSIs, namely, MSIs are assumed to be obtained by horizontal and vertical translation of the original image on the subpixel scale [3]. First, suppose that the number of MSIs is V. The multi-shift value on the subpixel scale between the first multiple subpixel shifted

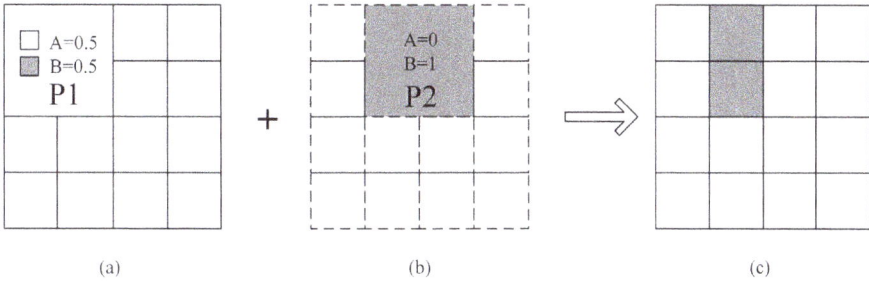

FIGURE 4.2 A simple example of subpixel mapping based on multi-shift images.

image and the vth (v = 1, 2, . . ., V) multiple subpixel shifted image is (x_v, y_v), which means to translate horizontally to the right x_v subpixels and vertically downward y_v subpixels, respectively. If the coordinate of a subpixel in the first reference image is (a_m, b_m), then the coordinate of the corresponding subpixel is ($a_m - x_v$, $b_m - y_v$) in the first MSI. As shown in Figure 4.2(a), the abundance value of class A in a mixed pixel P1 is 0.5, which is represented by white. The abundance value of class B is also 0.5, which is shown in gray. Assuming a scale S = 2, each mixed pixel will be divided into 2 × 2 = 4 subpixels because there will be 2 × 2 × 0.5 = 2 subpixels belonging to class B. In the absence of sufficient prior information, according to the principle of spatial correlation, the obtained subpixel mapping results are very uncertain; if another MSI as shown in Figure 4.2(b) is added at this time, this uncertainty can be reduced. The shifted value between pixel P2 and mixed P1 is x_v = 0.5, y_v = 0, so there are exactly two subpixels overlapping between pixel P1 and pixel P2. In pixel P2, the abundance value of class B is 1, and the abundance value of class A is 0. Therefore, because pixel P2 is a pure pixel belonging to class B, and the number of subpixels belonging to class B in mixed pixel P1 is equal to 2, it is possible to accurately determine subpixels belonging to class B in mixed pixel P1. The location of the subpixels is shown in Figure 4.2(c). Using the complementary information between the MSIs can effectively improve the subpixel mapping results.

4.2.2 Existing Subpixel Mapping Method Based on Multi-Shift Images

The basic principle of existing subpixel mapping based on MSIs is shown in Figure 4.3. First, each coarse MSI g_v^1 (v = 1, 2, . . . V, V is the number of MSIs) is obtained by the spectral unmixing method to obtain K low-resolution abundance images (K representing the number of classes contained in the image). Under a scale S, each pixel in multiple subpixel shifted image P_j^v (J = 1, 2, . . ., M, M is the number of pixels in the low-resolution abundance image) will be divided into S^2 subpixels. Then using a suitable subpixel sharpening method, K fine MSIs can be obtained; each MSI contains the abundance value $H_k^{spa}(p_j^v)$ (k = 1, 2, . . ., K) of each subpixel p_j^v (j = 1, 2, . . ., MS^2, MS^2 is the total number of subpixels) belonging to each class. The abundance value $H_k(p_j)$ of each subpixel p_j in the finer MSI can be obtained by formula (4.1). Finally, according to these more accurate subpixel abundance values

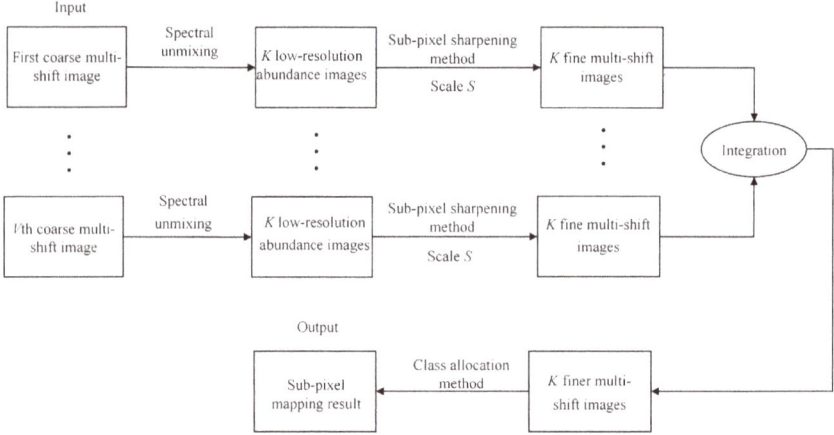

FIGURE 4.3   Existing subpixel mapping based on multi-shift images.

$H_k\left(p_j\right)$, the class labels are assigned to subpixels to obtain the final subpixel mapping result by class allocation method.

$$H_k(p_j)=\frac{1}{V}\sum_{v=1}^{V}H_k^{\mathrm{spa}}(p_j^v) \tag{4.1}$$

In particular, unlike the existing subpixel mapping methods based on MSIs, complex iterative operations and parameter selection are required. When the MSIs are used in interpolation-based subpixel mapping, not only are the mapping results better than the subpixel mapping result based on a single image, but also the physical meaning of the method at this time is simple, and the calculation speed is fast.

4.3   SUBPIXEL MAPPING METHOD BASED ON MULTI-SHIFT WITH SPATIAL-SPECTRAL INFORMATION

As shown in Figure 4.3, in the existing subpixel mapping method based on MSIs, all fine MSIs are generated through the path of spectral unmixing then subpixel sharpening. Since the path is obtained by a suitable spatial subpixel sharpening method, the fine MSIs obtained on the path will carry a large amount of spatial information, so this path can be called a spatial path. However, due to the potential uncertainties of the original coarse MSIs and the limitations of the existing spectral unmixing technology, it is difficult to extract all the supervision information of the original coarse MSIs, especially for the spectral information, through the single path. In order to solve this problem and make fine MSIs with more spatial-spectral information, a subpixel mapping method based on MSIs with spatial-spectral information (MSI-SS) is proposed.

4.3.1   MULTI-SHIFT IMAGE WITH MORE SPATIAL-SPECTRAL INFORMATION

Figure 4.4 shows the flowchart of the proposed MSI-SS method. First, a new path is added on the basis of the original spatial path $H_k^{\mathrm{spa}}(p_j^v)$, namely, named as the

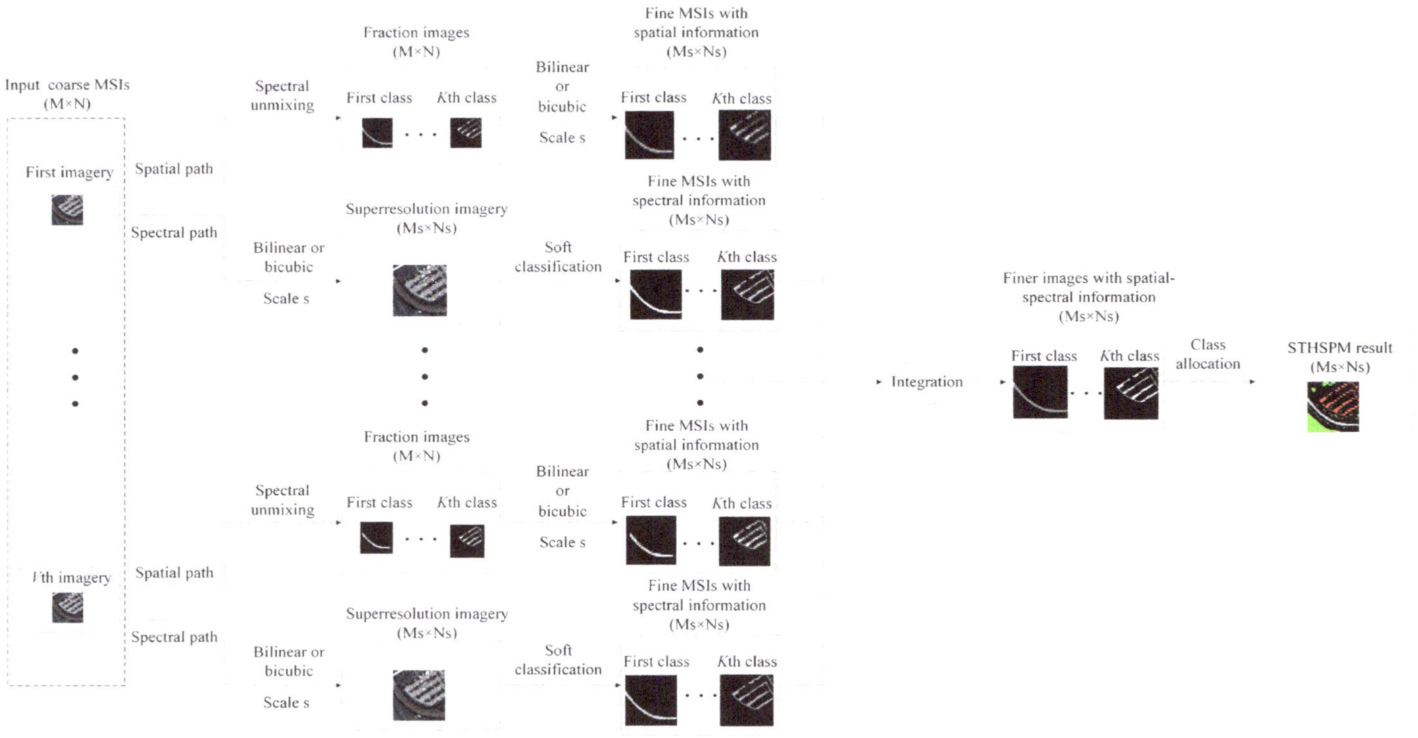

FIGURE 4.4   Multi-shift image with more spatial-spectral information.

spectral path. In the spectral path, each coarse MSI g_v^l is interpolated to obtain a high-resolution MSI g_v^h ($v = 1, 2, \ldots, V$, V is the number of MSIs). Due to the improvement of image resolution, the spectral information in the original image will be more easily extracted. The fine MSIs produced by spectral unmixing will contain MS^2 abundance values $H_k^{spe}(p_j^v)$ ($k = 1, 2, \ldots, K$; $j = 1, 2, \ldots, MS^2$), which $H_k^{spe}(p_j^v)$ have rich spectral information.

Next, in the finer MSIs, the abundance value $H_k(p_j)$ of the p_j subpixel belonging to the kth class can be obtained by integrating the previous two abundance values $H_k^{spa}(p_j^v)$ and $H_k^{spe}(p_j^v)$ with the use of formula (4.2) through appropriate weight parameter λ. In the formula (4.2), the weight parameter λ is used to weigh the influence of the spatial path $H_k^{spa}(p_j^v)$ and the spectral path $H_k^{spe}(p_j^v)$ on the final subpixel mapping result. Finally, the linear optimization technique class allocation method uses the abundance value $H_k(p_j)$ to assign class labels to subpixels to complete the final subpixel mapping:

$$H_k(p_j) = \frac{1}{2V}\left[\sum_{v=1}^{V} wH_k^{spe}(p_j^v) + \sum_{v=1}^{V}(1-w)H_k^{spa}(p_j^v)\right] \tag{4.2}$$

From the formula (4.2), it can be observed that two kinds of fine MSIs with different information and different abundance values are integrated with each other and complement each other to produce a finer MSI with more spatial-spectral information. This process is equivalent to adding more auxiliary information, and the accuracy of subpixel mapping will be further improved. In this chapter, the bilinear interpolation algorithm is selected in the MSI-SS method as the interpolation algorithm for two paths at the same time, marked as MSI-SS-BI, or bicubic interpolation is used as the interpolation algorithm for both paths at the same time, marked as MSI-SS-BIC.

The proposed MSI-SS method includes five steps in total:

Step 1. We utilize the phase correlation method to estimate appropriate subpixel shifts (x_v, y_v) ($v = 1, 2, \ldots, V$).

Step 2. In the spatial path, K abundance images $L_1^l, L_2^l, \ldots, L_K^l$ are derived by spectral unmixing each coarse MSI g_v^h. At the same time, in the spectral path, the high-resolution hyperspectral remote sensing imagery g_v^l is derived from each coarse MSI g_v^l by spectral interpolation.

Step 3. The fine MSIs that carry abundant spatial information are derived from the abundance images by spatial interpolation. The other fine MSIs with more spectral information are derived by unmixing the high-resolution imagery g_v^h.

Step 4. All the predicted values $H_k^{spa}(p_j^v)$ and $H_k^{spa}(p_j^v)$ from MSI-SS are integrated (see formula [4.2]). In this way, finer MSIs with more spatial-spectral information are generated.

Step 5. The soft attribute value $H_k(p_j)$ generated in step 3 is used for allocation of hard class labels, and subpixels for each class were allocated in turn.

4.3.2 Experiment Content and Result Analysis

Three sets of real hyperspectral remote sensing data are used to test the performance of the proposed method. To get closer to the real experimental environment, this chapter chooses the experimental method based on downsampling the original high-resolution image. The original high-resolution remote sensing image is downsampled to simulate the low-resolution coarse image through the mean filter. Because the linear unmixing model has simple physical meaning and convenient application, this chapter chooses this model as the spectral unmixing model. In addition, because the precise subpixel multi-shift in low-resolution images is difficult to estimate, it is difficult to evaluate the impact of image registration on real remote sensing data. To reduce the uncertainty of subpixel multi-shift estimation, the image registration method in literature [4] can be selected to obtain a suitable MSI. To simulate the real environment, in each experiment, all the original remote sensing images are firstly shifted and then downsampled to obtain coarse MSIs. Four sets of appropriate MSIs are selected in the three sets of experiments, and their subpixel multi-shifts are (0, 0), (0.6, 0), (0, 0.6), and (0.6, 0.6), respectively.

To obtain a fair comparison, only the influence of the newly added spectral path on the experimental results is considered. We select bilinear interpolation as the interpolation method of the spatial path and the spectral path at the same time or bicubic interpolation as the interpolation method of the spatial path and the spectral path at the same time. Six subpixel mapping methods based on interpolation are tested and compared. They are subpixel mapping based on bilinear interpolation (BI) [5], subpixel mapping based on bicubic interpolation (BIC) [5], subpixel mapping based on bilinear interpolation of multi-shift images (MSI-BI) [4], subpixel mapping based on bicubic interpolation of multiple multi-shift images (MSI-BIC)[4], the proposed subpixel mapping based on bilinear interpolation of multiple multi-shift images with spatial-spectral information (MSI-SS-BI), and the proposed subpixel mapping based on bicubic interpolation of multi-shift images with spatial-spectral information (MSI-SS-BIC). The mapping accuracy OA_i of each class and the overall accuracy PCC (%) and Kappa coefficient are used as the accuracy evaluation criteria for the three sets of experimental results.

4.3.2.1 Experiment 1

In experiment 1, the data set of the University of Pavia was selected as the research data for this experiment. The data were collected with the ROSIS spectrometer at the Faculty of Engineering of the University of Pavia. We chose an area with a size of 100×100 pixels as the test object, as shown in Figure 4.5(a). The area mainly includes four main classes: asphalt, grass, trees, and bricks. Figure 4.5(b) is used as a reference image, and the unmarked classes are replaced by black. In order to simulate the real situation, the original high-resolution remote sensing image is downsampled through an average filter of the scale $S = 4$ to produce a simulated coarse low-resolution image. The weight parameter $w = 0.6$ in MSI-SS-BI; the weight parameter $w = 0.4$ in MSI-SS-BIC.

The mapping results of the six subpixel mapping are shown in Figure 4.6. Through visual comparison, it can be found that in Figure 4.6(a)–(b), because there is no MSI

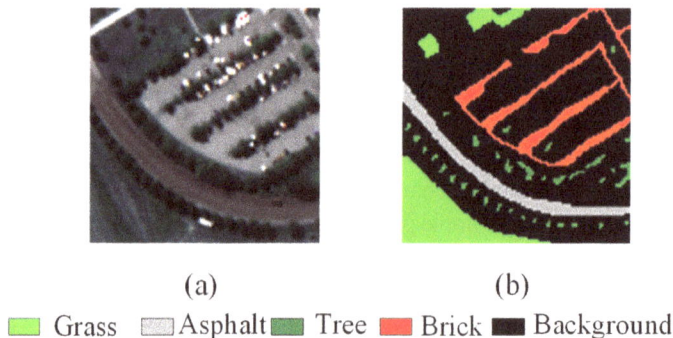

(a)   (b)

☐ Grass ▨ Asphalt ▬ Tree ▬ Brick ▬ Background

FIGURE 4.5 Pavia University data set: (a) false color composite image (RGB band: 19, 30, 44) and (b) reference image.

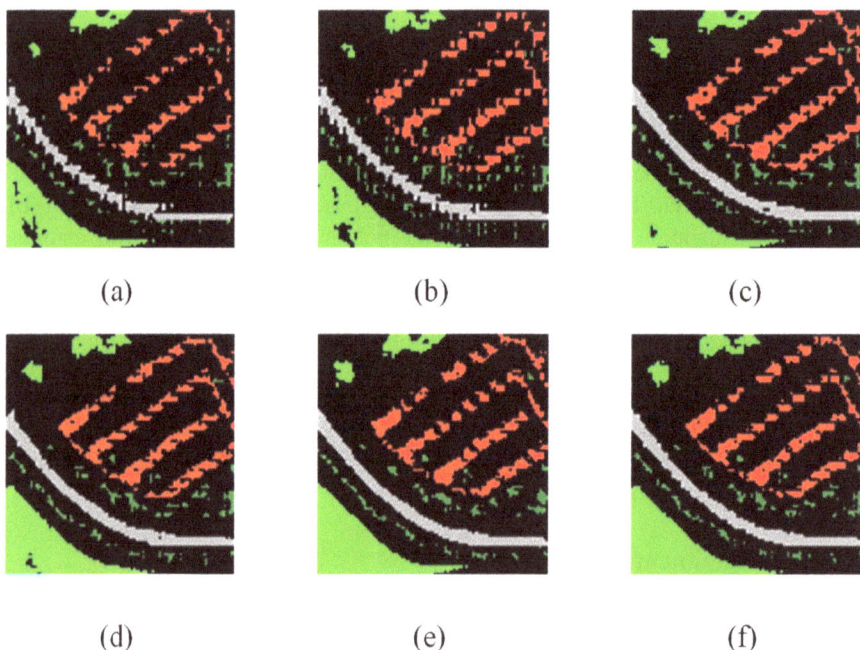

(a)   (b)   (c)

(d)   (e)   (f)

FIGURE 4.6 Subpixel mapping results in experiment 1 ($S = 4$): (a) BI, (b) BIC, (c) MSI-BI, (d) MSI-BIC, (e) MSI-SS-BI, and (f) MSI-SS-BIC.

as auxiliary information, the final subpixel mapping result is not ideal. For example, there are many burrs on the edges of asphalt and brick, which look very coarse. These phenomena are obviously alleviated in Figure 4.6(c)–(f). This is because the multiple MSIs complement each other as auxiliary information, which improves the accuracy of the mapping results. In particular, the MSIs in the proposed method have

TABLE 4.1
Mapping Accuracy of the Six Methods in Experiment 1 ($S = 4$)

Land Cover Class	BI	BIC	MSI-BI	MSI-BIC	MSI-SS-BI	MSI-SS-BIC
Grass (%)	90.04	90.86	92.38	92.92	93.47	94.07
Asphalt (%)	78.84	87.80	95.56	97.47	98.53	99.23
Tree (%)	30.02	33.94	38.27	41.88	47.07	51.16
Brick (%)	60.36	62.63	69.84	70.79	73.79	75.06
PCC (%)	73.66	75.46	80.64	81.89	83.74	84.99

more information than the existing MSIs, and the final mapping result is more similar to the reference image. In addition, because BIC has a better subpixel sharpening effect than BI, the result produced by MSI-SS-BIC is visually closest to the reference image.

In addition to intuitive visual comparison, the mapping accuracy (%) and overall mapping accuracy PCC (%) of each class in the six subpixel mapping results in experiment 1 were also tested, as shown in Table 4.1. Comparing the accuracy evaluation indicators in Table 4.1, it can be found that the accuracy of the subpixel mapping results has been significantly improved under the help of the MSIs. Compared with the existing subpixel mapping based on MSIs, the subpixel mapping based on MSIs with spatial-spectral information proposed in this chapter can obtain higher accuracy results. For types of grass, asphalt, tree, and brick, the mapping accuracy of the class has been improved by 1.2%, 1.8%, 9.2%, and 4.3%, respectively. In addition, MSI-SS-BIC can get the highest PCC value of 84.99%.

4.3.2.2 Experiment 2

Experiment 2 tests the performance of the proposed method in an environment with a larger number of classes. The data of experiment 2 still chose the Pavia city center data set obtained by ROSIS spectrometer. Although this data set and the data set of the University of Pavia in experiment 1 are collected by ROSIS spectrometer, the areas captured by the two are different. In addition, the data set of experiment 2 contains more classes and more complex distributions. Figure 4.7(a) shows the selected research area for this experiment with a size of 360×360 pixels. This area contains six classes, namely, background, water, street, trees, grass, and roof. Figure 4.7(b) is the reference image obtained by the classification method. Since the scale S has a great influence on the final subpixel mapping result, it is necessary to consider the performance of the proposed method at multiple scales. This experiment chooses three scales, namely, $S = 2$, $S = 4$, and $S = 8$. The weight parameter w is set to 0.7 in MSI-SS-BI and 0.6 in MSI-SS-BIC.

From the six subpixel mapping results shown in Figure 4.8, it can be seen that Figure 4.8(a)–(b) do not have multiple MSIs as auxiliary information, so there are many broken hole patches in the results; there is a lot of salt and pepper noise in the water. However, in Figure 4.8(c)–(f), these errors in mapping results have been significantly improved. Since the MSIs in the proposed method have more spatial-spectral

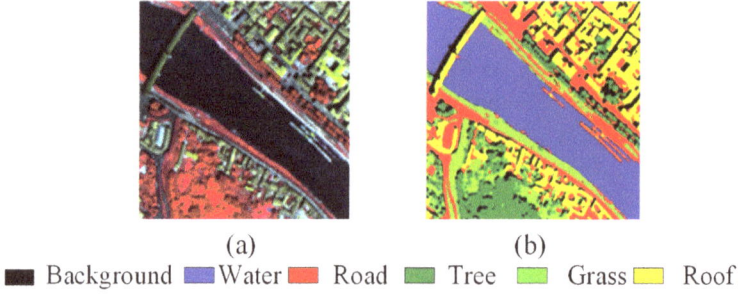

(a)   (b)

■ Background ■ Water ■ Road ■ Tree ■ Grass □ Roof

FIGURE 4.7   Pavia city center data set: (a) false color composite image (RGB band: 102, 56, 31) and (b) reference image.

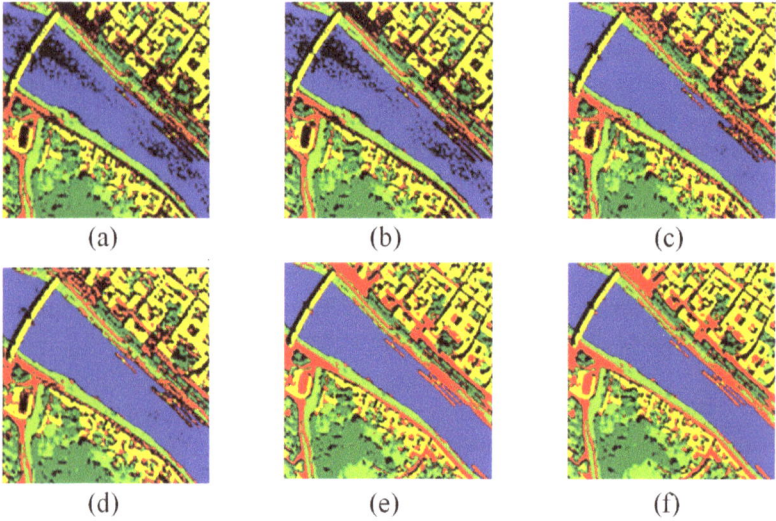

(a)   (b)   (c)

(d)   (e)   (f)

FIGURE 4.8   Subpixel mapping results in experiment 2 ($S = 4$): (a) BI, (b) BIC, (c) MSI-BI, (d) MSI-BIC, (e) MSI-SS-BI, and (f) MSI-SS-BIC.

information, Figure 4.8(e)–(f) are closer to the reference image than Figure 4.8(c)–(d). Figure 4.9(a)–(b) show the PCC (%) and Kappa of the six subpixel mapping methods at three scales. It can be observed from the results that as the scale S increases, the PCC (%) and Kappa coefficients of the six subpixel mapping methods are all decreasing. But similar to the conclusion of experiment 1, the proposed method can produce the highest PCC (%) value and Kappa coefficient at any scale.

4.3.2.3   Experiment 3

In experiment 3, the imagery is performed on an urban site of the airborne HYDICE from the mall in Washington, DC. We selected 240 × 300 pixels as the research area

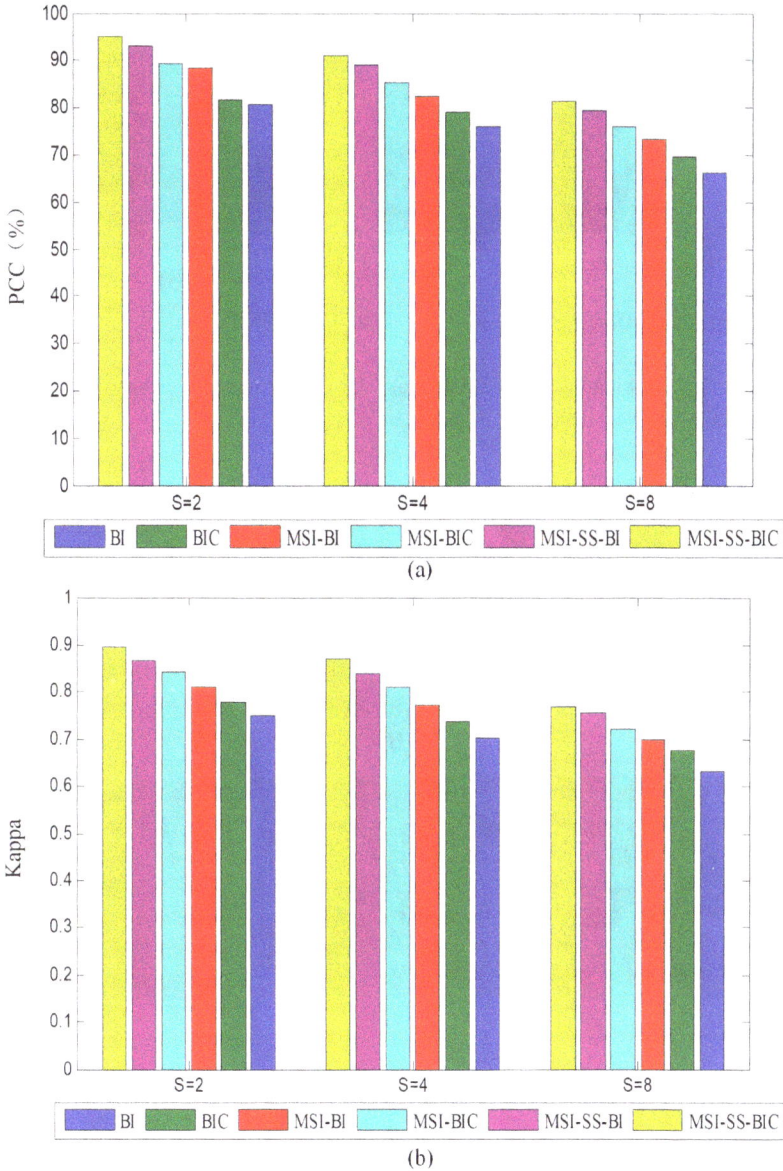

FIGURE 4.9   (a), (b) The relationship between PCC (%) and Kappa of the six methods and scale *S*.

of this experiment, as shown in Figure 4.10(a). The distribution of this area is more complicated than the previous two experiments, which mainly include seven classes: background, water, road, trees, grass, roof, and trail. The reference image shown in Figure 4.10(b) is obtained through the classification algorithm. The performance of the proposed method is also considered under multiple scales; this experiment

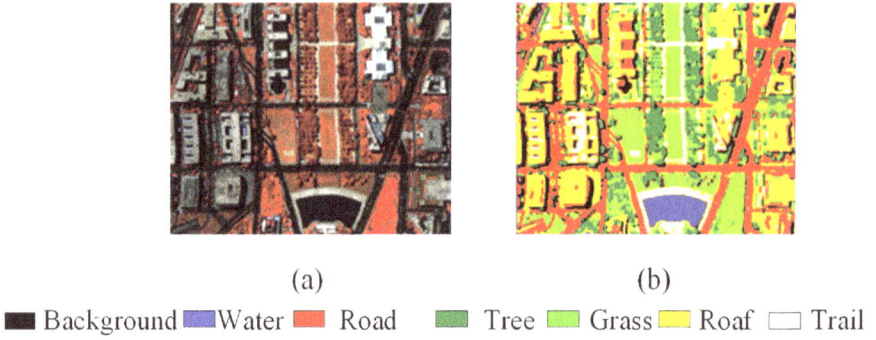

(a)   (b)

■ Background ■ Water ■ Road   ■ Tree □ Grass □ Roof □ Trail

FIGURE 4.10   Washington, DC, data set: (a) false color composite image (RGB band: 65, 52, 36) and (b) reference image.

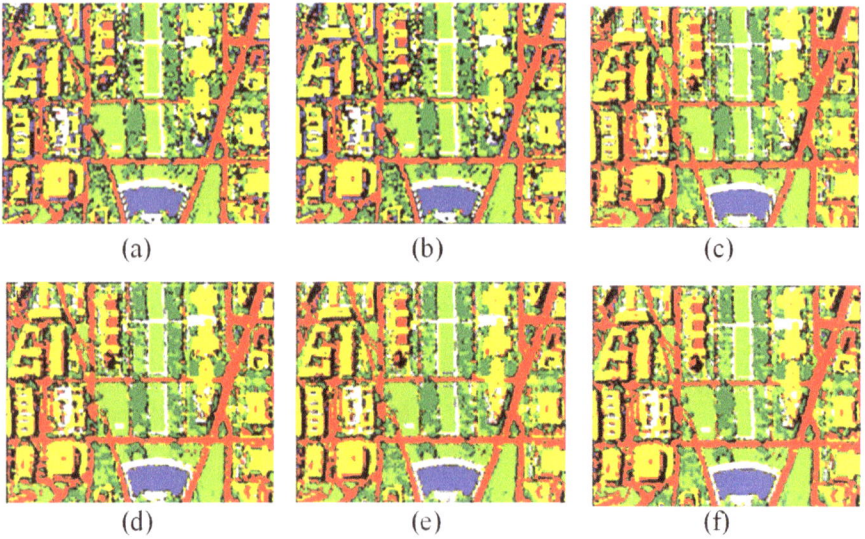

(a)   (b)   (c)

(d)   (e)   (f)

FIGURE 4.11   Subpixel mapping results in experiment 3 ($S = 3$): (a) BI, (b) BIC, (c) MSI-BI, (d) MSI-BIC, (e) MSI-SS-BI, and (f) MSI-SS-BIC.

chooses three scales $S = 3$, $S = 4$, and $S = 6$ to simulate the low-resolution images. The weight parameter is set to 0.5 in both MSI-SS-BI and MSI-SS-BIC.

Similar to the conclusions obtained in experiments 1 and 2, when there are no MSIs as auxiliary information, there is a lot of speckle effect in Figure 4.11(a)–(b). But with the help of MSIs, the mapping effect of Figure 4.11(c)–(f) is significantly better than in Figure 4.11(a)–(b). Since more space-spectrum information is utilized, MSI-SS-BI or MSI-SS-BIC is closer to the original reference image than MSI-BI or MSI-BIC. Figure 4.12(a)–(b) shows the PCC (%) and Kappa coefficients of the six methods at three scales, which are consistent with the aforementioned experimental

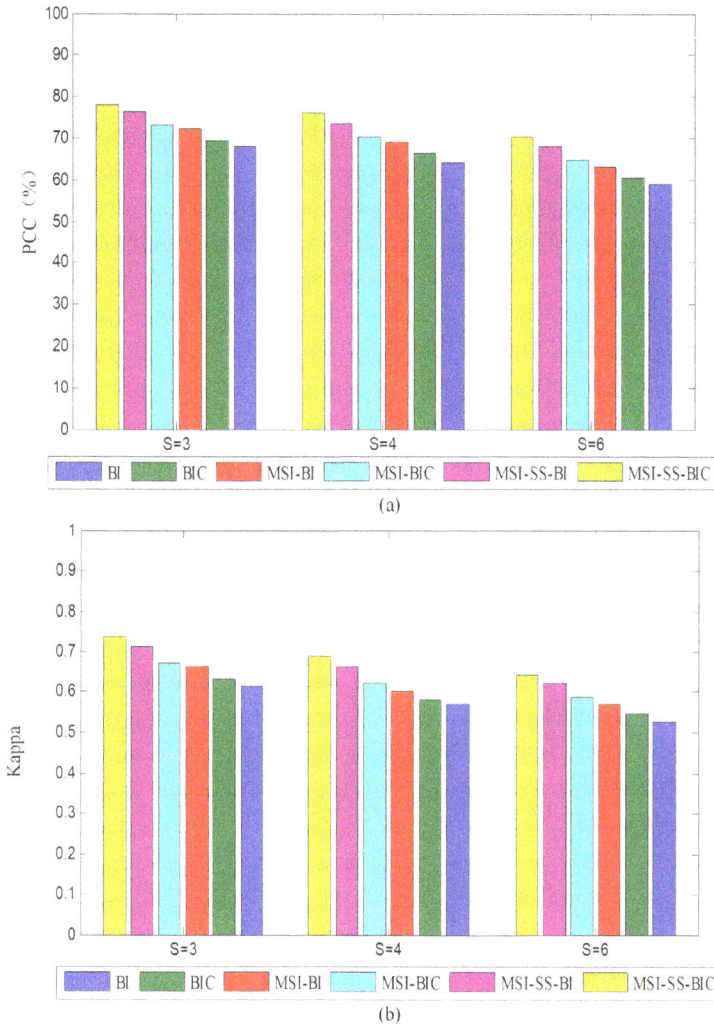

FIGURE 4.12   (a), (b) The relationship between PCC (%) and Kappa of the six methods and scale *S*.

results. The method proposed in this chapter is better than the existing MSI based method as the subpixel mapping method can obtain higher mapping accuracy.

4.3.2.4   Discussion

The method proposed in this chapter is to obtain the ideal subpixel mapping result by testing different weight parameters *w*. To obtain the appropriate weight parameter *w*, ten sets of values of the weight parameters are tested in the MSI-SS-BI method and the MSI-SS-BIC method. These values are between [0, 0.9], with an increment of 0.1 each time.

Figure 4.13(a)–(b) shows the relationship between the overall mapping accuracy PCC (%) produced by the MSI-SS-BIC method in experiment 2 ($S = 4$) and experiment 3 ($S = 4$) and different parameter values w. It can be seen from the result that when $w = 0$, the spectral path did not play any role, namely, the existing MSI-BIC method can be obtained. When the value w increases subsequently, the value of PCC (%) in Figure 4.13(a)–(b) also increases. This is because the newly added spectral path supplements the original spatial path with more spectral information of the original image, and the final subpixel mapping result obtained is improved. When w is 0.6 in experiment 2

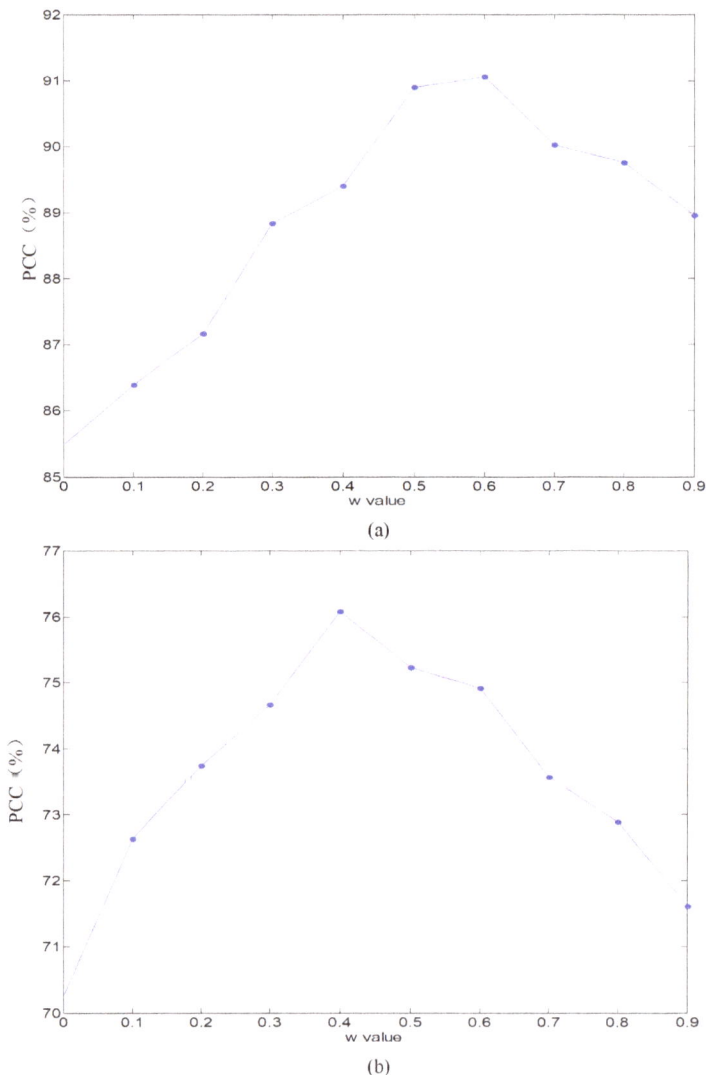

(a)

(b)

FIGURE 4.13   (a), (b) The relationship λ between the PCC (%) of MSI-SS-BIC in experiments 2 and 3.

and w is 0.5 in experiment 3, PCC (%) can get the highest value. However, when the w continues increasing, the spatial path will gradually lose its effect, the MSI will lose some of the spatial information of the original image, and the accuracy of the subpixel mapping result will be reduced again. Therefore, the choice of parameters has a very important impact on the performance of the proposed method.

This section analyzes the influence of the number of MSIs with spatial-spectral information on the final subpixel mapping results. We use four sets of suitable auxiliary MSIs to test experiment 2 ($S = 8$) and experiment 3 ($S = 3$) to obtain the corresponding overall mapping accuracy index PCC (%). Four sets of suitable MSIs are estimated by the method in [4], and their corresponding subpixel multi-shifts are (0.3, 0), (0, 0.3), (0.5, 0), and (0, 0.5). As shown in Figure 4.14(a)–(b), it can be

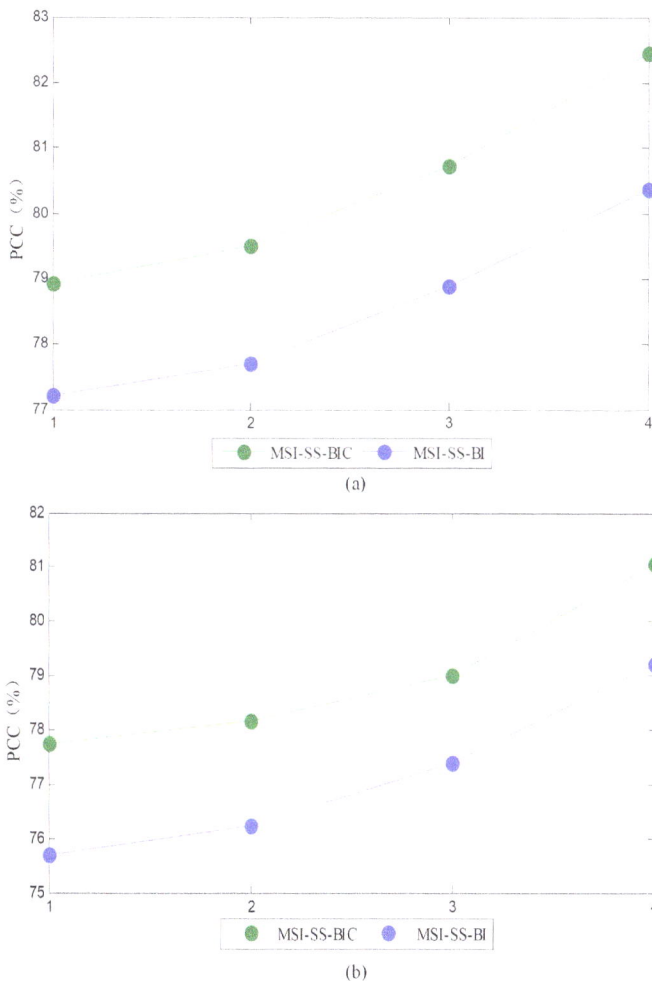

FIGURE 4.14   The influence of the number of multiple MSIs on MSI-SS: (a) experiment 2 and (b) experiment 3.

clearly seen that when the number of suitable MSIs increases from one to four in the above order, the accuracy of the subpixel mapping results is significantly improved. Therefore, using a larger number of MSIs in the proposed method, the subpixel mapping results obtained will be more accurate.

4.4 SUBPIXEL MAPPING BASED ON THE SPATIAL ATTRACTION MODEL WITH MULTI-SCALE SUBPIXEL SHIFTED IMAGES

Some existing subpixel mapping methods based on multiple MSIs [6–7] not only have a large computational burden, but also most of the methods require the assistance of prior structure information. In addition, in other subpixel mapping methods based on MSIs [4], although the rapid subpixel sharpening methods such as BI, BIC, and subpixel-pixel spatial attraction models (SPSAM) are used, the calculation speed of the method is improved, but because these methods only use the information of a single scale, unclear textures or blurry artifacts at the edges of the image will be produced in the image.

To solve these problems, this chapter uses the spatial attraction model to obtain MSIs with multiple scale information and uses it as auxiliary data to improve the subpixel mapping results. In the proposed method, the dominant smoothness item with the fine-scale information is obtained through the subpixel-subpixel spatial attraction model, and the SPSAM is used to generate an auxiliary penalty item with the coarse-scale information. Two items with multiple scale information for each class are integrated through appropriate parameters to obtain a finer subpixel sharpening result. Finally, according to the subpixel abundance value from the finer subpixel sharpening result, the class labels are assigned to subpixels to achieve the final subpixel mapping. The proposed method inherits the advantages of the spatial attraction model, namely, the calculation speed is fast and does not require any prior structure information, and the multi-shift images have multiple scale information, which makes the final subpixel mapping result more accurate. Compared with the existing subpixel mapping methods based on single-scale MSIs, the effectiveness of the proposed method is proved, and the subpixel mapping result with higher accuracy can be obtained. The following introduces the two spatial attraction models used in the proposed method, as well as the specific implementation process of the proposed method and analysis of experimental results.

4.4.1 SUBPIXEL-PIXEL SPATIAL ATTRACTION MODEL

Mertens first introduced the spatial attraction model [8] into the subpixel mapping method and obtained relatively ideal results. In the spatial attraction model, the attraction of each subpixel in the mixed pixel and its neighboring pixels is used to express the spatial correlation, and the class labels are assigned to subpixels according to the magnitude of the attraction value.

It is assumed that each mixed pixel will be decomposed into $S \times S$ subpixels with a scale S; a coarse original remote sensing image is unmixed to obtain K low-resolution abundance images L_k ($k = 1, 2, \ldots, K$, K represents the number of feature categories); $L_k (P_J)$ represents the abundance values of mixed pixels P_J ($J = 1, 2, \ldots,$ M, M is the total number of mixed pixels) belonging to the kth class; $H_k (p_j)$ is the abundance value of subpixels p_j ($j = 1, 2, \ldots, MS^2$, MS^2 is the total number of sub-pixels) belonging to the kth class. O_{kj} is defined as

$$O_{kj} = \begin{cases} 1, & \text{if the subpixel } p_j \text{ belongs to } k\text{th class} \\ 0, & \text{otherwise} \end{cases} \tag{4.3}$$

Subpixel mapping is then transformed into a problem of maximizing spatial correlation, and its mathematical model can be expressed by formula (4.4):

$$\max \quad SPA = \sum_{k=1}^{K} \sum_{j=1}^{MS^2} O_{kj} \cdot SD_{kj}$$

$$s.t. \quad \sum_{k=1}^{K} O_{kj} = 1, \quad j = 1, 2, \ldots, MS^2 \tag{4.4}$$

$$\sum_{j=1}^{MS^2} O_{kj} = K_k, \quad k = 1, 2, \ldots, K$$

where SPA represents the numerical value of spatial correlation; SD_{kj} is a measure of the spatial correlation of the p_j subpixel belonging to the kth class, which can be expressed by the following formula:

$$SD_{kj} = \sum_{C=1}^{M_n} w_j \cdot L_k(P_c) \tag{4.5}$$

where w_j represents the spatial correlation parameter; $L_k (P_C)$ represents the abundance value of the Cth neighborhood pixel P_C belonging to the kth class; and M_n is the total number of pixels in the neighborhood. This chapter chooses the eight neighborhoods as the neighborhood model of the proposed method.

According to formulas (4.4) and (4.5), the SPSAM can be obtained as shown in formula (4.6):

$$Attraction_{\text{subpixel/pixel}} = \max \sum_{k=1}^{K} \sum_{j=1}^{MS^2} \sum_{C=1}^{M_n} O_{jk} \cdot w_j \cdot L_k\left(P_c\right) \tag{4.6}$$

where w_j represents the spatial correlation between the central subpixel p_j and the adjacent pixels P_C:

$$w_j = \exp\left(-d\left(p_j, P_C\right)^2 / r_1\right) \tag{4.7}$$

where r_1 is the non-linear parameter of the exponential model; $d (p_j, P_C)$ is the Euclidean distance from the central subpixel p_j to the adjacent pixel P_C, as shown in Figure 4.15(a).

As shown in Figure 4.15(a), it can be found that the subpixel-pixel attraction model only considers the spatial correlation between the subpixel and the adjacent mixed

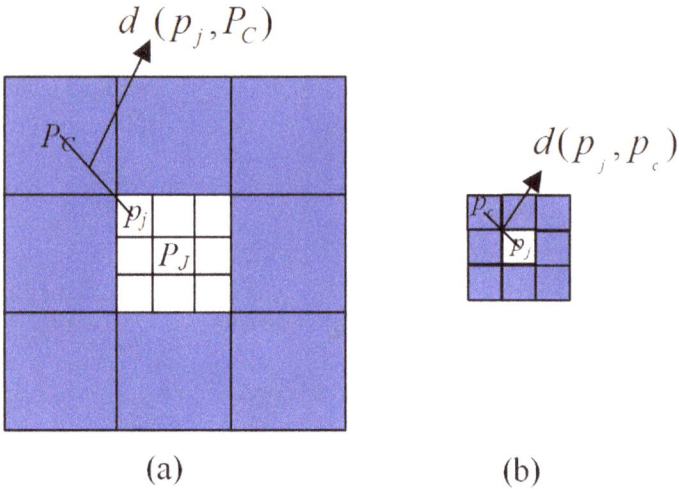

FIGURE 4.15   (a) A pixel P_J and its neighboring eight coarse-scale pixels p_j. (b) A subpixel and its neighboring eight fine-scale subpixels.

pixels. However, compared with the area of the subpixel, the adjacent area of the mixed pixel is much larger. Obviously, this kind of spatial scale is relatively coarse. It is difficult to gather all the subpixels belonging to the same class to form uniformly distributed patches, which leads to a lot of salt and pepper noise in the final subpixel mapping result.

4.4.2   SUBPIXEL-SUBPIXEL SPATIAL ATTRACTION MODEL

In response to this shortcoming, Wang et al proposed a subpixel-subpixel spatial attraction model 1 [9], which considers the spatial correlation between the central subpixel p_j and the neighboring subpixels p_c to make the spatial scale finer, thereby improving the final subpixel mapping result. At this time, the following formula is re-expressed in the fine-scale spatial attraction model SD_{kj}:

$$SD_{kj} = \sum_{C=1}^{M_n} w_j \cdot L_k(P_C)$$

$$= \sum_{C=1}^{M_n} w_j \cdot \frac{L_k(P_C) \cdot S^2}{S^2} \tag{4.8}$$

$$= \frac{1}{S^2} \sum_{c=1}^{M_n S^2} w_j \cdot x_{jc}$$

Assume that the principle x_{jc} of spatial correlation should satisfy formula (4.9). Then the subpixel-subpixel spatial attraction model can be obtained as shown in formula (4.10). At this time, the correlation weight parameter w_j between the center subpixel p_j and the neighboring subpixel p_c can be expressed by formula (4.11):

$$x_{jc} = \begin{cases} 1, \text{if the subpixel } p_j \text{ and subpixel } p_c \text{ belongs to the same class} \\ 0, \text{otherwise} \end{cases} \tag{4.9}$$

$$Attraction_{\text{subpixel/subpixel}} = \max \sum_{j=1}^{MS^2} \sum_{c=1}^{M_nS^2} w_j x_{jc} \tag{4.10}$$

$$w_j = \exp(-d(p_j, p_c)^2 / r_2) \tag{4.11}$$

where r_2 is the non-linear parameter of the exponential model and $d(p_j, p_c)$ represents the Euclidean distance from the central subpixel p_j to the neighboring subpixel p_c. Figure 4.15(b) shows the fine-scale spatial correlation intention.

Although the SPSAM has made the spatial scale more refined, which is conducive to the formation of clusters of neighboring subpixels belonging to the same class, if only the correlation between the subpixels is used to determine the classification of each class edge, this may result in excessively smooth edge.

4.4.3 Spatial Attraction Model With Multi-Scale Subpixel Shifted Image

In this chapter, two spatial attraction models are applied to MSIs to generate multi-scale subpixel shifted images (MSSIs) with multiple scale information to replace single-scale subpixel shifted images (SSIs) with the single-scale information to improve the accuracy of subpixel mapping results.

In the vth (v = 1, 2, . . ., V, V is the number of MSIs) MSI, the subpixel p_j^v (j = 1, 2, . . ., MS^2, MS^2 is the total number of subpixels in the vth MSI) is utilized by the subpixel-subpixel spatial attraction model to generate the abundance value $H_k^{\text{fs}}(p_j^v)$ with fine-scale information as the dominant smooth term, as shown in formula (4.12). At the same time, the abundance value $H_k^{\text{cs}}(p_j^v)$ with coarse-scale information is generated by the SPSAM, which is an auxiliary penalty item, as shown in formula (4.13):

$$H_k^{\text{fs}}(p_j^v) = \max \sum_{j=1}^{MS^2} \sum_{c=1}^{M_nS^2} \exp(-d(p_j^v, p_c^v)^2 / r_1) \cdot x_{jc} \tag{4.12}$$

$$H_k^{\text{cs}}(p_j^v) = \max \sum_{k=1}^{K} \sum_{j=1}^{MS^2} \sum_{C=1}^{M_n} o_{jk} \cdot \exp(-d(p_j^v, P_c^v)^2 / r_2) \cdot L_k(P_c^v) \tag{4.13}$$

Through these two formulas, it can be found that the difference between the smooth term $H_k^{\text{fs}}(p_j^v)$ with fine-scale information and the penalty term $H_k^{\text{cs}}(p_j^v)$ with coarse-scale information is obvious. The smooth term $H_k^{\text{fs}}(p_j^v)$ fully considers the homogeneity between the central subpixel p_j^v and the adjacent subpixels p_c^v in the vth MSI, so that the subpixels belonging to the same class are better gathered together to form uniformly distributed patches. However, only the homogeneity between pixels is considered, which may make the edges of some classes appear excessively smooth, or the small local subpixels are homogenized by adjacent subpixels belonging to most classes. The penalty term $H_k^{cs}(p_j^v)$ is just the opposite. $H_k^{cs}(p_j^v)$ fully considers the heterogeneity between the center subpixel p_j^v and the adjacent mixed pixels P_c^v in the vth MSI, which can effectively alleviate the problems caused by the smooth term $H_k^{fs}(p_j^v)$. However, because this spatial scale is relatively coarse, the penalty term $H_k^{cs}(p_j^v)$ makes the final result have a lot of salt and pepper noise. Therefore, for each MSI, a more reasonable model is that the smooth term $H_k^{fs}(p_j^v)$ is selected as the dominant term, and the penalty term $H_k^{cs}(p_j^v)$ is selected as the auxiliary term. The smooth term $H_k^{fs}(p_j^v)$ and the penalty term $H_k^{cs}(p_j^v)$ are then integrated through appropriate parameters w to obtain the target term $H_k(p_j^v)$ with multiple scale information at the same time, as shown in (4.14):

$$H_k(p_j^v) = H_k^{fs}(p_j^v) + w\, H_k^{cs}(p_j^v) \qquad (4.14)$$

Figure 4.16 shows the specific operational steps of the subpixel mapping based on the spatial attraction model with multi-scale subpixel shifted images (SAM-MSSI) proposed in this chapter.

Step 1. We utilized phase correlation to estimate subpixel shifts (x_v, y_v) ($v = 1, 2, \ldots, V$).

Step 2. V coarse images can be derived by spectral unmixing of V SSI. The results for each coarse image are K-class abundance images.

Step 3. The target term $H_k(p_j^v)$ in the multi-scale MSIs will be obtained by formula (4.14) by using the smooth term $H_k^{fs}(p_j^v)$ and coarse penalty term $H_k^{cs}(p_j^v)$.

Step 4. For each land cover class, V multi-scale MSIs will be integrated by formula (4.15) to obtain a more refined subpixel sharpening result with more spatial scale information:

$$H_k(p_j) = \frac{1}{V} \sum_{v=1}^{V} H_k(p_j^v) \qquad (4.15)$$

Step 5. Under the fixed limit of the number of subpixels in each class, according to the subpixel sharpening result produced in step 4, we use the linear optimization technique class allocation method to assign class labels to subpixels to achieve the final subpixel mapping.

4.4.4 EXPERIMENT CONTENT AND RESULT ANALYSIS

To verify the performance of the proposed method, two synthetic remote sensing images and a hyperspectral remote sensing image are selected as experimental data. The experimental method based on downsampling the reference image is used for the two synthetic remote sensing images. This simulation experimental method can effectively avoid the influence of spectral unmixing errors on the experimental results and only consider the performance of the subpixel mapping method. In the method based on downsampling the reference image, the synthesized fine remote sensing image is degraded by $s \times s$ mean filter to simulate the low-resolution abundance images after spectral unmixing. The experimental results produced by this

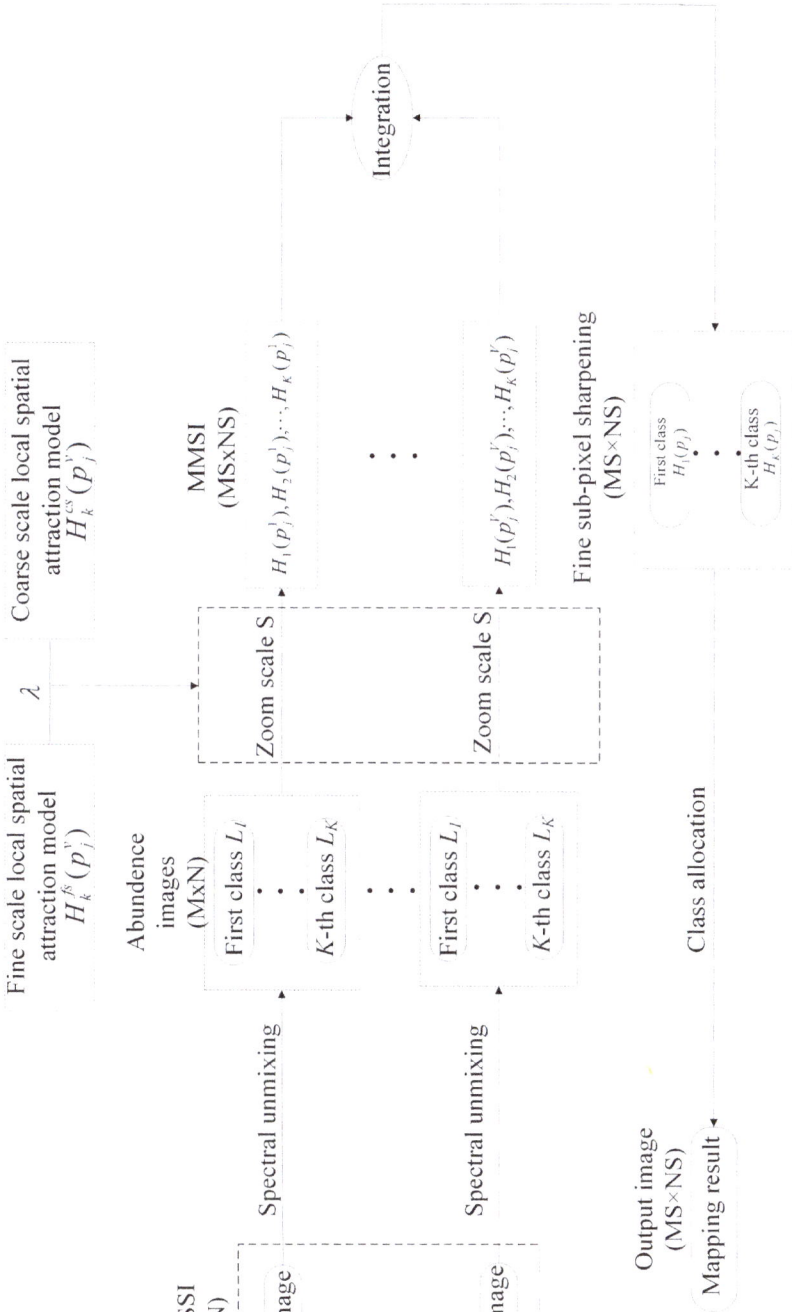

FIGURE 4.16 SAM-MSSI.

experimental method often need to eliminate the influence of pure pixels on the results. Therefore, the two evaluation accuracy indicators PCC (%) and Kappa will become two new evaluations by removing pure pixels, namely, PCC' (%) and Kappa'. For the real hyperspectral remote sensing image, the experimental method based on downsampling the original high-resolution image is used to simulate the real experiment. In the real hyperspectral remote sensing image experiment, the spectral unmixing method based on support vector machine is used to generate the low-resolution abundance images of each class, and the two evaluation indicators of PCC (%) and Kappa are used to evaluate the final subpixel mapping results. In addition, in order to reduce the uncertainty of subpixel multi-shift, this chapter adopts the commonly used method in the existing subpixel multi-shift estimation, namely, the original high-resolution remote sensing image is first shifted and then downsampled [4] to obtain the simulated low-resolution MSIs. In the three sets of experiments, the three suitable subpixel multi-shifts are (0, 0), (1, 0), and (0, 1).

Five subpixel mapping methods were used for testing and comparison. They are BI based on a single image [5], spatial attraction model based on single image (SPSAM) [8], bilinear interpolation based on multiple multi-shift images (BI-SSI) [4], spatial attraction model based on multi-shift images (SPSAM-SSI) [3], and the proposed spatial attraction model based on multi-scale multi-shift images (SAM-MSSI).

4.4.4.1 Experiment 1

In experiment 1, the synthetic multispectral image obtained in September 2008 was used as the experimental data, as shown in Figure 4.17(a). The study area is located in Nanjing

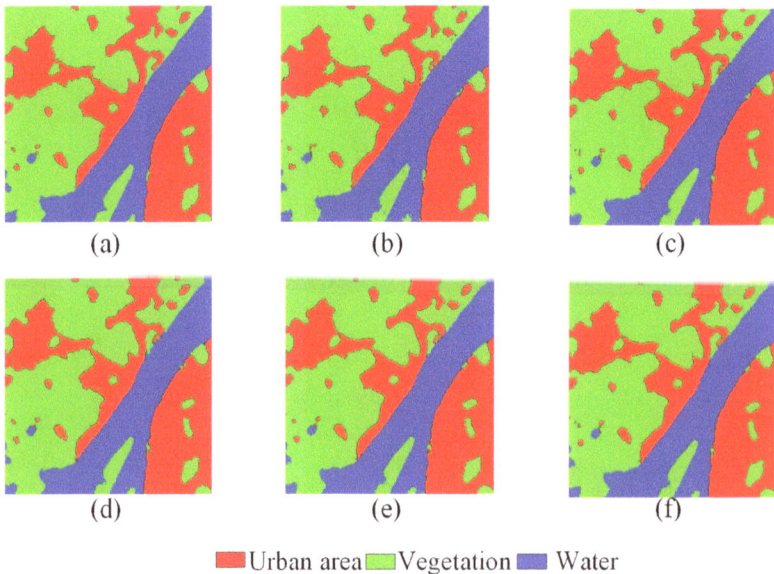

FIGURE 4.17   Subpixel mapping results in experiment 1 ($S = 4$): (a) synthetic multispectral image, (b) BI, (c) SPSAM, (d) BI-SSI, (e) SPSAM-SSI, and (f) SAM-MSSI.

City, Jiangsu Province, China, with a size of 200 × 200 pixels, including three classes: vegetation, water, and urban area. In this experiment, the parameter w is set to 0.6.

To simulate the abundance images produced by spectral unmixing, the scale S is set to 4. The corresponding five subpixel mapping results are shown in Figure 4.17(b)–(f). Through intuitive visual comparison, the subpixel mapping results based on multiple MSIs shown in Figure 4.17(d)–(f) are significantly better than the subpixel results based on a single image shown in Figure 4.17(b)–(c). For example, in Figure 4.17(b)–(c), there are many burrs at the boundary between the urban area and the water, and salt and pepper noise also appears in some areas. These erroneous mapping phenomena are alleviated with the help of MSIs. In particular, the proposed SAM-MSSI can make the subpixels belonging to the same class cluster together under the action of the fine-scale term, and at the same time avoid excessive smoothing of the boundary under the action of the coarse-scale term. The subpixel mapping result obtained by SAM-MSSI is closer to the reference image than the other subpixel mapping results based on multiple MSIs.

In addition to the intuitive visual comparison, Table 4.2 lists the indicators of the mapping accuracy of each class OA_i and the overall mapping accuracy PCC' (%) after removing the pure pixels for the five subpixel mapping results. Compared with the other four subpixel mapping methods, in the subpixel mapping results obtained by the proposed SAM-MSSI method, the mapping accuracy of the three classes is the highest at 94.66%, 96.19%, and 94.57%, respectively. In addition, when using multi-shift images, the PCC'(%) based on the bilinear interpolation algorithm increased from 91.66% to 93.03%, and the PCC'(%) based on the spatial attraction model algorithm increased from 92.63% to 93.78%. Since the multi-shift images obtained in the SAM-MSSI method have multi-scale information, the proposed method can obtain the highest PCC'(%) value of 94.89%.

In addition, the five subpixel mapping methods are also tested with two other scales $S = 2$ and $S = 8$ scales. The results of PCC' (%) and Kappa' are shown in Figure 4.18. Similar to the conclusion in Table 4.2, the proposed method can still obtain the highest PCC' (%) value and Kappa' value with the help of the two scales. It is proved that under different scales S, the proposed method can still get better results.

4.4.4.2 Experiment 2

A reference image with a size of 300 × 300 pixels in the central Pavia data set is selected as the experimental data. As shown in Figure 4.19(a), the data include

---

TABLE 4.2
Mapping Accuracy of the Five Methods in Experiment 1 ($S = 4$)

Land Cover Class	BI	SPSAM	BI-SSI	SPSAM-SSI	SAM-MSSI
Urban area (%)	91.20	92.45	93.21	93.87	94.66
Water (%)	92.53	94.65	94.10	95.20	96.19
Vegetation (%)	90.60	92.03	92.47	93.16	94.57
PCC' (%)	91.16	92.63	93.03	93.78	94.89

(a)

(b)

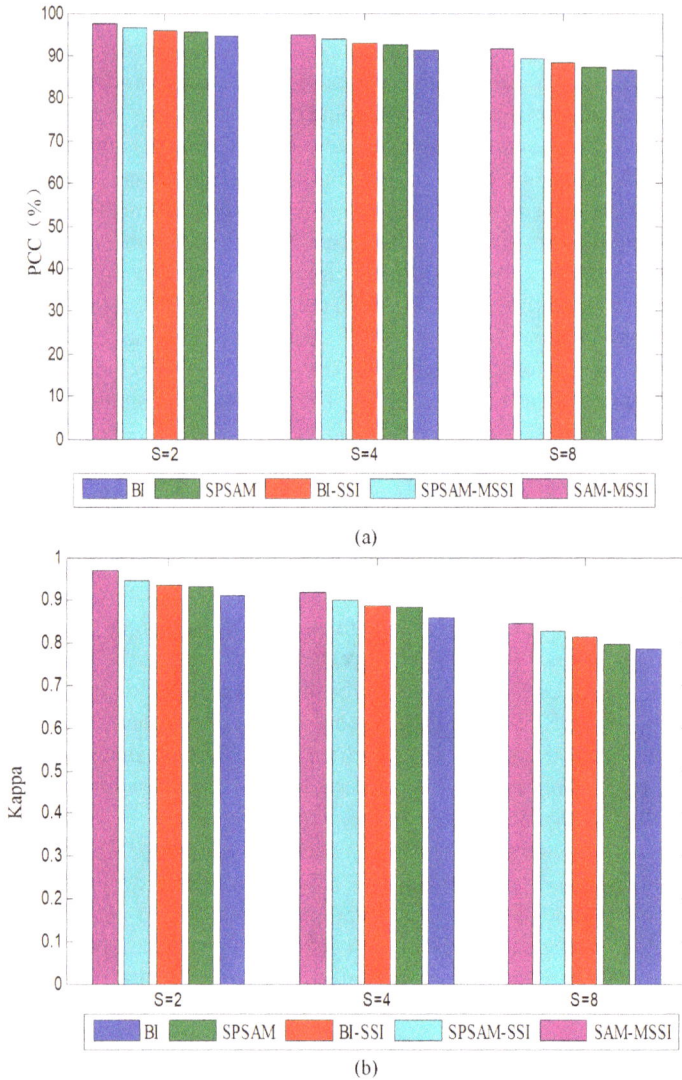

FIGURE 4.18   (a), (b) The relationship between PCC' (%) and Kappa' of the five methods and scale S.

six land cover classes: background, water, road, tree, grass, and roof. The data are downsampled by mean filter with a scale $S = 6$ to generate the simulated low-resolution abundance images. The subpixel mapping method uses these low-resolution abundance images to generate a high-resolution image with the spatial distribution information of land cover classes. In experiment 2, the parameter w was set to 0.7.

Figure 4.19(b)–(f) show the results of five subpixel mapping methods. As shown in Figure 4.19(b)–(c), due to the conflicting features of each class in the

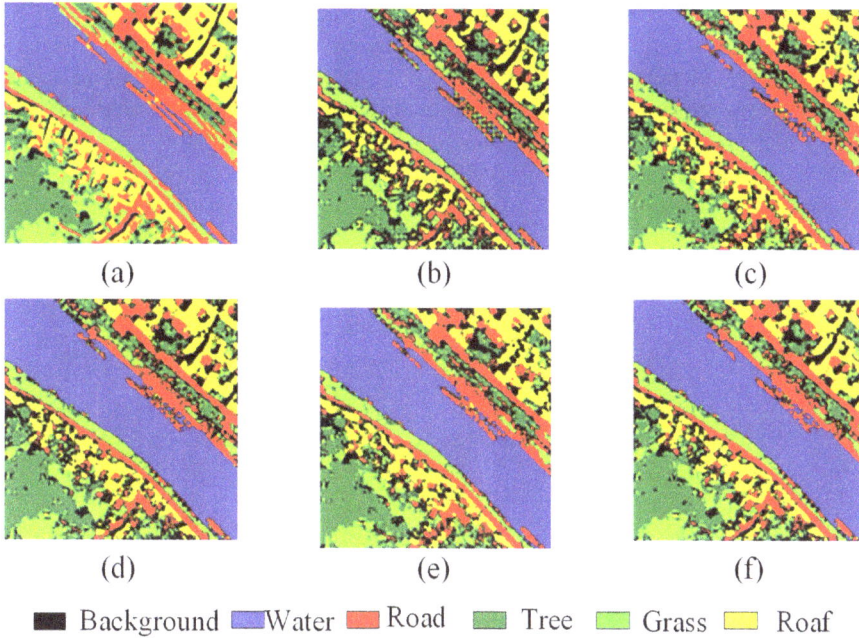

■ Background ▨Water ▨Road ▨ Tree ▨ Grass ▨ Roaf

FIGURE 4.19 Subpixel mapping results in experiment 2 (S = 6): (a) synthetic hyperspectral image, (b) BI, (c) SPSAM, (d) BI-SSI, (e) SPSAM-SSI, and (f) SAM-MSSI.

TABLE 4.3

Mapping Accuracy of the Five Methods in Experiment 2 ($S = 6$)

Land Cover Class	BI	SPSAM	BI-SSI	SPSAM-SSI	SAM-MSSI
Background (%)	62.81	66.59	70.94	71.74	73.26
Water (%)	95.98	96.67	96.89	97.15	97.31
Road (%)	62.76	67.32	68.42	70.99	71.96
Tree (%)	70.76	72.94	74.06	75.59	76.64
Grass (%)	65.95	68.58	70.55	72.61	74.16
Road (%)	70.32	71.76	74.31	74.40	76.60
PCC' (%)	77.39	78.27	79.98	80.84	82.28

space, there are many disconnected cone-shaped patches in the subpixel mapping results based on a single image. In the subpixel mapping results based on MSIs, these phenomena are obviously effectively alleviated. In particular, comparing the subpixel mapping results based on a single-scale MSI, with the help of multi-scale MSIs, the subpixel mapping result generated by SAM-MSSI are closer to the reference image.

In the five subpixel mapping results, the mapping accuracy (%) of each class and the overall mapping accuracy of PCC' (%) are shown in Table 4.3. Similar to the conclusion of experiment 1, the mapping accuracy of the proposed SAM-MSSI method is significantly higher than the other four subpixel mapping methods. This reason is that the fine-scale smoothing term makes the subpixels belonging to the same class cluster together to form a uniformly distributed local area, and the coarse-scale penalty term prevents the edges being over-smoothed.

Next, we tested the PCC' (%) and Kappa' accuracy of the experimental data on three scales $S = 4$, $S = 6$, $S = 10$. As shown in Figure 4.20, although the accuracy of

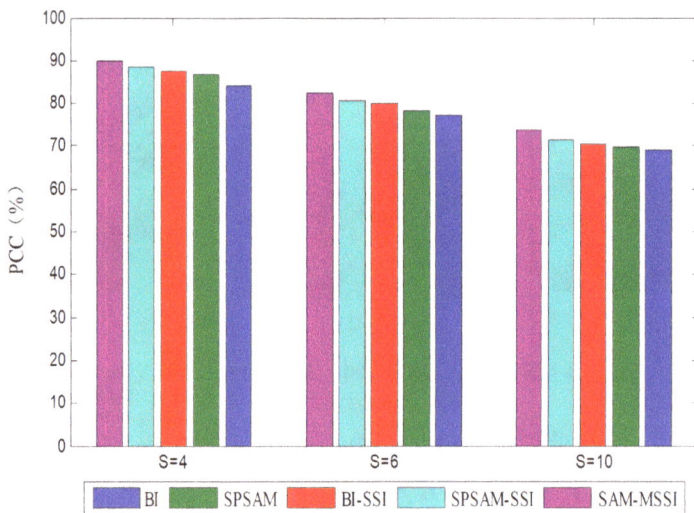

(a)

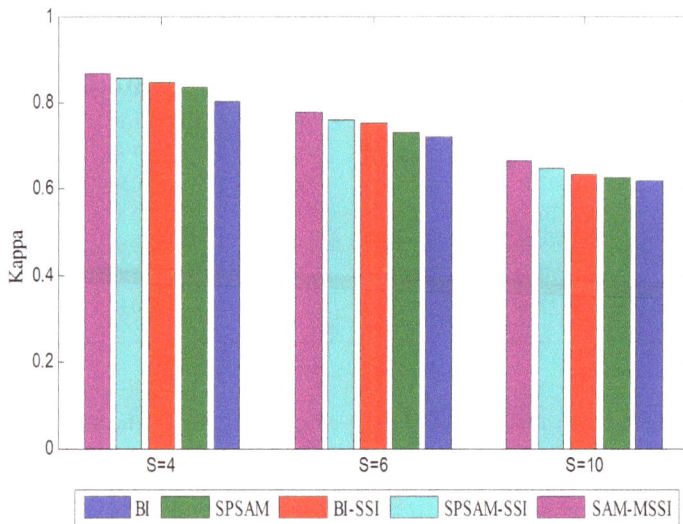

(b)

FIGURE 4.20   (a), (b) The relationship between PCC' (%) and Kappa' of five methods and scale S.

the mapping results of the five subpixel mapping methods decreases as the scale S increases, it is similar to the conclusion of experiment 1. Compared with the other four methods, the proposed method still obtains the highest mapping accuracy.

4.4.4.3 Experiment 3

In experiment 3, a set of real hyperspectral remote sensing data was used for research. As shown in Figure 4.21(a), this real hyperspectral experimental data still selects the study area of 240×240 pixels in the Washington, DC, data set. The original image is downsampled to simulate a low-resolution image with the scale $S = 2$. The reference image is shown in Figure 4.21(b). The parameter w is set to 0.5 here.

Figure 4.22 shows the results of five subpixel mapping methods. Due to the influence of spectral unmixing errors, some pixels are inevitably misclassified in the

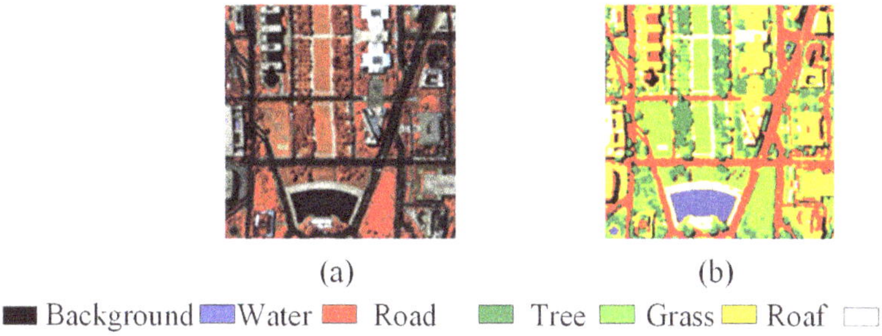

(a)   (b)

■ Background ■ Water ■ Road   ■ Tree ■ Grass ■ Roaf ☐

FIGURE 4.21   Washington, DC, data set: (a) false color composite image (RGB band: 65, 52, 36) and (b) reference image.

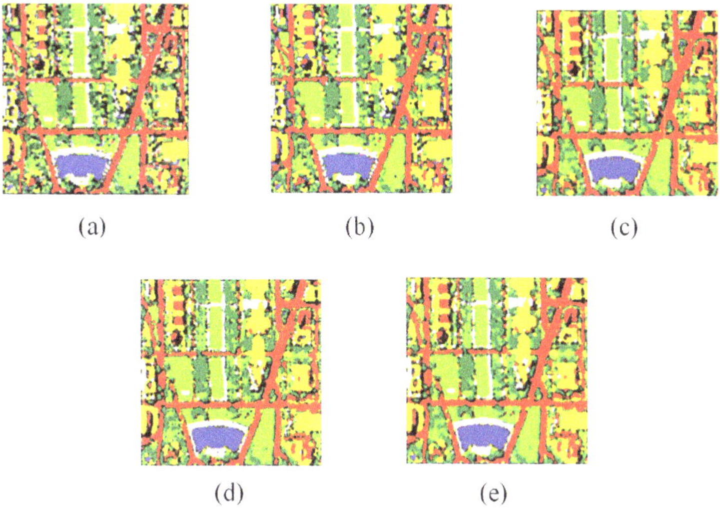

(a)   (b)   (c)

(d)   (e)

FIGURE 4.22   Subpixel mapping results in experiment 3 ($S = 2$): (a) BI, (b) SPSAM, (c) BI-SSI, (d) SPSAM-SSI, and (e) SAM-MSSI.

results. For example, in the five images, some pixels in the grass area are erroneously classified as trees due to the conflicts between the classes in the space. Therefore, the spectral unmixing error seriously affects the overall performance of the subpixel mapping method. However, similar to the conclusions of experiments 1 and 2, when there is no MSI, the speckle effect of the subpixel mapping result is very serious. From the visual point of view, the speckle effect is well alleviated under the action of the MSI. In particular, due to the special effects of the multi-scale MSIs, the final subpixel mapping results produced by SAM-MSSI are more accurate than the sub-pixel mapping results produced by BI-SSI and SPSAM-SSI and are closer to the reference image.

Since it is impossible to determine whether a pixel is a pure pixel by using SVM-based spectral unmixing technology, the pure pixels in the abundance image cannot be excluded when performing accuracy statistics for each class in experiment 3. The performance of spectral unmixing also needs to be considered. This is different from experiments 1 and 2, because in the previous two experiments, only synthesized images were studied, and the step of spectral unmixing were not considered. Therefore, in experiment 3, two accuracy indicators PCC and Kappa are needed to evaluate the five methods. Observing the mapping accuracy (%) in Table 4.4, the BI-SSI, SPSAM-SSI, and SAM-MSSI with multiple MSIs are better than the BI and SPSAM without multiple MSIs. Since the MSI in the proposed method has the multi-scale information, in terms of overall mapping accuracy evaluation PCC (%) and Kappa, SAM-MSSI is higher than BI-SSI and SPSAM-SSI.

4.4.4.4 Discussion

This section analyzes the influence of the number of suitable multi-scale multi-shift images on the proposed SAM-MSSI method. Experiment 1 ($S = 4$) and experiment 2 ($S = 6$) are tested by four sets of multi-scale multi-shift images with different numbers. The corresponding subpixel multi-shift is shown in Table 4.5. As shown in Figure 4.23, when the number of suitable multi-scale multi-shift images increases from 3 to 12, the corresponding PCC' (%) value is also increased. According to the

TABLE 4.4
Mapping Accuracy of Five Methods in Experiment 3 ($S = 2$)

Land Cover Class	BI	SPSAM	BI-SSI	SPSAM-SSI	SAM-MSSI
Background (%)	73.44	74.97	77.31	78.26	79.38
Water (%)	85.56	88.65	90.89	92.67	94.83
Road (%)	70.55	72.74	75.63	77.04	79.11
Tree (%)	72.45	74.84	76.53	78.08	79.96
Grass (%)	74.70	77.86	80.72	82.27	84.46
Road (%)	70.67	73.09	75.40	77.68	79.83
Trail (%)	73.88	75.35	76.10	81.98	83.10
PCC (%)	76.82	79.76	81.81	83.08	84.71
Kappa	0.7356	0.7588	0.7825	0.8002	0.8174

TABLE 4.5
Subpixel Multi-Shift of Multi-Scale Multi-Shift Image

Number				MSSI								
3				(0,0)	(0,1)	(0,1)						
6			(0,0)	(0,1)	(0,1)	(1,1)	(0,2)	(2,0)				
9		(0,0)	(0,1)	(0,1)	(1,1)	(0,2)	(2,0)	(2,2)	(4,0)	(0,4)		
12	(0,0)	(0,1)	(0,1)	(1,1)	(0,2)	(2,0)	(2,2)	(4,0)	(0,4)	(4,4)	(2,4)	(4,2)

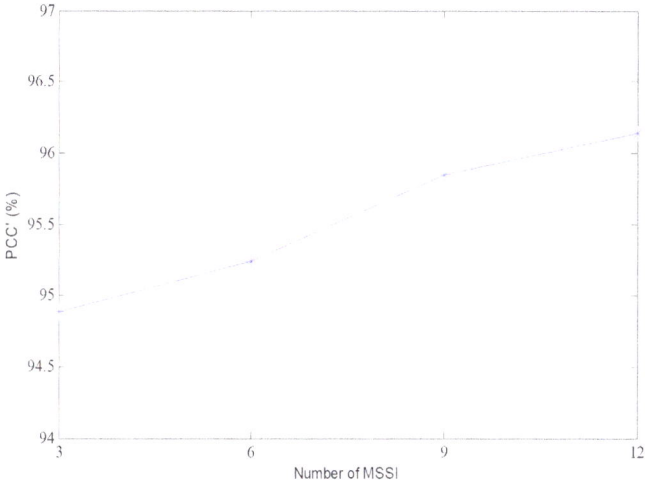

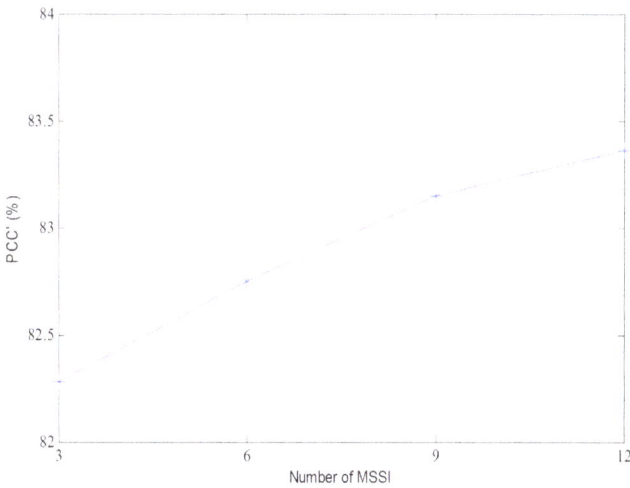

FIGURE 4.23   The influence of the number of multi-scale multi-shift images on SAM-MSSI: (a) experiment 1 ($S = 4$) and (b) experiment 2 ($S = 6$).

above experimental results, the more suitable multi-shift images mean the higher accuracy of the subpixel mapping results can be obtained. However, when inaccurate multi-shift information is generated, the accuracy of subpixel mapping results is often affected.

The registration error reflects the accuracy of subpixel multi-shift estimation, and it is one of the main factors affecting the performance of the method. In this section, the simulated registration error ranges from −0.4 to 0.4 pixels, with an interval of 0.1 pixels. Experiment 1 ($S = 4$) and experiment 2 ($S = 6$) are tested by different registration errors. If the registration error is 0.1, the original subpixel shift (0, 0), (0, 1), and (1, 0) will be adjusted to (0.1, 0.1), (0.1, 1.1), and (1.1, 0.1). Figure 4.24 shows the relationship between the PCC' (%) of the proposed method and the registration error. It can be seen from Figure 4.24 that the registration error has a negative impact

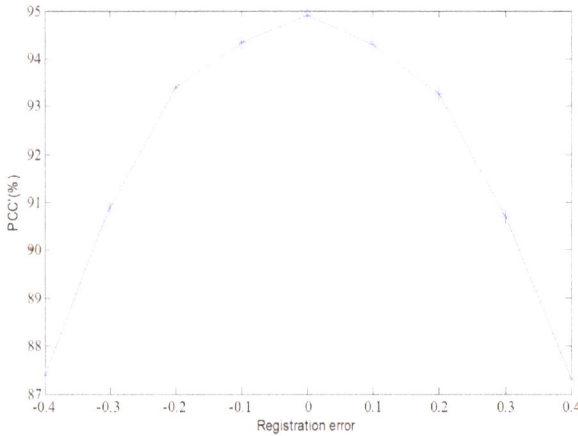

(a)

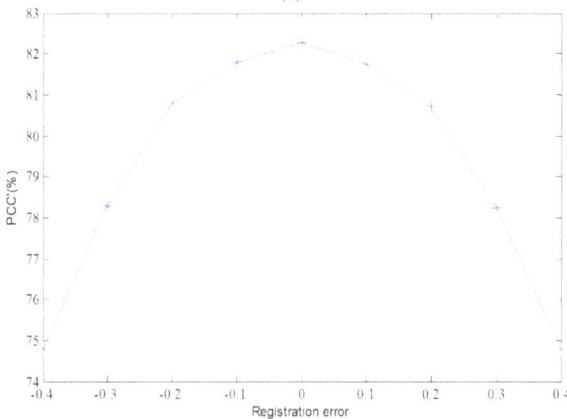

(b)

FIGURE 4.24   The relationship between PCC' (%) of SAM-MSSI and registration error: (a) experiment 1 ($S = 4$) and (b) experiment 2 ($S = 6$).

on the proposed method. As the registration error increases, the accuracy of the proposed method decreases. Therefore, accurate subpixel multi-shift estimation is very important to the performance of the proposed method.

4.5   UTILIZING PARALLEL NETWORKS TO PRODUCE SUBPIXEL SHIFTED IMAGES WITH MULTI-SCALE SPATIAL-SPECTRAL INFORMATION FOR SUBPIXEL MAPPING

Through the research of the existing subpixel mapping based on the MSIs, it is found that the fine MSIs are usually generated by a single subpixel sharpening method. In this way, the type of information carried by the fine MSIs is always single. To enrich the types of information carried by the fine MSIs, this section combines the methods proposed in the previous sections 4.3 and 4.4 and utilizes parallel networks to produce subpixel shifted images with multi-scale spatial-spectral information for subpixel mapping (SSI-MSSI).

4.5.1   MULTI-SCALE NETWORK AND SPATIAL-SPECTRAL NETWORK

In the multi-scale network, the previously proposed mixed spatial attraction model (MSAM) is still used to generate $H_k^{fs}(p_j^v)$, which is the smooth term with fine-scale information, and to generate $H_k^{cs}(p_j^v)$, which is the penalty term with coarse-scale information. Then according to formula (4.14), $H_k(p_j^v)$, which are the fine shifted images with multi-scale information, are obtained.

In the spatial-spectral network, to improve the resolution of the original coarse multi-shift images, projection onto convex set (POCS) is a more effective algorithm instead of the simple interpolation algorithm. Then the fine MSIs $H_k^{ss}(p_j^v)$ with richer spatial-spectral information can be obtained by unmixing the improved images.

The super-resolution reconstruction model of POCS can be obtained by the following steps. First assume that $f[n_1, n_2, \lambda_0]$, where (n_1, n_2) represents the high-resolution discrete spatial coordinates, is a two-dimensional high-resolution target image with a fixed wavelength λ_0. Then the continuous spatial image $f(x_1, x_2, \lambda_0)$, where (x_1, x_2) represents the continuous spatial coordinates, will be recovered from the target image by the Nyquist sampling theorem [10], where the sampling pulse sequence of the continuous spatial image can be expressed by the following formula:

$$f_s(x_1, x_2, \lambda_0) = \sum_{n_1=0}^{N_1-1} \sum_{n_2=0}^{N_2-1} f[n_1, n_2, \lambda_0] \times \delta\left(x_1 - \frac{n_1}{L_1}, x_2 - \frac{n_2}{L_2}\right) \qquad (4.16)$$

where N_1 stands for the horizontal range of the high-resolution target image; N_2 is the vertical range of the high-resolution target image; $\delta(\bullet)$ is the unit impulse function; L_1 and L_2 denote the sampling density relationship between the low-resolution image and the high-resolution image. When the number of samples per unit area of the low-resolution image is 1, the number of samples per unit area of the high-resolution image is $L_1 \times L_2$. The restored continuous spatial image $f(x_1, x_2, \lambda_0)$ is illustrated in formula (4.17):

$$f_c(x_1,x_2,\lambda_0) = \iint f(v_1,v_2,\lambda_0) \times h(x_1-v_1,x_2-v_2) dv_1 dv_2$$

$$= \sum_{n_1=0}^{N_1-1}\sum_{n_2=0}^{N_2-1} f[n_1,n_2,\lambda_0] \times \iint h_r\left(x_1-\frac{n_1}{L_1},x_2-\frac{n_2}{L_2}\right) \times h(x_1-v_1,x_2-v_2) dv_1 dv_2 \quad (4.17)$$

$$= \sum_{n_1=0}^{N_1-1}\sum_{n_2=0}^{N_2-1} f[n_1,n_2,\lambda_0] \times h_b(x_1,x_2,n_1,n_2)$$

where $h_r(\bullet)$ stands for reconstruction filter function.

Spatial domain filtering means the inevitable optical blur due to the limitations of sensors. The image $f_c(x_1, x_2, \lambda_0)$ can be obtained from the continuous spatial image $f(x_1, x_2, \lambda_0)$ through spatial domain filtering. Then $h(\bullet)$ represents the spatial domain filtering response function.

$$g_c(x_1,x_2,\lambda_i) = \int_0^\infty f_c\left(x_1,x_2,\lambda\right) r_i(\lambda) d\lambda$$

$$= h_b\left(x_1,x_2,n_1,n_2\right) \times \sum_{n_1=0}^{N_1-1}\sum_{n_2=0}^{N_2-1}\int_0^\infty f[n_1,n_2,\lambda] r_i(\lambda) d\lambda \quad (4.18)$$

$$= \sum_{n_1=0}^{N_1-1}\sum_{n_2=0}^{N_2-1} \Psi_{i,n_1,n_2}\left\{f\left[n_1,n_2,\lambda\right]\right\} \times h_b\left(x_1,x_2,n_1,n_2\right)$$

Spectral domain filtering represents the influence of external factors such as band selection and atmospheric disturbance on the hyperspectral band. Because the wavelength λ_0 is fixed in the previous description, the spectral domain filter function $r_1(\lambda), r_2(\lambda), \ldots, r_Q(\lambda)$ is introduced, where Q represents the number of bands. If λ_0 is seen as a continuous variable λ, $g_c(x_1, x_2, \lambda_i)$, will be obtained in any fixed space (x_1, x_2) or (n_1, n_2) by using the image $f_c(x_1, x_2, \lambda_0)$ through spectral domain filtering.

$$g_c(x_1,x_2,\lambda_i) = \int_0^\infty f_c\left(x_1,x_2,\lambda\right) r_i(\lambda) d\lambda$$

$$= h_b\left(x_1,x_2,n_1,n_2\right) \times \sum_{n_1=0}^{N_1-1}\sum_{n_2=0}^{N_2-1}\int_0^\infty f[n_1,n_2,\lambda] r_i(\lambda) d\lambda \quad (4.19)$$

$$= \sum_{n_1=0}^{N_1-1}\sum_{n_2=0}^{N_2-1} \Psi_{i,n_1,n_2}\left\{f\left[n_1,n_2,\lambda\right]\right\} \times h_b\left(x_1,x_2,n_1,n_2\right)$$

where $r_i(\lambda)$ represents a spectral domain filter response function, which is used to obtain an image $g_c(x_1, x_2, \lambda_i)$ with a continuously variable wavelength λ and to constitute an operator Ψ_{i,n_1,n_2}. When the wavelength and spatial position change, $g_c(x_1, x_2, \lambda_i)$ can be regarded as a two-dimensional continuous image.

To obtain the feasible operational results, the previous results should be spatially discretized. Therefore, the low-resolution discrete observation image $g(m_1, m_2, \lambda_i)$ ($[m_1, m_2]$ represents the low-resolution discrete spatial coordinates) can be obtained through the spatial sampling process.

$$g(m_1, m_2, \lambda_i) = g_c(x_1, x_2, \lambda_i)\big|_{x_1=m_1, x_2=m_2}$$

$$= \sum_{n_1=0}^{N_1-1} \sum_{n_2=0}^{N_2-1} \Psi_{i,n_1,n_2} \left\{ f\left[n_1, n_2, \lambda\right] \right\} \times h_b\left(m_1, m_2, n_1, n_2\right) \qquad (4.20)$$

In actual operation, for the operator Ψ_{i,n_1,n_2}, it can be regarded as a matrix with a size of $1 \times \lambda$, then this matrix is dot multiplied by $f[n_1, n_2, \lambda]$ to obtain $g(m_1, m_2, \lambda_i)$.

For example, when $Q = 100$, the previous model can be expressed as

$$
\begin{bmatrix} g(m_1, m_2, \lambda_1) \\ g(m_1, m_2, \lambda_2) \\ . \\ . \\ . \\ g(m_1, m_2, \lambda_{100}) \end{bmatrix} = \begin{bmatrix} \Psi_{1,n_1,n_2} \\ \Psi_{2,n_1,n_2} \\ . \\ . \\ . \\ \Psi_{100,n_1,n_2} \end{bmatrix} \times \begin{bmatrix} f(n_1, n_2, \lambda_1) \cdot h_b(m_1, m_2, n_1, n_2) \\ f(n_1, n_2, \lambda_2) \cdot h_b(m_1, m_2, n_1, n_2) \\ . \\ . \\ . \\ f(n_1, n_2, \lambda_{100}) \cdot h_b(m_1, m_2, n_1, n_2) \end{bmatrix} \qquad (4.21)
$$

Additive noise $v(m_1, m_2, \lambda_i)$ is added to the low-resolution discrete observation image $g(m_1, m_2, \lambda_i)$. Then $v(m_1, m_2, \lambda_i)$ represents the superposition of multiple noise such as sensor sampling. The final super-resolution reconstruction model of the hyperspectral remote sensing image can be expressed as

$$g_f(m_1, m_2, \lambda_i) = g(m_1, m_2, \lambda_i) + v(m_1, m_2, \lambda_i)$$

$$= g_c(x_1, x_2, \lambda_i)\big|_{x_1=m_1, x_2=m_2} + v(m_1, m_2, \lambda_i)$$

$$= \sum_{n_1=0}^{N_1-1} \sum_{n_2=0}^{N_2-1} \Psi_{i,n_1,n_2} \left\{ f\left[n_1, n_2, \lambda\right] \right\} \times h_b\left(m_1, m_2, n_1, n_2\right) \qquad (4.22)$$

$$+ v(m_1, m_2, \lambda_i) \qquad (i = 1, 2, ..., Q)$$

The spatial domain filter function H_{spa} represents the spatial domain filtering process, and the spectral domain filter function H_{spe} represents the spectral domain filtering process. The super-resolution reconstruction model mentioned can be abbreviated as formula (4.23):

$$g^H = H_{spa} H_{spe} f^H + v \qquad (4.23)$$

where f^H represents a high-resolution target image, g^H denotes a low-resolution observation image, and v is noise. Figure 4.25 shows the previous super-resolution reconstruction model.

In the actual operation process, the original high-dimensional data can be first mapped to a low-dimensional transformation space through the endmember of interest (EOI), and then the previously mentioned super-resolution reconstruction model is applied. Such a process can reduce the complexity of the super-resolution algorithm and protect the class of interest (COI). The selection principle of EOI is

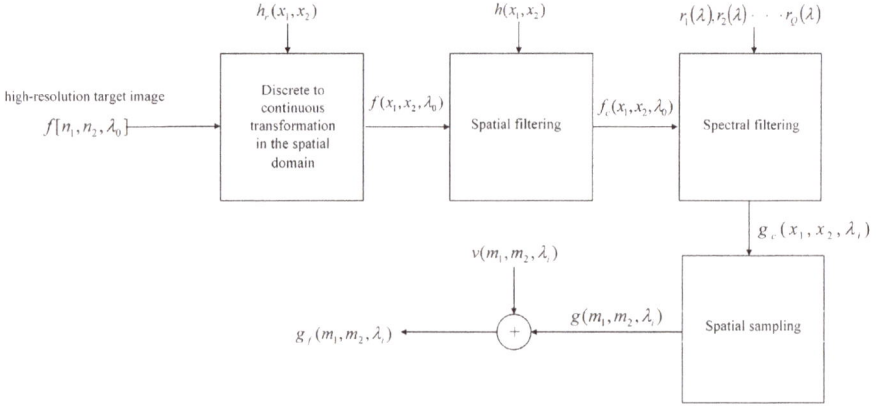

FIGURE 4.25   Super-resolution algorithm imaging model using POCS reconstruction.

discussed in detail in Chapter 6. It is possible to add a function Φ_{inv} that maps the original high-dimensional data to low-dimensional data and a function Φ that maps the low-dimensional data to the original high-dimensional data to achieve this goal.

Spectral endmembers can be obtained by the N-FINDR endmember extraction method. EOI can be formed by COI spectral endmembers. Then Φ is a column vector composed of EOI. $\Phi_{inv} = (\Phi^T\Phi)^{-1}\, \Phi^T\, \Phi$, and Φ_{inv} is equivalent to the left multiplication of the corresponding matrix. The relationship between the reduced-dimensional high-resolution target image f^L and the reduced-dimensional low-resolution observation image g^L can be expressed by the following formula:

$$g^{\mathrm{L}} = \Phi_{inv} H_{\mathrm{spa}} H_{\mathrm{spe}} \Phi f^{\mathrm{L}} + v \tag{4.24}$$

It can be noted from formula (4.24) that the reducing dimensionality of the input data can simplify the super-resolution reconstruction process. The formula (4.24) expresses the relationship between the low-dimensional transform domain f^L and g^L, and the function of Φ_{inv} is to return the output data to the low-dimensional transform domain. Further, the $\Phi_{inv}H_{\mathrm{spa}}H_{\mathrm{spe}}\Phi$ is replaced with the synthesis operator H, and formula (4.24) can be abbreviated into the following form:

$$g^{\mathrm{L}} = Hf^{\mathrm{L}} + v \tag{4.25}$$

Suppose that the previously mentioned super-resolution reconstruction process is implemented in a sliding area window, H_{spa} and H_{spe} do not change with the sliding of the area window due to the time-invariant characteristics. H is derived from the deduction of Φ_{inv}, H_{spa}, H_{spe}, and Φ, but this derivation process is very complicated. Since all input data are linearly operated, the resulting data can be expressed as the weighted sum of all discrete data. Therefore, each element in the input data can be 1 in turn, and the remaining elements are 0, and they can be placed in the super-resolution reconstruction formula. Then the result derived from the super-resolution reconstruction formula is the corresponding weighted value of the position element $H_{i,j}$:

$$H_{i,j} = H_{spa} H_{spe} \Phi_{inv} a_{i,j} \tag{4.26}$$

After all weighted values are obtained, the operator H_o of the same size as the input data can be obtained. When H_o is used in the input data, it is equivalent to the process of point-by-point corresponding multiplication and then summation, namely, the synthetic operator H can be derived. As mentioned earlier, the synthesis operator H does not change with the spatial position transformation; therefore, it can be applied to the entire iterative operation process, and this method will greatly improve the speed of the algorithm. In POCS, the feasibility problem of convex optimization is utilized to optimize the super-resolution process. The closed convex set contains various constraints, and the intersection is formed by iteratively mapping the initial value step by step. The optimal reconstruction solution is obtained through the final mapping point. Finally, by unmixing the improved image, a fine shifted image $H_k^{ss}(p_j^v)$ with richer spatial-spectral information is obtained.

4.5.2 MULTI-SCALE SPATIAL-SPECTRAL INFORMATION

The fine MSI $H_k(p_j^v)$ with multi-scale information and the fine MSI $H_k^{ss}(p_j^v)$ with richer space-spectral information are combined by formula (4.27) to obtain the MSI $H_k(p_j)$ with multi-scale spatial-spectral information.

$$H_k(p_j) = \frac{1}{2V}\left[\sum_{v=1}^{V} wH_k^{ss}(p_j^v) + \sum_{v=1}^{V}(1-w)H_k(p_j^v)\right] \tag{4.27}$$

where w represents the weight parameter, and v represents the number of MSIs.

The SSI-MSSI flowchart is shown in Figure 4.26. The proposed SSI-MSSI method includes five steps:

Step 1. Using phase correlation to estimate subpixel shifts (x_v, y_v) ($v = 1, 2, \ldots, V$).

Step 2. In the multi-scale network, K low-resolution abundance images are obtained by unmixing the original coarse MSIs. At the same time, the POCS super-resolution reconstruction algorithm in the spatial-spectral network is used to obtain the improved images from the original coarse MSIs.

Step 3. MSAM is used to obtain the fine MSIs $H_k(p_j^v)$ with multi-scale information from coarse abundance images. The fine MSIs $H_k^{ss}(p_j^v)$ with richer spatial-spectral information are obtained by unmixing the improved images.

Step 4. The fine MSIs $H_k(p_j^v)$ with multi-scale information and the fine MSIs $H_k^{ss}(p_j^v)$ with richer spatial-spectral information will be integrated according to formula (4.27) to produce the MSIs with multi-scale-scale spatial-spectral information.

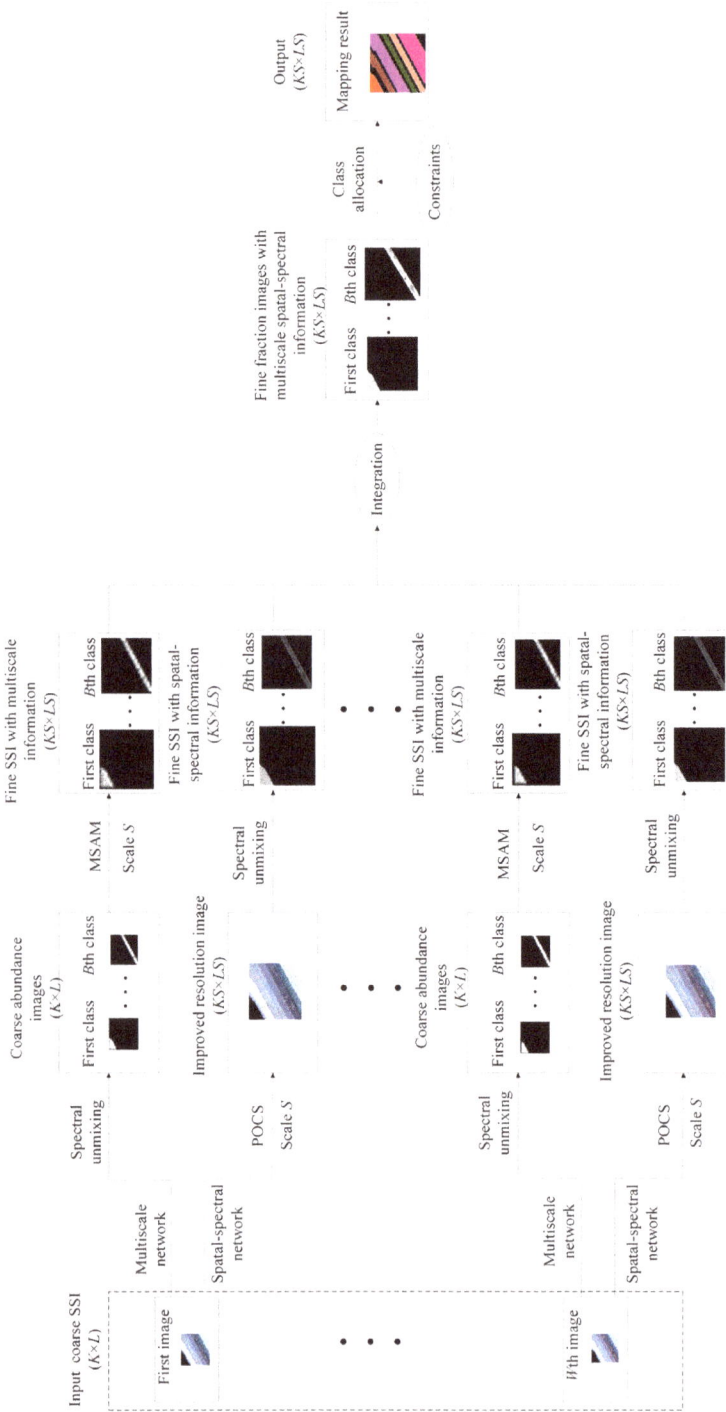

FIGURE 4.26 Spatial-spectral interpolation with multi-scale spatial-spectral information.

Step 5. According to the abundance value of $H_k(p_j)$, the class labels are assigned to subpixels by using the class allocation method to obtain the final subpixel mapping result.

As shown in Figure 4.26, when compared with the existing subpixel mapping method based on MSIs, the MSI has more information types in the proposed method. The SSI-MSSI provides the more multi-scale spatial-spectral information, producing the more ideal subpixel mapping result.

4.5.3 EXPERIMENTAL CONTENT AND RESULT ANALYSIS

Experiments based on three remote sensing images are tested to evaluate the SSI-MSSI. In the quantitative evaluation, the fine original remote sensing image is used for downsampling processing to obtain a simulated coarse MSI. In such an operational case, we can know the land cover classes at the subpixel level in order to directly evaluate the influence of image registration error on the proposed method. This chapter adopts the principle of the highest soft attribute values assigned first (HAVF) as the class allocation method. The least squares linear mixing model (LSLMM) has a simple physical meaning and is used for spectral unmixing here [11, 12]. The appropriate MSIs are derived by the method in [4]. The original multispectral or hyperspectral remote sensing image is first degraded and then shifted to generate the coarse MSIs. Four MSIs are considered, and the subpixel shifts are defined as (0, 0), (0.5, 0), (0, 0.5), and (0.5, 0.5).

Five subpixel mapping methods are compared: subpixel mapping based on a single image by bicubic interpolation (BIC) [5], subpixel mapping based on a single image by mixed space attraction model (MSAM) [9], subpixel mapping based on MSIs by bicubic interpolation (SSI-BIC) [4], subpixel mapping based on MSIs with multi-scale spatial-spectral information (SSI-MSAM) proposed in section 4.3 [13], and the proposed SSI-MSSI. PCC (%) and Kappa were used to quantitatively evaluate the accuracy of subpixel mapping results.

4.5.3.1 Experiment 1

As shown in Figure 4.27(a), a real hyperspectral data set is obtained by AVIRIS over Salinas Valley in 1998. The 80 × 80 size with 204 spectral bands is selected as the tested region. Figure 4.27(a) is first downsampled by the scale $S = 3$ and then shifted to produce the coarse MSIs. The simulated coarse MSI with a subpixel offset of (0, 0) is shown in Figure 4.27(b). Due to the low resolution, it is difficult to obtain the spatial distribution information of land cover classes from Figure 4.27(b). Subpixel mapping technology is utilized to derive the fine distribution information at finer spatial resolution. The weight parameter w is chosen to be 0.7.

Figure 4.27 (c)–(d) present the results of POCS and POCS with EOI. For visual comparison, POCS with EOI is closer to the original image than POCS. For further quantitative comparison, the relative error of super-resolution reconstruction is used to evaluate the performance of POCS with EOI. Super-resolution reconstruction relative error is defined as the sum of all reconstruction pixels absolute errors and the sum of all original image pixels ratio. Table 4.6 shows the reconstruction errors of

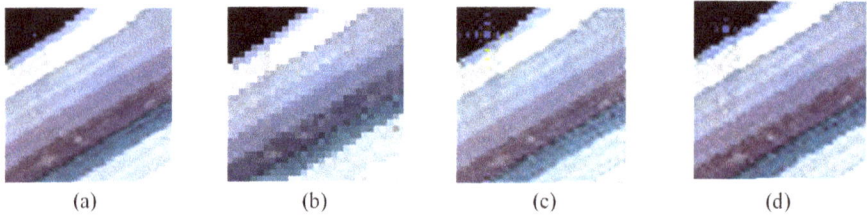

(a)   (b)   (c)   (d)

FIGURE 4.27   Salinas Valley data: (a) false color composite image (RGB band: 16, 27, 145); (b) coarse simulated image ($S = 3$); (c) POCS; and (d) POCS with EOI.

TABLE 4.6
Super-Resolution Reconstruction Error

Land Cover Class	POCS	POCS With EOI
Broccoli green 1 week	8.85%	4.14%
Corn senesced green weeds	6.51%	3.29%
Lettuce romaine 4 weeks	6.60%	3.76%
Lettuce romaine 5 weeks	7.52%	4.03%
Lettuce romaine 6 weeks	8.36%	4.40%
Lettuce romaine 7 weeks	8.08%	4.24%

different classes of POCS and POCS with EOI. As shown in Table 4.6, the reconstruction error of each class of POCS with EOI is lower than that of POCS. Compared with POCS, the operation time of POCS with EOI is increased by more than three times. This is because that EOI can reduce complexity and protect the land cover classes. To obtain the better subpixel mapping result, POCS with EOI is selected as the super-resolution reconstruction method in the spatial-spectral network.

The reference image is shown in Figure 4.28(a), which contains six land cover classes. The results of five subpixel mapping methods are presented in Figure 4.28(d)–(f). Through visual comparison, the subpixel mapping based on MSIs in Figure 4.28(d)–(f) obtains better mapping results than the subpixel mapping based on a single image in Figure 4.28(b)–(c). It can be observed in Figure 4.28(b)–(c) that there are many broken patches and obvious burrs. As shown in Figure 4.28(d)–(f), this phenomenon is alleviated with the help of MSIs. Compared with the other two subpixel mapping methods based on MSIs, the proposed SSI-MSSI is closer to the reference image.

Quantitatively assess the performance of the five methods through the mapping accuracy (%) and PCC (%) of each category. As shown in Table 4.7, the mapping accuracy (%) of each category of SSI-MSSI is higher than the other four methods. When using MSIs, the PCC (%) value of BIC increases from 90.80% to 96.22%. For the MSAM method, when the MSI is utilized, the PCC (%) value increases from 93.20% to 98.49%. Since the MSI in the proposed method can provide multi-scale spatial-spectral information, SSI-MSSI has the highest PCC (%) value of 99.09%.

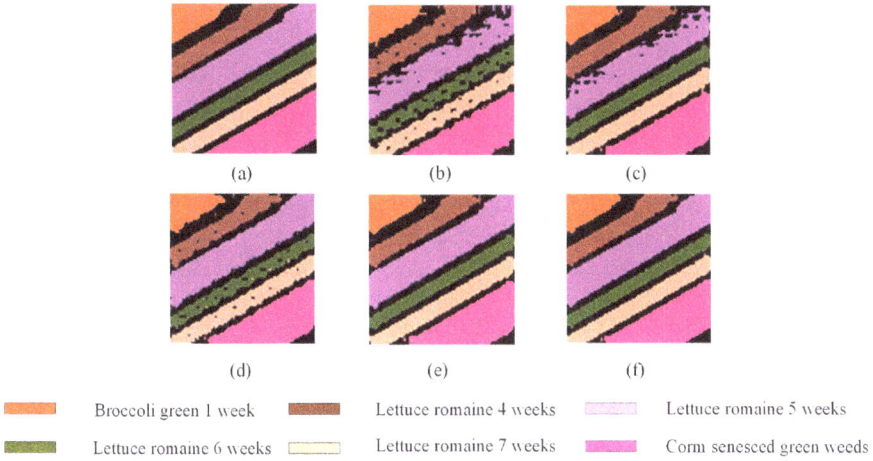

🟧 Broccoli green 1 week	🟫 Lettuce romaine 4 weeks	▦ Lettuce romaine 5 weeks
🟩 Lettuce romaine 6 weeks	▢ Lettuce romaine 7 weeks	🟪 Corn senesced green weeds

FIGURE 4.28 Subpixel mapping results in experiment 1 ($S = 3$): (a) reference image, (b) BIC, (c) MSAM, (d) SSI-BIC, (e) SSI-MSAM, and (f) SSI-MSSI.

TABLE 4.7
Mapping Accuracy of the Five Methods in Experiment 1 ($S = 3$)

Land Cover Class	BIC	MSAM	SSI-BIC	SSI-MSAM	SSI-MSSI
Broccoli green 1 week (%)	96.93	97.70	97.76	97.95	98.21
Corn senesced green weeds (%)	98.46	98.27	98.57	98.75	99.42
Lettuce romaine 4 weeks (%)	93.34	96.43	94.32	97.40	98.21
Lettuce romaine 5 weeks (%)	81.58	82.62	97.78	99.03	99.52
Lettuce romaine 6 weeks (%)	89.94	97.97	92.01	98.24	98.56
Lettuce romaine 7 weeks (%)	93.48	97.78	95.28	98.47	99.45
PCC (%)	90.80	93.20	96.22	98.49	99.09

4.5.3.2 Experiment 2

A real multispectral image with a larger region is selected to test in experiment 2. The multispectral images were taken by Landsat 8 satellite over Rome. In Figure 4.2(a), the 300×300 area is selected as the test data. Red, blue, near-infrared, short-wavelength infrared 1, and short-wavelength infrared 2 constitute the six bands of the test data. In Figure 4.29(b), the original multispectral image is downsampled by $S = 6$ and then translated to generate the coarse MSIs. The weight parameter w is set to 0.7 here.

There are four land cover classes in the reference image in Figure 4.30(a), including water, vegetation, building, and soil. Figure 4.30(b)–(f) show the subpixel mapping results of the five methods. For the subpixel mapping results based on a single image, some broken cone-shaped patches are shown in Figure 4.30(b)–(c). In the subpixel mapping results based on MSIs, as shown in Figure 4.30(d)–(f), this phenomenon is

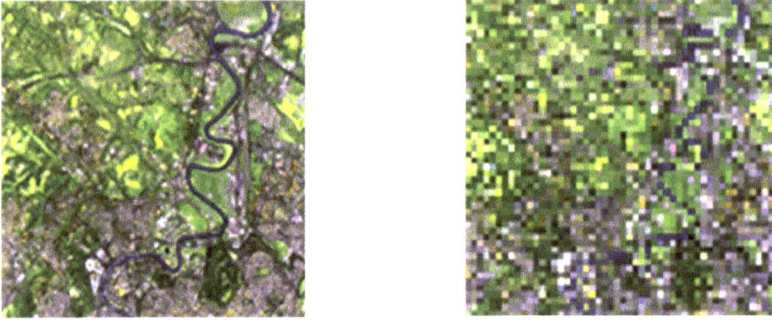

FIGURE 4.29 Rome data: (a) false color composite image (RGB band: 6, 5, 2) and (b) coarse simulated image ($S = 6$).

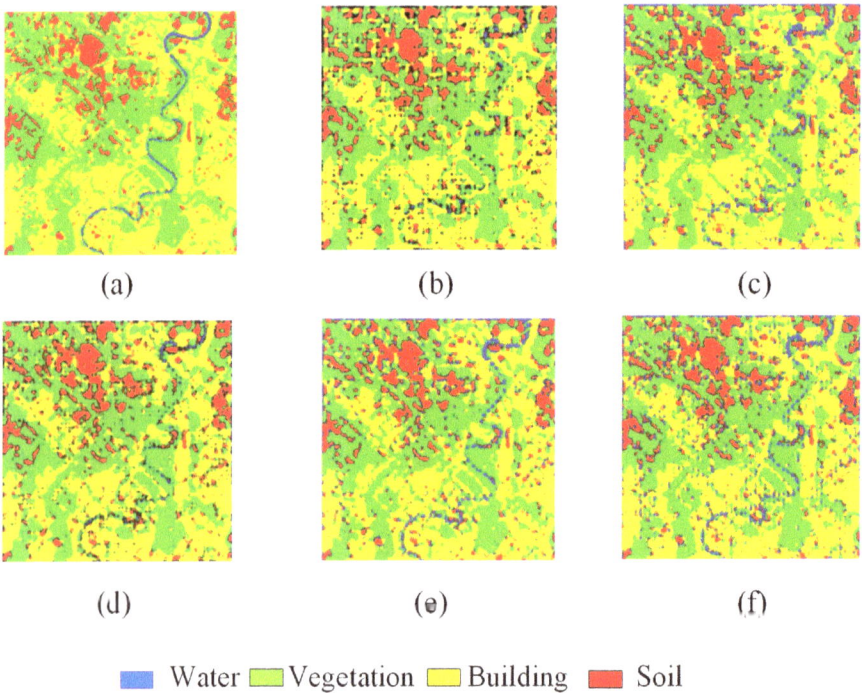

(a)   (b)   (c)

(d)   (e)   (f)

■ Water ▢Vegetation ▢Building ■ Soil

FIGURE 4.30 Subpixel mapping results in experiment 2 ($S = 6$): (a) reference image, (b) BIC, (c) MSAM, (d) SSI-BIC, (e) SSI-MSAM, and (f) SSI-MSSI.

alleviated. Compared with SSI-BIC and SSI-MSAM, the proposed SSI-MSSI is visually closer to the reference image.

The mapping accuracy (%) of each class and PCC (%) of the five methods are shown in Table 4.8. The mapping accuracy (%) of each class of SSI-MSSI is higher than the other four methods. To test the influence of the scale S on the subpixel

TABLE 4.8
Mapping Accuracy of the Five Methods in Experiment 2 (S = 3)

Land Cover Class	BIC	MSAM	SSI-BIC	SSI-MSAM	SSI-MSSI
Water (%)	19.43	42.26	26.13	50.75	60.36
Vegetation (%)	65.07	66.81	67.46	69.77	75.93
Building (%)	72.55	72.64	74.15	74.28	80.65
Soil (%)	59.86	57.58	62.66	61.46	70.37
PCC (%)	67.12	68.84	69.26	70.41	74.07

mapping results, the other two scales $S = 3$ and $S = 10$ were tested. Figures 4.31(a)–(b) show the PCC (%) and Kappa values of five subpixel mapping methods with three scales S. Compared with the other methods, the SSI-MSSI proposed in this chapter still obtains the highest PCC (%) and Kappa values.

4.5.3.3 Experiment 3

In experiment 3, a real hyperspectral image is selected with more classes, which is obtained by ROSIS in Pavia, northern Italy. The region of 360×360 pixels as the tested region is as shown in Figure 4.32(a). The coarse MSIs are produced by downsampling Figure 4.32(a) with $S = 3$, 5, and 9. In Figure 4.32(b), the original fine hyperspectral image is degraded with $S = 5$ to produce coarse MSIs. The weight w is defined as 0.6.

There are six land cover classes in reference image shown in Figure 4.33(a), including shadow, water, road, tree, grass, and roof. The results of the five subpixel mapping methods are shown in Figure 4.33(b)–(f). When the MSIs are not applied, Figure 4.33(b)–(c) will generate speckle artifacts. With the help of MSIs, these phenomena are alleviated in Figure 4.33(d)–(f). Because the MSIs in the proposed method have various types of information and provide more multi-scale spatial-spectral information, the proposed SSI-MSSI yields a more accurate subpixel mapping result than SSI-BIC and SSI-MSSI.

Figure 4.34(a)–(b) describes the PCC (%) and Kappa values of the five subpixel mapping methods with $S = 3$, 5, and 9. Similar to the previous experimental results, the PCC (%) and Kappa values in SSI-MSSI are higher than those in the other four subpixel mapping methods.

4.5.4.4 Discussion

This section discusses some factors affecting the proposed SSI-MSSI. First, the weight parameter w affects the influence of the multi-scale term $H_k(p_j^v)$ and the spatial-spectral term $H_k^{ss}(p_j^v)$ on the formula (4.27). To analyze the appropriate weight parameter w of SSI-MSSI, experiment 2 ($S = 6$) and experiment 3 ($S = 5$) calculated the PCC (%) values of ten combinations with an interval of 0.1 in the interval of [0, 0.9]. The results are shown in Figure 4.35(a)–(b). When $w = 0$, the spatial-spectral term $H_k^{ss}(p_j^v)$ has no effect, and only the existing SSI-MSSI results can be obtained. When the value of w increases, the PCC (%) value is noted to increase, because the

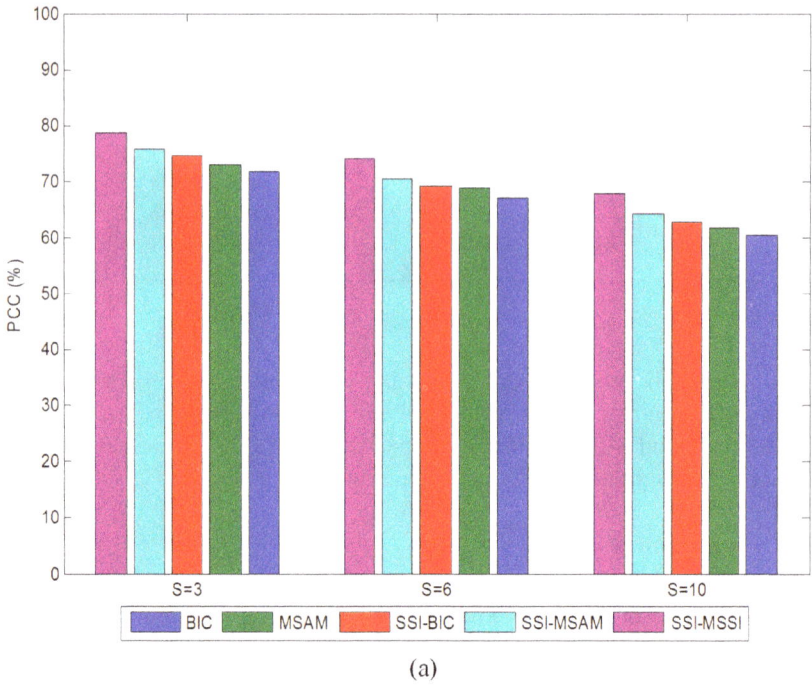

(a)

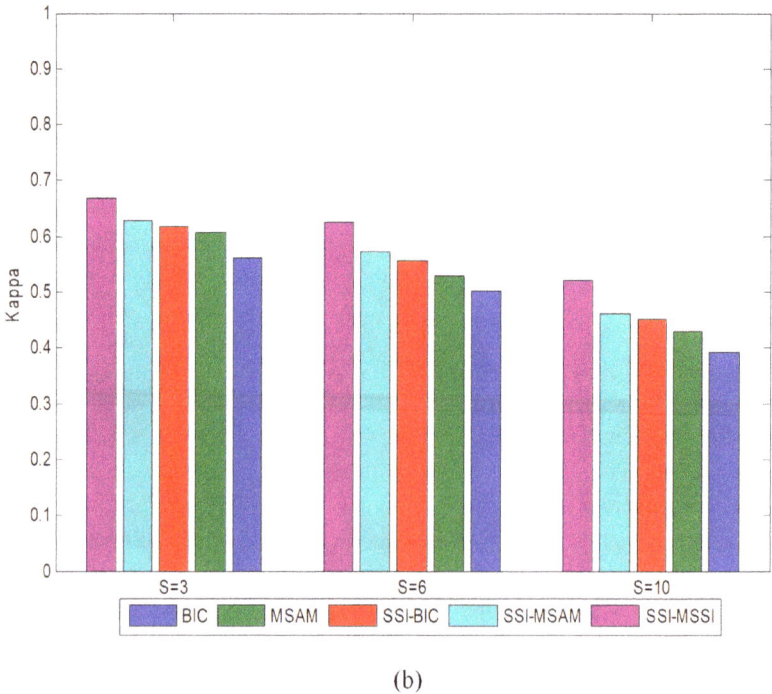

(b)

FIGURE 4.31   (a) PCC value (%) in relation to factor *S*. (b) Kappa value in relation to factor *S*.

(a)   (b)

FIGURE 4.32   Pavia city center composite data: (a) false color composite image (RGB band: 102, 56, 31) and (b) simulated coarse image ($S = 5$).

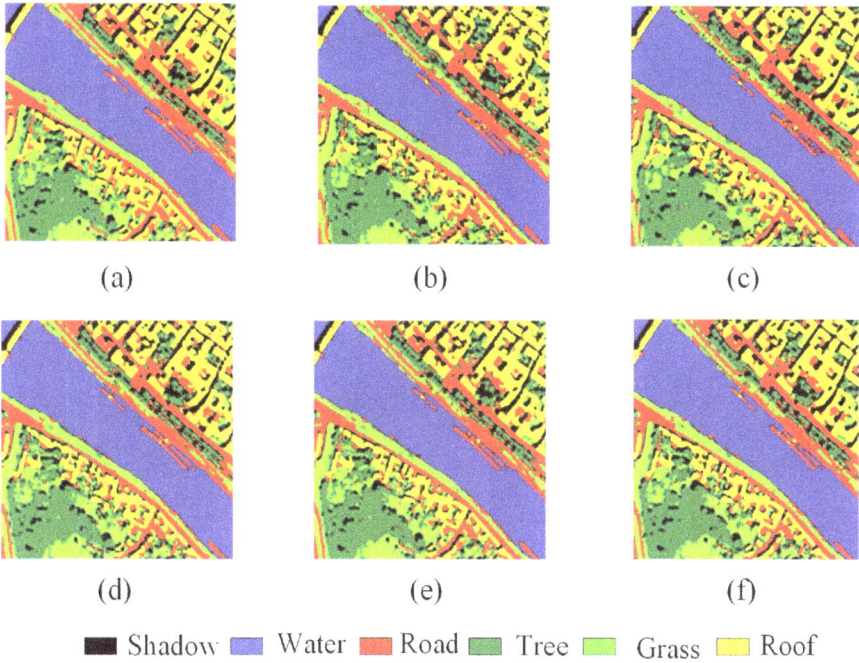

(a)   (b)   (c)

(d)   (e)   (f)

■ Shadow ■ Water ■ Road ■ Tree ■ Grass ■ Roof

FIGURE 4.33   Subpixel mapping results in experiment 3 ($S = 5$): (a) reference image, (b) BIC, (c) MSAM, (d) SSI-BIC, (e) SSI-MSAM, and (f) SSI-MSSI.

MSIs will contain more types of information at this time. With $w = 0.7$ in experiment 2 and $w = 0.6$ in experiment 3, the PCC (%) value is the highest. When w continues to rise, the contribution of the multi-scale term $H_k(p_j^v)$ to formula (4.27) decreases. The reduction of multi-scale information will affect the final subpixel mapping results.

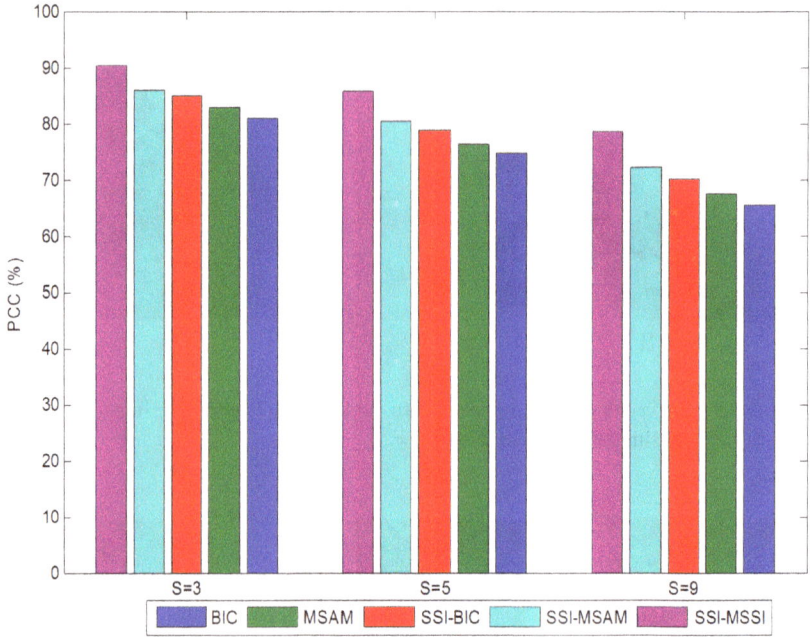

(a)

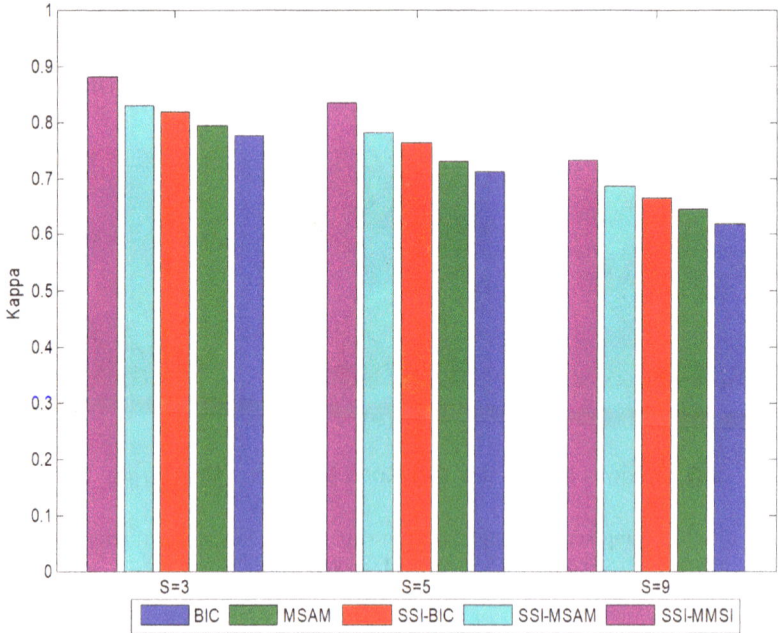

(b)

FIGURE 4.34   (a) PCC value (%) in relation to factor *S*. (b) Kappa value in relation to factor *S*.

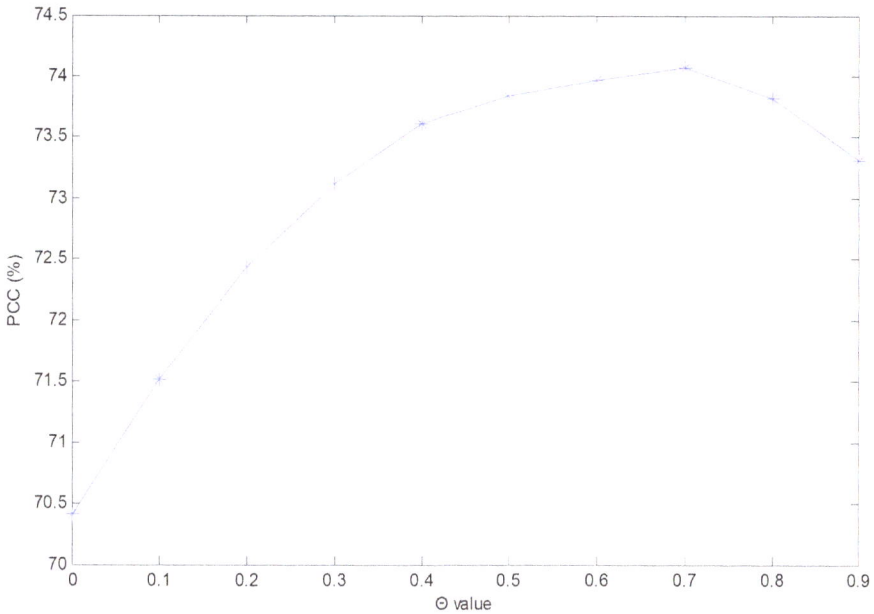

(a)

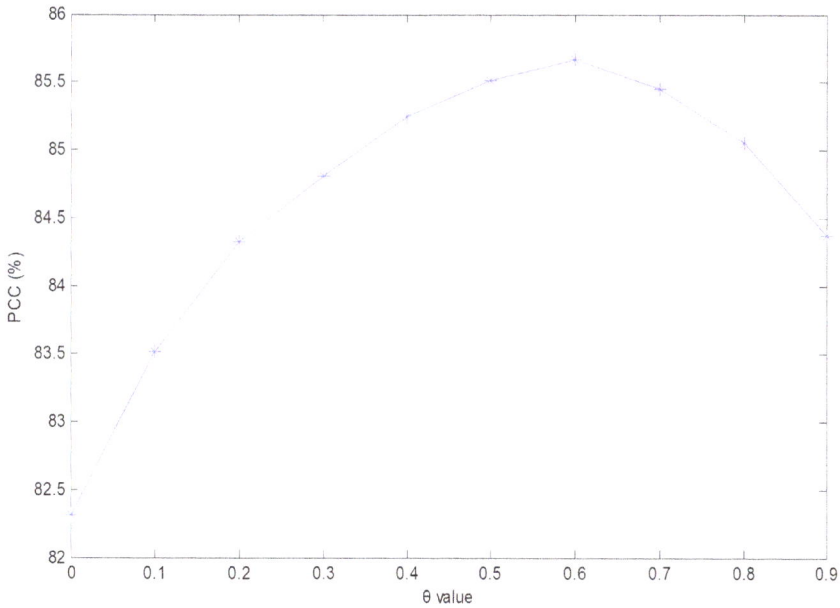

(b)

FIGURE 4.35   (a) PCC value (%) of SSI-MSSI in relation to weight parameter w in experiment 2. (b) PCC value (%) of SSI-MSSI in relation to weight parameter w in experiment 3.

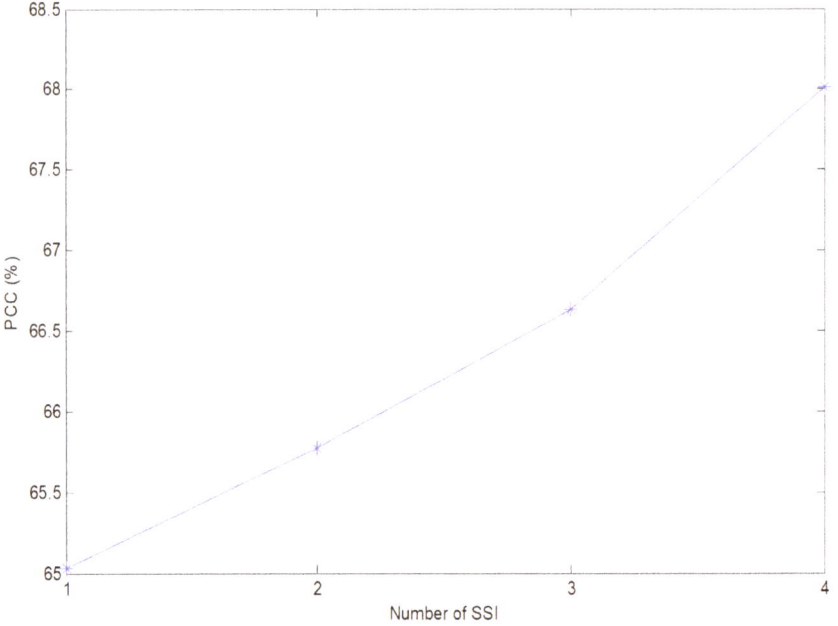

(a)

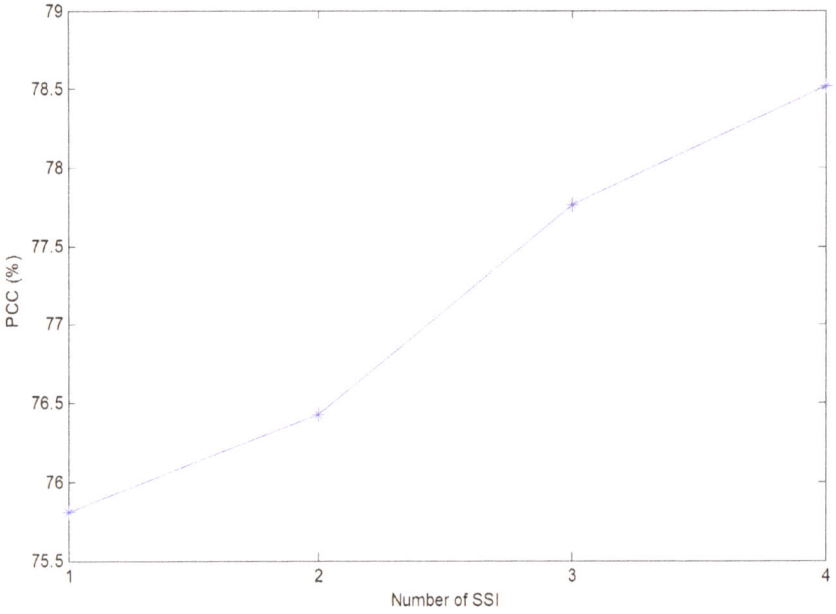

(b)

FIGURE 4.36   Influence of the number of multi-shift images for SSI-MSSI: (a) experiment 2 and (b) experiment 3.

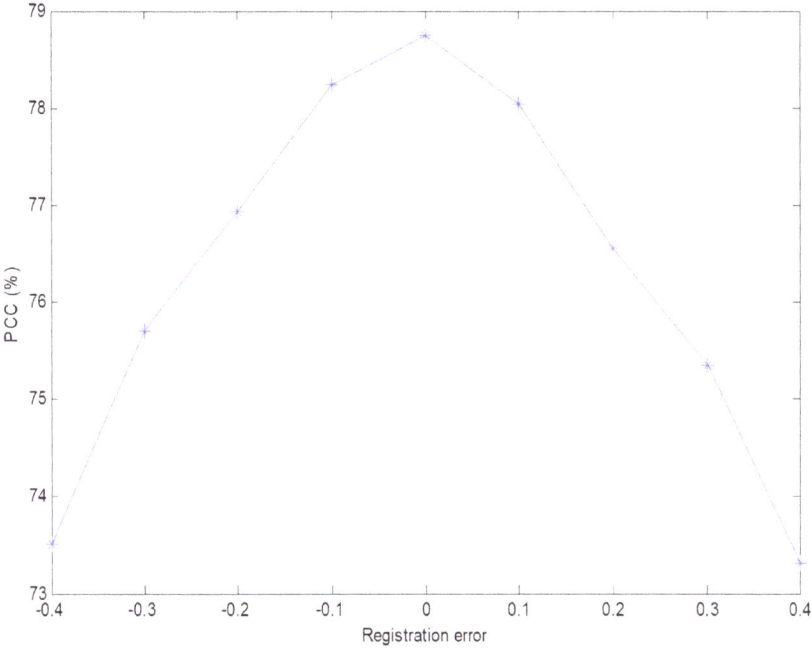

(a)

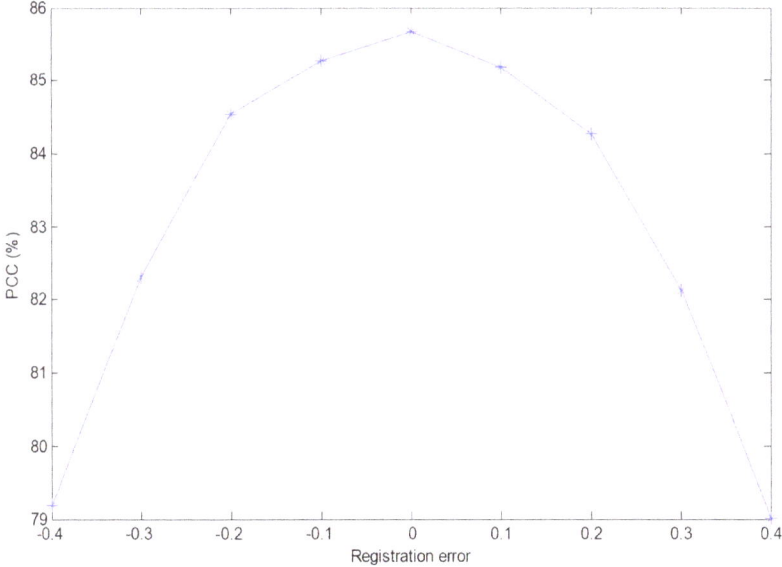

(b)

FIGURE 4.37   PCC value (%) of SSI-MSSI in relation to the registration error: (a) experiment 2 and (b) experiment 3.

OCSI
$(L \times W)$

PFSI
$(LS \times WS)$

Spectral
unmixing

Coarse abundance
images with $F_m(Q_j^{OCSI})$
$(L \times W)$

Spectral
unmixing

ATPK

Fine abundance
images with $F_m(q_j^{OCSI})$ ⟶ images with $F_m(q_j^{PFSI})$ ⟶
$(LS \times WS)$

Fine abundance

$(LS \times WS)$

Fine scale temporal
relationship
$D_a^{temporal}(q_j^{OCSI}; q_j^{PFSI})$

Temporal dependence
$O^{temporal}$

Ideal square
wave filter

PSF filter

Improved coarse
abundance images with
$\hat{F}_m(Q_j^{OCSI})$
$(L \times W)$

⟶

Simulated coarse
abundance images with
$F_m(Q_j^{PFSI})$
$(L \times W)$

⟶

Coarse scale temporal
relationship
$D_a^{temporal}(Q_j^{OCSI}; Q_j^{PFSI})$

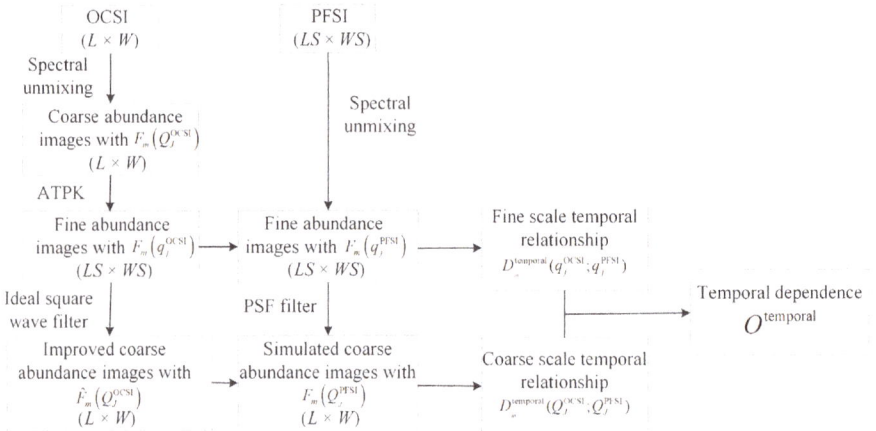

FIGURE 4.38   The flowchart of obtaining temporal dependence.

Secondly, the influence of the appropriate number of MSIs on the proposed method is analyzed. For experiment 2 ($S = 10$) and experiment 3 ($S = 9$), four MSIs are used for testing. Through image registration, the appropriate subpixel offsets are (0, 0), (0.5, 0), (0, 0.5), (0.5, 0.5). As shown in Figure 4.38(a)–(b), when the number of MSIs increases from one to four in the previous order, the accuracy of the subpixel mapping results will be improved.

Third, the previous experimental results show that a large number of appropriate MSIs can obtain better subpixel mapping results. However, when there is an inaccurate offset in the MSIs, it will influence the final accuracy. Registration error is a major factor affecting the performance of SSI-MSSI. It is the accuracy of subpixel offset estimation. Here, the simulated registration error range is −0.4 to 0.4 mixed pixels, with an interval of 0.1 pixels. Different registration errors were tested in experiment 2 ($S = 3$) and experiment 3 ($S = 5$). If the registration error is 0.1, the original shifts (0, 0), (0, 0.5), (0.5, 0), and (0.5, 0.5) are converted into (0.1, 0.1), (0.1, 0.6), (0.6, 0.1), and (0.6, 0.6). Figure 4.37(a)–(b) show the PCC (%) value related to the registration error. It can be seen from Figure 4.37(a)–(b) that the registration error has a negative impact on the proposed method. When the registration error increases, the accuracy of this method decreases. Therefore, it is extremely important to accurately estimate the subpixel offset in this method.

Finally, computational efficiency is also an important factor in evaluating the method because a parallel network is added to subpixel mapping based on MSIs, and more information types are used. The proposed SSI-MSSI takes longer to obtain subpixel mapping results than the other four methods.

4.6   SPATIOTEMPORAL SUBPIXEL MAPPING BY CONSIDERING THE POINT SPREAD FUNCTION EFFECT

Spatiotemporal subpixel mapping as a new subpixel mapping model that uses the appropriate prior fine spectral image (PFSI) as prior knowledge constraints shows great potential in recent years. Ling et al. [14] first propose the concept of spatiotemporal subpixel mapping. Spatiotemporal subpixel mapping is then further developed and applied in the traditional spatial attraction algorithm [15] and Hopfield neural network [16]. Subsequently, temporal dependence is introduced into spatiotemporal subpixel mapping to form a spatial-temporal dependence model to improve the mapping result [17–19].

However, the existing spatiotemporal subpixel mapping models usually describe the temporal dependence that is the relationship between the coarse abundance images from the original coarse spectral image (OCSI) and the fine abundance images from the PFSI, the accuracy of temporal dependence information could be affected due to the different scales and properties of two abundance images. In addition, Wang et al. [20] proved that when reducing the point spread function (PSF) effect, the performance of the traditional subpixel mapping models based on monotemporal image is improved, but the PSF effect is still not considered in the existing spatiotemporal subpixel mapping models. A general spatiotemporal subpixel mapping model based on fine and coarse scales temporal dependence by considering PSF effect (FCSTD) is proposed here.

Suppose S is the scale factor between OCSI and PFSI, namely, each mixed pixel in OCSI is considered to divide into S^2 subpixels in PFSI. Q_J^{OCSI} ($J = 1, 2, \ldots, N$, N is the number of mixed pixels) is a mixed pixel in OCSI, and q_j^{PFSI} ($j = 1, 2, \ldots, N \times S^2$, $N \times S^2$ is the number of subpixels) is a subpixel in PFSI. The coarse abundance images from the spectral unmixing of OCSI contain the coarse proportion $F_m\left(Q_J^{OCSI}\right)$ of the mth ($m = 1, 2, \ldots, M$, M is the number of land cover classes) class for mixed pixel Q_J^{OCSI}, and the fine abundance images from the spectral unmixing of PFSI contain the fine proportion $F_m\left(q_j^{PFSI}\right)$ of the mth class for subpixel q_j^{PFSI}. The motivation of spectral unmixing of PFSI is to obtain the abundance images with the same scale and properties as spectral unmixing of OCSI in the next processing.

In the FCSTD model, first, the coarse abundance images of OCSI and the fine abundance images of PFSI are derived by spectral unmixing, respectively. Second, considering the PSF effect, the fine abundance images of OCSI and the improved coarse abundance images of OCSI are obtained in turn from the coarse abundance images of OCSI by utilizing area-to-point kriging (ATPK) and then ideal square wave filter, and at the same time the simulated coarse abundance images of PFSI are obtained from the fine abundance images of PFSI by PSF filter. Third, the spatial dependence is derived by describing the relationship between central subpixel and neighbor pixels or subpixels in the improved coarse abundance images of OCSI. Fourth, as shown in Figure 4.38, the temporal dependence includes the fine scale described by the relationship between the fine abundance images of OCSI and the fine abundance images of PFSI, and the coarse scale described by the relationship between the improved coarse abundance images of OCSI and the simulated coarse abundance images of PFSI. Finally, according to maximizing the spatiotemporal dependence combined spatial dependence and temporal dependence, the mapping result is produced. Next, each module is introduced in detail.

4.6.1 Spatial Dependence

Inspired by [20], the PSF effect is considered in spatial dependence of FCSTD. The coarse abundance images of OCSI are first improved by the ATPK method which can describe the semi-variogram at different scales using the PSF filter to produce the fine abundance images with the fine proportion $F_m\left(q_j^{OCSI}\right)$ of the mth class for subpixel q_j^{OCSI}. The fine abundance images are then convolved with the ideal square wave filter to obtain the improved coarse abundance images with the enhanced coarse proportion $\hat{F}_m\left(Q_J^{OCSI}\right)$. More details can be found in [20].

Next, spatial dependence is derived by utilizing the improved coarse abundance images of OCSI. According to [17], spatial dependence could be described in two ways as formulas (4.28) through (4.32). One is based on the relationship between central subpixel and neighbor subpixels, and the other is based on the relationship between central subpixel and neighbor pixels. According to maximizing spatial dependence, the mathematical model of both types could be quantified by

$$\text{Max } O^{\text{spatial}} = \sum_{m=1}^{M}\sum_{j=1}^{S^2} x_{j,m} \times D_m^{\text{spatial}}(q_j^{OCSI}) \tag{4.28}$$

$$x_{j,m} = \begin{cases} 1, \text{ if subpixel } q_j^{OCSI} \text{ belongs to class } m \\ 0, \text{ otherwise} \end{cases} \tag{4.29}$$

where O^{spatial} is the objective function of spatial dependence, q_j^{OCSI} is central subpixel within mixed pixel Q_J^{OCSI}, $x_{j,m}$ is a class indicator, and $D_m^{\text{spatial}}(q_j^{OCSI})$ is the described spatial relationship between q_j^{OCSI} and neighbor pixels or subpixels for class m. As shown in Figure 4.39(a), when describing the relationship between q_j^{OCSI} and neighbor pixels, $D_m^{\text{spatial}}(q_j^{OCSI})$ is defined as

$$D_m^{\text{spatial}}(q_j^{OCSI}) = \sum_{n=1}^{N^*} \hat{F}_m(Q_n^{OCSI})\big/d^E(q_j^{OCSI},Q_n^{OCSI})^\gamma \tag{4.30}$$

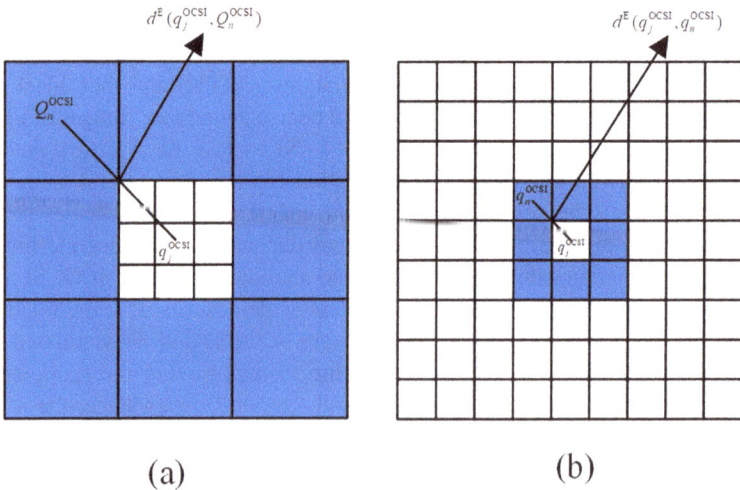

(a)   (b)

FIGURE 4.39 Spatial dependence between central subpixel q_j and (a) its eight neighboring pixels Q_n, and (b) its eight neighboring subpixels q_n.

As shown in Figure 4.2(b), when describing the relationship between q_j^{OCSI} and neighbor subpixels, $D_m^{\text{spatial}}(q_j^{\text{OCSI}})$ is described as

$$D_m^{\text{spatial}}(q_j^{\text{OCSI}}) = \sum_{n=1}^{N^*} B\left(q_j^{\text{OCSI}};q_n^{\text{OCSI}}\right)\Big/ d^{\text{E}}(q_j^{\text{OCSI}},q_n^{\text{OCSI}})^{\gamma} \tag{4.31}$$

where $d^{\text{E}}(q_j^{\text{OCSI}},Q_n^{\text{OCSI}})$ is the spatial distance from q_j^{OCSI} to the nth neighboring pixel Q_n^{OCSI}; $d^{\text{E}}(q_j^{\text{OCSI}},q_n^{\text{OCSI}})$ is the spatial distance from q_j^{OCSI} to the nth neighboring subpixel q_n^{OCSI}; N^* is the number of neighboring pixels or subpixels, which is set to eight to confirm the position of the central subpixel; γ is a non-linear parameter; and $B\left(q_j^{\text{C}};q_n^{\text{C}}\right)$ is defined as

$$B\left(q_j^{\text{OCSI}};q_n^{\text{OCSI}}\right) = \begin{cases} 1, \text{ if } q_j^{\text{OCSI}} \text{ and } q_n^{\text{OCSI}} \text{are the same class} \\ 0, \text{ otherwise} \end{cases} \tag{4.32}$$

4.6.2 TEMPORAL DEPENDENCE

The fine-scale temporal dependence $D_m^{\text{temporal}}(q_j^{\text{OCSI}};q_j^{\text{PFSI}})$ describes the relationship between the fine abundance images of OCSI with the fine proportion $F_m\left(q_j^{\text{OCSI}}\right)$ and the fine abundance images of PFSI with the fine proportion $F_m\left(q_j^{\text{PFSI}}\right)$, which is defined as

$$D_m^{\text{temporal}}(q_j^{\text{OCSI}};q_j^{\text{PFSI}}) = B\left(q_j^{\text{OCSI}};q_j^{\text{PFSI}}\right)\Big/ d^{\text{T}}(q_j^{\text{OCSI}},q_j^{\text{PFSI}})^{\mu} \tag{4.33}$$

where $d^{\text{T}}(q_j^{\text{OCSI}},q_j^{\text{PFSI}})$ is the time interval between subpixel q_j^{OCSI} of OCSI and subpixel q_j^{PFSI} of PFSI, which is calculated by the collection time intervals between OCSI and PFSI. Then μ is a free non-linear parameter that is usually set to 1 [17]. By comparing $F_m\left(q_j^{\text{OCSI}}\right)$ with $F_m\left(q_j^{\text{PFSI}}\right)$, $B\left(q_j^{\text{OCSI}};q_j^{\text{PFSI}}\right)$ is defined as

$$B\left(q_j^{\text{OCSI}};q_j^{\text{PFSI}}\right) = \begin{cases} 1, \text{ if } q_j^{\text{OCSI}} \text{ and } q_j^{\text{PFSI}} \text{ are the same class} \\ 0, \text{ otherwise} \end{cases} \tag{4.34}$$

To describe the coarse-scale temporal dependence, the fine abundance images of PFSI with the fine proportion $F_m\left(q_j^{\text{PFSI}}\right)$ are first convolved with the PSF filter to obtain the simulated coarse abundance images with the simulated coarse proportion $F_m\left(Q_J^{\text{PFSI}}\right)$. The coarse-scale temporal dependence $D_m^{\text{temporal}}(Q_J^{\text{OCSI}};Q_J^{\text{PFSI}})$ describes the relationship between the improved coarse abundance images of OCSI with the enhanced coarse proportion $\hat{F}_m\left(Q_J^{\text{OCSI}}\right)$ and the simulated coarse abundance images with the simulated coarse proportion $F_m\left(Q_J^{\text{PFSI}}\right)$, which is described by

$$D_m^{\text{temporal}}(Q_J^{\text{OCSI}};Q_J^{\text{PFSI}}) = B\left(Q_J^{\text{OCSI}};Q_J^{\text{PFSI}}\right)\Big/ d^{\text{T}}(Q_J^{\text{OCSI}},Q_J^{\text{PFSI}})^{\mu} \tag{4.35}$$

where $d^{\text{T}}(Q_J^{\text{OCSI}},Q_J^{\text{PFSI}})$ is the time interval between pixel Q_J^{OCSI} of OCSI and pixel Q_J^{PFSI} of PFSI, and also measured by the collection time intervals between OCSI and PFSI. By comparing $\hat{F}_m\left(Q_J^{\text{OCSI}}\right)$ with $F_m\left(Q_J^{\text{PFSI}}\right)$, $B\left(Q_J^{\text{OCSI}};Q_J^{\text{PFSI}}\right)$ is described as

$$B\left(Q_J^{\text{OCSI}};Q_J^{\text{PFSI}}\right)=\begin{cases}1,\text{ if they are with the same proportion}\\0,\text{ otherwise}\end{cases} \tag{4.36}$$

Next, according to maximizing temporal dependence, the objective function of temporal dependence O^{temporal} is obtained by the linear combination of $D_m^{\text{temporal}}(q_j^{\text{OCSI}};q_j^{\text{PFSI}})$ and $D_m^{\text{temporal}}(Q_J^{\text{OCSI}};Q_J^{\text{PFSI}})$ and is defined as

$$\text{Max } O^{\text{temporal}}=\sum_{m=1}^{M}\left(\begin{array}{c}\sum_{J=1}^{N}\left(1-\lambda\right)\times D_m^{\text{temporal}}(Q_J^{\text{OCSI}};Q_J^{\text{PFSI}})\\+\sum_{j=1}^{N\times S^2}\lambda\times D_m^{\text{temporal}}(q_j^{\text{OCSI}};q_j^{\text{PFSI}})\end{array}\right) \tag{4.37}$$

where λ is a weight parameter.

4.6.3 SPATIOTEMPORAL DEPENDENCE

The spatiotemporal dependence O could be derived by combining O^{spatial} and O^{temporal}:

$$\text{Max } O=\beta\times O^{\text{spatial}}+\left(1-\phi\right)O^{\text{temporal}} \tag{4.38}$$

where β is a weight parameter.

Finally, the mapping result is obtained according to maximizing spatiotemporal dependence. In addition, the two constraints in (4.39) need to be met:

$$\text{s. t. }\sum_{m=1}^{M}x_{j,m}=1,\ j=1,\ 2,\ ...,\ S^2$$

$$\sum_{j=1}^{S^2}x_{j,m}=\text{round}\left(\hat{F}_m\left(Q_J^{\text{OCSI}}\right)\times S^2\right)1,\ m=1,\ 2,\ ...,\ M \tag{4.39}$$

where round(•) means the integer nearest to $\hat{F}_m\left(Q_J^{\text{OCSI}}\right)\times S^2$. The first formula in (4.39) means each subpixel belongs to only one class, and the second formula means the subpixels for each class need to meet the constraint provided by the enhanced coarse proportion $\hat{F}_m\left(Q_J^{\text{OCSI}}\right)$.

4.6.4 EXPERIMENTAL CONTENT AND RESULT ANALYSIS

One synthetic data set and one real spectral data set are tested in the experiments. According to the general experimental process of subpixel mapping, the fine synthetic data or real spectral data are downsampled by $S \times S$ mean filter to obtain the simulated coarse image. A mixed pixel in the simulated coarse image is considered as S^2 subpixels in the fine data. In this case, the true spatial distribution of all subpixels in each mixed pixel in the simulated coarse image is known by referring to the fine data. This experimental processing method avoids the image registration error when manually adding class labels to obtain the reference image. For real spectral data, the abundance images are derived from real spectral data by spectral unmixing based on least squares support vector machine.

Because both subpixel mapping based on considering PSF effect model and spatiotemporal subpixel mapping model have been proved they are superior to the traditional subpixel mapping models [16, 20], we compare the proposed FCSTD model with these two advanced models. The pixel swapping-algorithm (PSA) method belongs to the spatial relationship between central subpixel and neighbor subpixels, and radial basis function interpolation (RBF) method belongs to the spatial relationship between central subpixel and neighbor pixels. Here, the three general models of six methods by utilizing PSA and RBF were compared: 1) The first model is subpixel mapping based on considering the PSF effect [20], namely, PSA-PSF and RBF-PSF. 2) The second model is spatiotemporal subpixel mapping proposed by Wang et al. [17], namely, PSA-SSM and RBF-SSM, and the spatial dependence of the two methods takes the PSF effect into account as well. 3) The third model is the proposed FCSTD, namely, PSA-FCSTD and RBF-FCSTD. In addition, the empirical values of the weight parameters λ and β are obtained by many experiments, λ is set to 0.7 and 0.6 in the synthetic data set and real data, and β is set to 0.6 and 0.5 in the synthetic data set and real data. Overall accuracy (OA) and kappa coefficient (Kappa) are used to evaluate the six subpixel mapping methods. All experiments are tested by MATLAB 2018a software package.

4.6.4.1 Synthetic Data

The data set contains three maps from National Land-Cover Database (NLCD), which were captured in 2001, 2006, and 2011. The data set with a 30-m spatial resolution is obtained by classification based on raster method. As shown in Figure 4.40, the tested area is with 1000 × 1000 pixels, which covers an area in Georgia, United States, and includes 15 land cover classes. In this experiment, the NLCD 2001 image in Figure 4.40(a) is selected as the PFSI. The NLCD 2006 image in Figure 4.40(b) and NLCD 2011 image in Figure 4.40(c) are downsampled by an 10 × 10 mean filter to produce the simulated OCSIs with a 300-m spatial resolution, respectively. Therefore, Figure 4.40(b) and Figure 4.40(c) are selected as reference images for the mapping results according to the general experimental process of subpixel mapping introduced before. The mapping results of six subpixel mapping methods are shown in Figure 4.41. To facilitate observation, as shown in Figure 4.42, a sub-region of the six subpixel mapping results for NLCD 2006 with 100 × 100 pixels marked in Figure 4.40(b) are magnified to facilitate observation.

By visual comparison in the marked area of Figure 4.42, the mapping results of the proposed FCSTD model in Figure 4.42(f)–(g) are closer to the reference image in Figure 4.42(a). We also use OA (%) and Kappa to evaluate the performance of six subpixel mapping methods for NLCD 2006 image and NLCD 2011 image. According to OA (%) and Kappa listed in Table 4.9, the PSA-FCSTD and RBF-FCSTD can obtain the higher OA (%) and Kappa than the other four methods.

In addition, the six subpixel mapping methods are tested with three S values (i.e., 5, 10, and 20) for NLCD 2006 and NLCD 2011. OA (%) values in relation to different S values are shown in Figure 4.43. The proposed PSA-FCSTD and RBF-FCSTD still could obtain the higher OA (%) values than the other four methods.

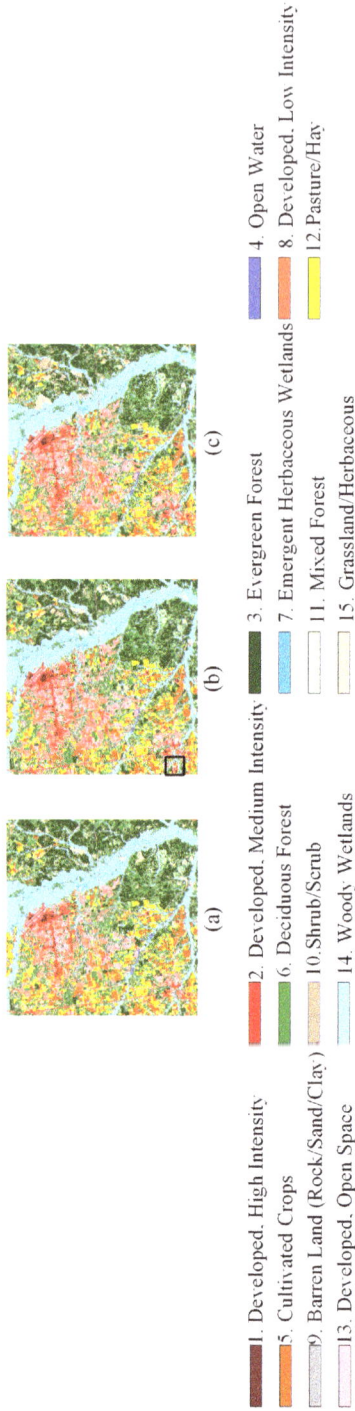

FIGURE 4.40 Synthetic data: (a) NLCD 2001, (b) NLCD 2006, and (c) NLCD 2011.

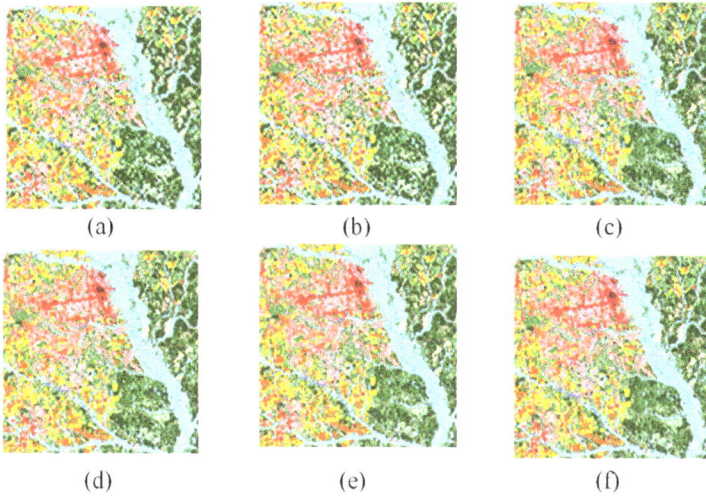

FIGURE 4.41 Subpixel mapping results of NLCD 2006 ($S = 10$): (a) PSA-PSF, (b) RBF-PSF, (c) PSA-SSM, (d) RBF-SSM, (e) PSA-FCSTD, and (f) RBF-FCSTD.

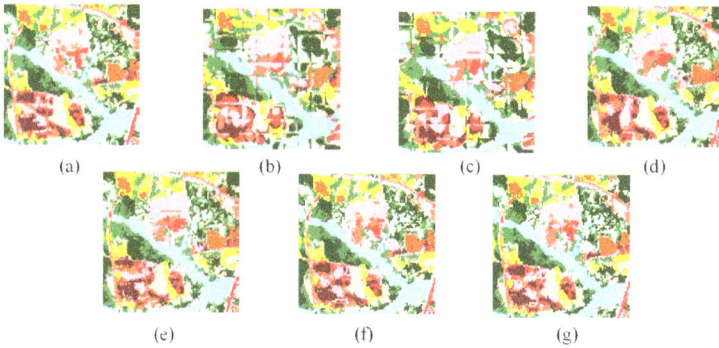

FIGURE 4.42 Sub-region of SM results for NLCD 2006 ($S = 10$): (a) reference image, (b) PSA-PSF, (c) RBF-PSF, (d) PSA-SSM, (e) RBF-SSM, (f) PSA-FCSTD, and (g) RBF-FCSTD.

TABLE 4.9
Quantitative Evaluation of the Different Methods ($S = 10$)

Prior Image	NLCD 2006		NLCD 2011	
Method	OA (%)	Kappa	OA (%)	Kappa
PSA-PSF	59.64	0.5355	58.96	0.5308
RBF-PSF	60.03	0.5415	59.39	0.5371
PSA-SSM	91.65	0.9043	89.05	0.8753
RBF-SSM	91.74	0.9055	89.25	0.8776
PSA-FCSTD	93.44	0.9249	90.55	0.8905
RBF-FCSTD	93.82	0.9293	91.17	0.8995

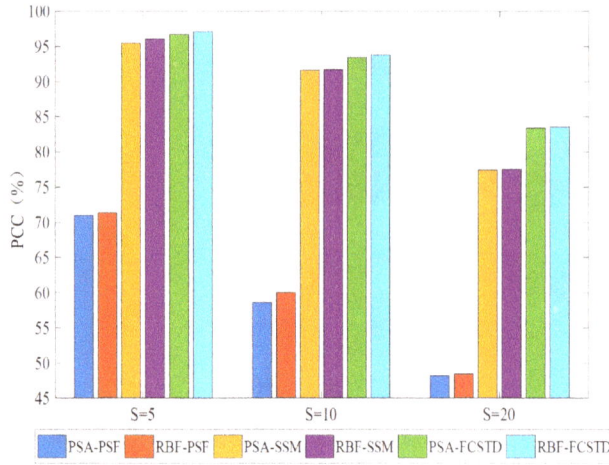

(a)

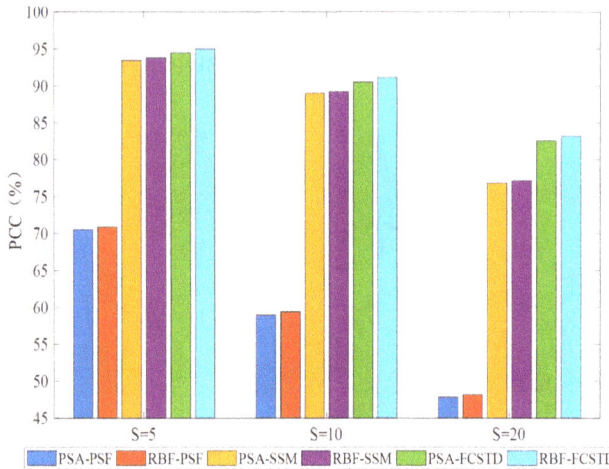

(b)

FIGURE 4.43 (a) OA (%) of the six subpixel mapping methods in relation to different S (a) for NLCD 2006 and (b) for NLCD 2011.

4.6.4.2 Real Data

This experiment aims to test the performance of the proposed FCSTD model for real spectral data, and also for OCSI and PFSI from different sensors. A multispectral image with 30-m spatial resolution and eight spectral bands is obtained by Landsat 8 operational land imager in 2014, and a hyperspectral image with 30-m spatial resolution and 198 spectral bands is obtained by Hyperion imaging spectrometer in 2002. As shown in Figure 4.44, the tested area has 200×160 pixels, which covers an area in Rome, Italy. The fine multispectral image in Figure 4.44(a) is selected as the PFSI. The fine hyperspectral image in Figure 4.44(b) is downsampled by a 5×5 mean filter to simulate OCSI with a 150-m spatial resolution. As shown in Figure 4.44(c),

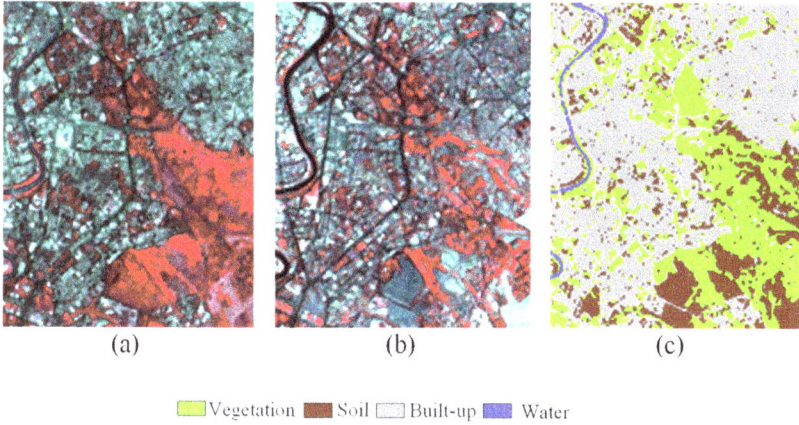

(a)   (b)   (c)

Vegetation   Soil   Built-up   Water

FIGURE 4.44   Real data: (a) multispectral image (bands 5, 2, and 3 for red, green, and blue); (b) hyperspectral image (bands 150, 10, and 24 for red, green, and blue); and (c) reference image.

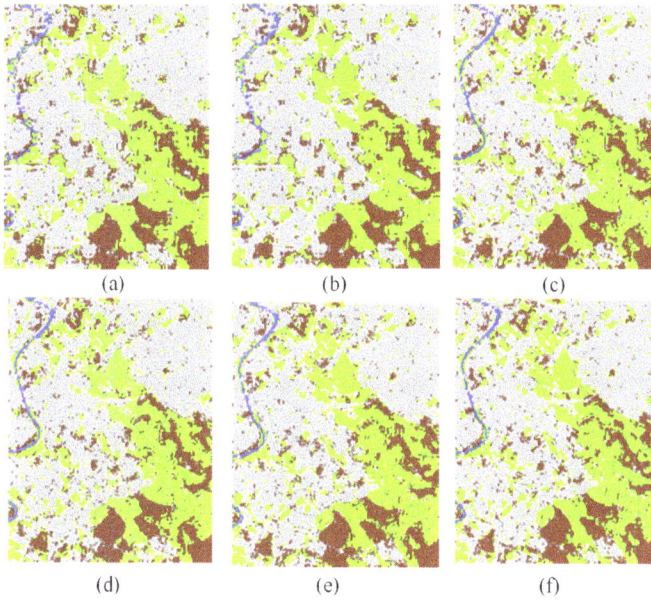

(a)   (b)   (c)

(d)   (e)   (f)

FIGURE 4.45   Subpixel mapping results of hyperspectral image ($S = 5$): (a) PSA-PSF, (b) RBF-PSF, (c) PSA-SSM, (d) RBF-SSM, (e) PSA-FCSTD, and (f) RBF-FCSTD.

the reference image is derived from the classification result of Figure 4.44(b), which includes four land cover classes.

The mapping results of six subpixel mapping methods are shown in Figure 4.45. The mapping accuracies of each land cover class, OA (%) and Kappa are listed in Table 4.10. Similar to the experimental results of synthetic data, both visual

TABLE 4.10
Quantitative Evaluation of the Different Methods ($S = 5$)

Method	Vegetation (%)	Soil (%)	Built-up (%)	Water (%)	OA (%)	Kappa
PSA-PSF	72.34	62.14	85.25	44.05	78.13	0.6332
RBF-PSF	73.88	62.92	85.88	44.65	78.42	0.6373
PSA-SSM	78.12	66.70	87.89	70.17	81.60	0.6892
RBF-SSM	78.43	66.87	87.95	70.44	81.75	0.6918
PSA-FCSTD	80.48	69.92	88.68	69.89	83.20	0.7162
RBF-FCSTD	80.52	70.42	88.95	71.27	83.45	0.7206

comparison and quantitative evaluation show that the proposed FCSTD model shows the better performance than the other two models.

4.7 SUMMARY

Since the existing subpixel mapping methods based on MSIs cannot make full use of spatial-spectral information of the original image, a subpixel mapping method based on MSIs with spatial-spectral information has the following advantages: (1) This method inherits the advantages of the existing interpolation-based MSI subpixel mapping method, that is, the calculation speed is fast, and the implementation is simple. In addition, the method can also be applied without any spatial prior structural information. (2) Compared with the existing subpixel mapping method based on MSIs, this method can more fully consider the spatial-spectral information of the original remote sensing image and then obtain more accurate subpixel mapping results.

Aiming at the problem of the single-scale information of MSIs in the existing subpixel mapping methods based on MSIs, subpixel mapping based on the spatial attraction model with multi-scale subpixel shifted images is proposed. In this method, the multi-scale shifted images are obtained by integrating the smooth term of fine-scale homogeneity and the penalty term of coarse-scale heterogeneity. The smoothing term can make the subpixels belonging to the same class gather to form a uniformly distributed local area, while the penalty term can effectively preserve the regional details and restrain the edges from being excessively over-smoothed. Therefore, the MSIs of the proposed method have more abundant scale information. Compared to the other four subpixel mapping methods, the proposed SAM-MSSI can provide the more ideal subpixel mapping result.

Combining the previous two methods proposed in this chapter, utilizing parallel networks to produce subpixel shifted images with multi-scale spatial-spectral information for subpixel mapping is proposed, which solves the problem of the single type of information contained in the MSIs in the existing subpixel mapping method. This method combines the MSIs with multi-scale information generated by the multi-scale network and the MSIs with spatial-spectral information generated by the spatial-spectral network to generate a MSI with multi-scale spatial-spectral information, and then enriches the type of information in the MSIs. Through visual and quantitative verification, when compared with the existing subpixel mapping methods based on MSIs, the method proposed in this chapter can obtain the better subpixel mapping results.

To improve the mapping result of spatiotemporal subpixel mapping, a general spatiotemporal subpixel mapping model named FCSTD is proposed. In the proposed FCSTD model, the scales and properties of the two abundance images from OCSI and PFSI are equal. By describing the fine-scale and coarse-scale temporal dependence, the temporal dependence information is more accurate and richer. In addition, the PSF effect is considered to improve the final mapping result. The PSA-FCSTD and RBF-FCSTD from the proposed model are evaluated and compared to four subpixel mapping methods from two state-of-the-art subpixel mapping models. The results indicated that the FCSTD achieved better mapping results.

REFERENCES

[1] Ling F, Wu S, Xiao F, Wu K, Li X. Summary of research on remote sensing image subpixel mapping[J]. Chinese Journal of Image and Graphics, 2011, 16(8): 1335–1345.

[2] Wang Q, Shi W, Wang L. Indicator cokriging-based subpixel land cover mapping with shifted images[J]. IEEE Journal of Selected Topics in Applied Earth Observations and Remote Sensing, 2014, 7(1): 327–339.

[3] Xu X, Zhong Y, Zhang L. A sub-pixel mapping based on an attraction model for multiple shifted remotely sensed images[J]. Neurocomputing, 2014, 134(9): 79–91.

[4] Wang Q, Shi W. Utilizing multiple subpixel shifted images in subpixel mapping with image interpolation[J]. IEEE Geoscience and Remote Sensing Letters, 2014, 11(4): 798–802.

[5] Wang L, Wang Z, Dou Z, Wang Y. Edge-directed interpolation-based sub-pixel mapping[J]. Remote Sensing Letters, 2013, 12(4): 1195–1203.

[6] Xu X, Zhong Y, Zhang L, Zhang H. Sub-pixel mapping based on a MAP model with multiple shifted hyperspectral imagery[J]. IEEE Journal of Selected Topics in Applied Earth Observations and Remote Sensing, 2013, 6(2): 580–593.

[7] Wang Q, Shi W, Wang L. Indicator cokriging-based subpixel land cover mapping with shifted images[J]. IEEE Journal of Selected Topics in Applied Earth Observations and Remote Sensing, 2014, 7(1): 327–339.

[8] Mertens K C, Basets B D, Verbeke L P C, Wulf R De. A sub-pixel mapping algorithm based on sub-pixel/pixel spatial attraction models[J]. International Journal of Remote Sensing, 2006, 27(15): 3293–3310.

[9] Chen Y, Ge Y, Wang Q, Jiang Y. A subpixel mapping algorithm combining pixel-level and subpixel-level spatial dependences with binary integer programming[J]. Remote Sensing Letters, 2014, 5(10): 902–911.

[10] Richards J A, Jia X. Using suitable neighbors to augment the training hyperspectral maximum likelihood classification[J]. IEEE Geoscience and Remote Sensing Letters, 2008, 5(4): 774–777.

[11] Park B, Windham W R, Lawrence K C, Smith D P. Contaminant classification of poultry hyperspectral imagery using a spectral angle mapper algorithm[J]. Biosystems Engineering, 2007, 96(3): 323–333.

[12] Jia S, Qian Y. Spectral and spatial complexity-based hyperspectral unmixing[J]. IEEE Transactions on Geoscience and Remote Sensing, 2006, 45(12): 3867–3879.

[13] Wang P, Wang L. Soft-then-hard super-resolution mapping based on a spatial attraction model with multiscale sub-pixel shifted images[J]. International Journal of Remote Sensing, 2017, 38(15): 4303–4326.

[14] Ling F, Li W, Du Y, Li X. Land cover change mapping at the subpixel scale with different spatial-resolution remotely sensed imagery[J]. IEEE Geoscience and Remote Sensing Letters, 2011, 8(1): 182–186.

[15] Wu K, Wang Y, Niu R Q, Wei L F. Subpixel land cover change mapping with multi-temporal remote-sensed images at different resolution[J]. Journal of Applied Remote Sensing, 2014, 1: 097299.

[16] Wang Q, Shi W, Atkinson P M, Li Z. Land cover change detection at subpixel resolution with a Hopfield neural network [J]. IEEE Journal of Selected Topics in Applied Earth Observations and Remote Sensing, 2015, 8(3): 1339–1352.

[17] Wang Q, Shi W, Atkinson P M. Spatiotemporal subpixel mapping of time-series images [J]. IEEE Transactions Geoscience and Remote Sensing, 54(9): 5397–5411.

[18] He D, Zhong Y, Zhang L. Spatiotemporal subpixel geographical evolution mapping [J]. IEEE Transactions on Geoscience and Remote Sensing, 2019, 57(4): 2198–2220.

[19] He D, Zhong Y, Zhang L. Spectral-spatial-temporal MAP-based sub-pixel mapping for land-cover change detection[J]. IEEE Transactions on Geoscience and Remote Sensing, 2020, 58(3): 1696–1717.

[20] Wang Q, Zhang C, Tong X, Atkinson P M. General solution to reduce the point spread function effect in subpixel mapping[J]. Remote Sensing of Environment, 2020, 251: 112054.

5 Subpixel Mapping of Remote Sensing Image Based on Fusion Technology

5.1 INTRODUCTION

Through the aforementioned introduction, we already know that the subpixel mapping technology is an ill-conditioned inverse problem [1]. It is often necessary to use auxiliary information to improve the technology. At the same time, considering that the coarse resolution of the original remote sensing image is one of the important factors affecting the accuracy of the final subpixel mapping result. In view of this, in this chapter, we use the already mature fusion technology in the field of remote sensing to improve the resolution of the original remote sensing image, supply more spatial-spectral information, and ultimately improve the accuracy of the subpixel mapping result. The effective combination of fusion technology and subpixel mapping technology also provides readers with a new way of handling remote sensing image subpixel mapping.

This chapter uses fusion technology to propose three algorithms. First, a soft-then-hard subpixel mapping based on pansharpening technique for remote sensing image (STHSRM-PAN) is proposed. Using the fusion method based on pansharpening technique [2–3], the original coarse remote sensing images are fused with the panchromatic images with high spatial resolution in the same area to obtain the improved remote sensing image. Experiments have proved that a higher subpixel mapping result can be obtained by the proposed STHSRM-PAN.

Secondly, especially for the subpixel mapping of a hyperspectral remote sensing image, a subpixel land cover mapping based on dual processing paths for hyperspectral image (DPP) is proposed. This algorithm consists of a fusion path and a deep path. After the parallel processing path, the abundance images with richer multi-scale spatial-spectral information are generated. Ultimately, due to the richness of the information types of abundance images, the subpixel mapping accuracy of the hyperspectral remote sensing image is improved.

Finally, subpixel mapping based on multi-source remote sensing fusion data for land cover classes (SPM-MRSFD) is proposed. First, the original hyperspectral image and the auxiliary panchromatic image are fused to produce the high spatial and spectral resolution fused image by pansharpening technology. Second, the fused image with spatial-spectral information and the auxiliary digital surface model (DSM) of light detection and ranging (LiDAR) with elevation information are fused to obtain the multi-source remote sensing fusion data (MRSFD) with spatial-spectral-elevation information by feature fusion. Finally, the abundance images with the proportions of subpixels belonging to land cover classes are derived by unmixing the MRSFD, and the class labels are allocated to subpixels to obtain the final subpixel mapping result according to these proportions' information. Experimental results show that SPM-MRSFD obtains the more accurate mapping results than state-of-the-art subpixel mapping methods.

5.2 SOFT-THEN-HARD SUBPIXEL MAPPING BASED ON PANSHARPENING TECHNOLOGY

STHSRM is a major type of subpixel mapping. However, due to the coarse resolution of the original remote sensing image, it is difficult to extract the full spatial-spectral information from the original image for the abundance images. The proposed algorithm applies the pansharpening technique for subpixel mapping to generate the abundance images with more spatial-spectral information, which improves the subpixel mapping result. First, the original low-resolution remote sensing image and the panchromatic images are fused by using pansharpening technology to obtain the improved image with high spectral resolution and high spatial resolution. Then, the high-resolution abundance images with more spatial-spectral information are obtained by unmixing the improved image. Finally, according to the abundance value of the high-resolution abundance images, the class labels are assigned to subpixels. Compared with the existing soft-then-hard subpixel mapping methods, the algorithm shows the best performance in PCC (%) and Kappa among the three experimental results.

5.2.1 Pansharpening Technology

Pansharpening technology refers to the fusion of the original coarse remote sensing image (multispectral image or hyperspectral image) and the high spatial resolution panchromatic image of the same area, so that the obtained image has the high spectral resolution of the former and the high spatial resolution of the latter [2]. This can be regarded as a special problem of image fusion, because the unique result contains the spatial details carried by the panchromatic image and the spectral band of the original remote sensing image.

Due to the advantages of high-fidelity rendering of spatial details and good robustness during registration, component substitution (CS) is as widely used as the pansharpening model. This algorithm uses pansharpening based on CS. CS projects the original remote sensing image to another space in order to separate the spatial structure and spectral information into different components. Subsequently, the high spatial resolution panchromatic image is used to replace the components containing the spatial structure, thereby enhancing the spatial resolution of the transform domain image. The greater the correlation is between the pansharpening image and the replaced component means the smaller the spectral distortion introduced by the fusion method. For this reason, the panchromatic image and the replaced component are histogram-matched before the replacement occurs, and the panchromatic image

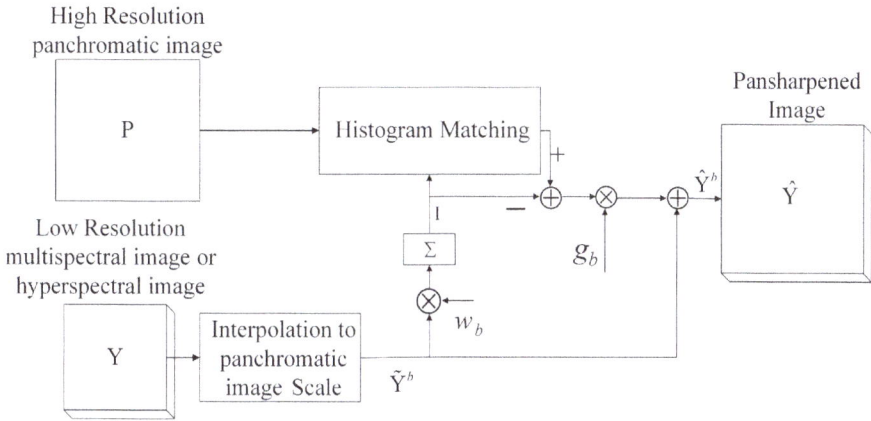

High Resolution
panchromatic image

P

Low Resolution
multispectral image or
hyperspectral image

Y

Interpolation to
panchromatic
image Scale $\tilde{Y}^b$

Histogram Matching

w_b

Σ

g_b

Pansharpened
Image

$\hat{Y}^b$ $\hat{Y}$

FIGURE 5.1   Flowchart of pansharpening model based on component substitution.

matched by the histogram will show the same mean and variance as the component
to be replaced. Finally, the inverse transform is applied to obtain the fused image,
and the pansharpening process based on CS is completed. Figure 5.1 shows the flowchart of the pansharpening model based on CS.

The pansharpening model based on CS can generally be expressed by the following formula:

$$\hat{Y}^b = \tilde{Y}^b + g_b\left(P - I\right) \tag{5.1}$$

where b ($b = 1, 2, \ldots, N$) represents the bth spectral band; Y represents the original
coarse remote sensing image; $\hat{Y}$ is the pansharpened image after pansharpening; $\hat{Y}^b$
represents the image after the bth spectral band in the pansharpened image; $\tilde{Y}^b$ represents the bth band of which the original remote sensing spectral image is interpolated
to the panchromatic image size; $g_b = [g_1, g_2, \ldots, g_N]$ is the gain vector; P represents
the high spatial resolution panchromatic image; and I is defined as

$$I = \sum_{b=1}^{N} w_b \tilde{Y}^b \tag{5.2}$$

where the weight vector $w_b = [w_1, w_2, \ldots, w_N]^T$ is a measure of the spectral overlap
between the spectral band and the panchromatic image. There are many commonly
used pansharpening methods that belong to the CS type, such as principal component analysis (PCA) [4], gram schmidt (GS) [5], and intensity-hue-saturation (HIS)
[6]. PCA is selected as the pansharpening technique of this algorithm with its simple
and fast implementation. PCA generates a set of scalar images called principal components by linearly transforming the original image. The basic assumption of PCA
is that the first principal component contains all spatial information, while the other
principal components contain the spectral information. Formula (5.1) and formula
(5.2) can describe the entire pansharpening process, where PCA acts on the remote
sensing image to obtain the vector and coefficient vector.

In particular, the complexity of the PCA algorithm is reduced by introducing the previously proposed endmember of interest (EOI). EOI is composed of spectral endmembers. The column vector of EOI is defined as an operator Φ that represents the low-dimensional data mapped to the original high-dimensional data; $\Phi_{inv} = \left(\Phi^T \Phi \right)^{-1} \Phi^T$ represents an operator that maps the original high-dimensional data to low-dimensional data. First, the mapping operator Φ_{inv} is used to map the original high-dimensional data to the low-dimensional transform space. Then the PCA process described earlier is applied. Finally, the operator Φ is utilized to map the pansharpening results to the original space. Since the dimensionality of the input data is decreased, the PCA process is simplified.

5.2.2 STHSRM-PAN

The proposed STHSRM-PAN algorithm improves the final subpixel mapping result by providing more spatial-spectral information. First, the pansharpening method based on PCA is used to improve the original coarse image Y, and the improved image $\hat{Y}$ is obtained by formula (5.1) and formula (5.2). Then, the high-resolution abundance images H_k with rich spatial-spectral information are generated by unmixing the image $\hat{Y}$, and the abundance images contain subpixel abundance values. Finally, the class allocation method based on the principle of units of class (UOC) is used to assign class labels to subpixels to obtain the final mapping result.

The flowchart of STHSRM-PAN is shown in Figure 5.2. Implementation consists of the following three steps:

Step 1. The original coarse remote sensing image and the high spatial resolution panchromatic image are fused through pansharpening based on PCA (see from formula [5.1] and formula [5.2]) to generate an improved image.

Step 2. By unmixing the improved image, the high-resolution abundance images with rich spatial-spectral information are obtained. This algorithm uses the spectral unmixing method based on the least squares linear mixing model [7] to predict the proportion of land cover classes in the image obtained by the pansharpening technique.

Step 3. According to the subpixel abundance values provided by the high-resolution abundance images, the UOC class allocation method is used to assign labels to obtain the final subpixel mapping result.

The abundance images are generated from the original low-resolution remote sensing image in the existing STHSRM, while the abundance images are obtained from the pansharpened result in the proposed STHSRM-PAN. Since the fusion image obtained by pansharpening technology combines the spatial details of the panchromatic image and the spectral details of the spectral image, the abundance images can extract more spatial-spectral information from the original image. Compared with the existing subpixel mapping methods based on the existing STHSRM, STHSRM-PAN can obtain the more ideal subpixel mapping result.

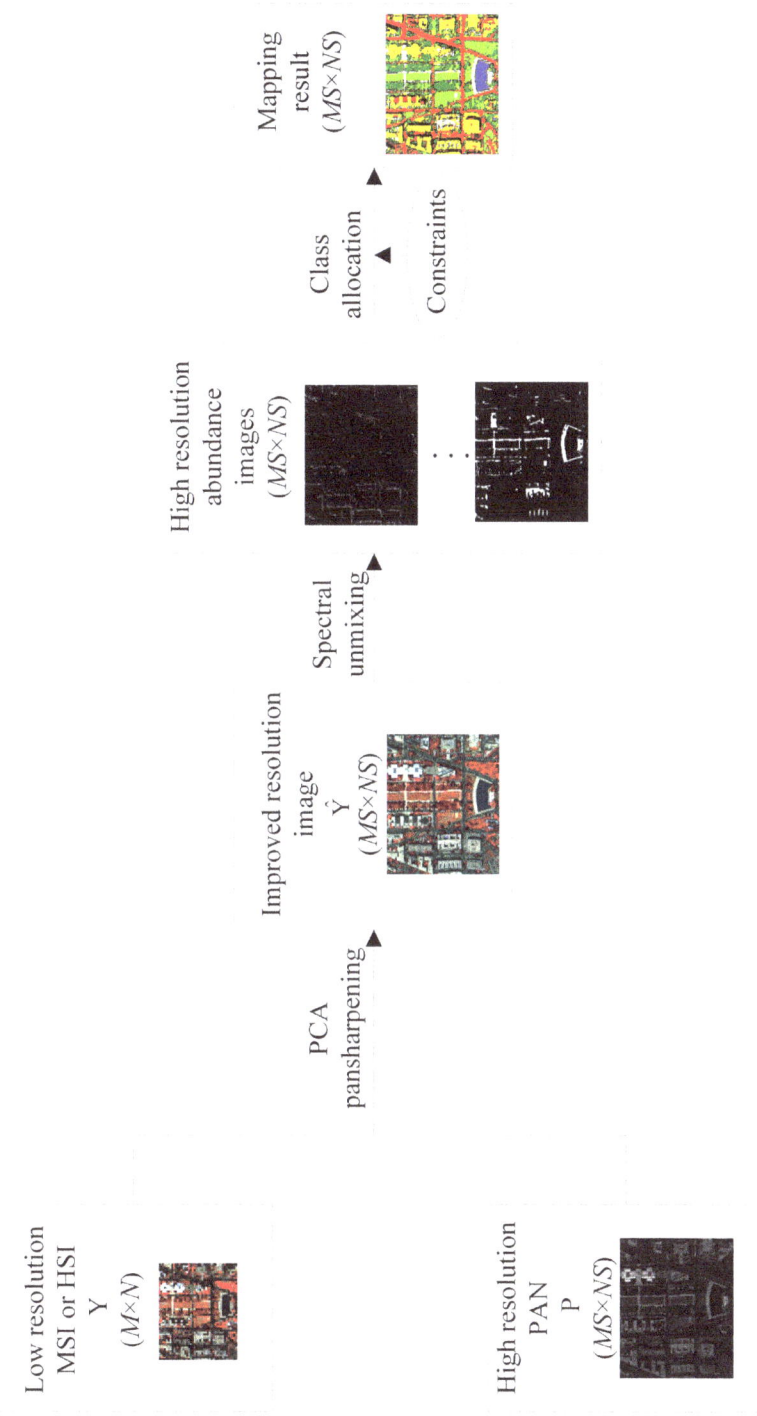

FIGURE 5.2 Flowchart of STHSRM-PAN.

5.2.3 Experimental Content and Result Analysis

Experiments on three remote sensing images have verified the performance of the proposed method. For quantitative evaluation, the low-resolution images obtained by downsampling the original fine remote sensing images are selected as experimental data. In addition, only considering the effect of the pansharpening technique and avoiding the impact of errors caused by the acquisition of the panchromatic image, the appropriate synthetic panchromatic image is created by the spectral response of the IKONOS satellite [8, 9]. The satellite captured four multispectral bands (0.45–0.52 μm, 0.52–0.60 μm, 0.63–0.69 μm, and 0.76–0.90 μm) and a panchromatic image (0.45–0.90 μm).

Six subpixel mapping methods are tested and compared: edge interpolation (EI) [10], spatial-spectral bicubic interpolation (SSBIC) [11], radial basis function (RBF) interpolation [12], high-accuracy surface modeling (HASM) [13], HNN based on fused image (HNNF) [14], and the proposed STHSRM-PAN. According to the relevant literature, the relevant parameters of the five algorithms are selected and compared. PCC (%) and Kappa are used to evaluate the performance of six subpixel mapping methods.

5.2.3.1 Experiment 1

Experiment 1 is to verify the performance of the proposed method against the multispectral image. Only considering the impact of pansharpening technology on the mapping result, this set of experimental data is semi-synthesized. It is based on a standard hyperspectral image data set, which has 610×340 pixels, 115 spectral bands, spanning a spectral range of 0.43–0.86 μm, and a spatial resolution of 1.3 m. It was acquired by the ROSIS sensor during the flight activity of the University of Pavia. The spectral response of the IKONOS satellite is used to create suitable multispectral and panchromatic images. The semi-synthetic multispectral image is shown in Figure 5.3(a), and the test area is 100×100 pixels. The reference image is shown in Figure 5.3(b), including asphalt, grass, tree, and brick.

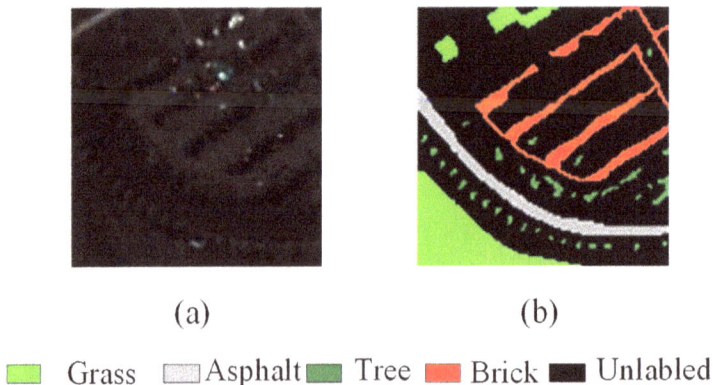

(a)   (b)

■ Grass ▢ Asphalt ■ Tree ■ Brick ■ Unlabled

FIGURE 5.3 Pavia University data set: (a) false color composite image (RGB band: 3, 2, 1) and (b) reference image.

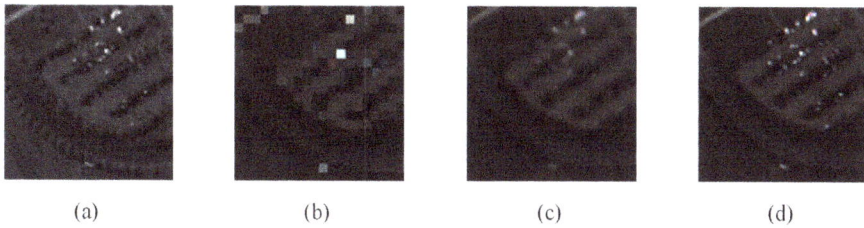

| (a) | (b) | (c) | (d) |

FIGURE 5.4 (a) Panchromatic image. (b) Low-resolution image ($S = 5$). (c) PCA. (d) PCA with EOI.

TABLE 5.1

Super-Resolution Reconstruction Errors for Different Classes

Land Cover Class	PCA (%)	PCA With EOI (%)
Grass	5.31	4.53
Asphalt	4.56	3.47
Tree	6.12	5.39
Brick	3.38	2.43

The semi-synthesized panchromatic image is shown in Figure 5.4(a). Figure 5.4(b) shows the low-resolution image generated by downsampling the high-resolution multispectral image in Figure 5.3(a). It can be found that due to the low resolution, it is difficult to obtain the spatial distribution information of the land cover classes from Figure 5.4(b). The purpose of pansharpening technology is to fuse a low-resolution spectral image and a high spatial resolution panchromatic image to obtain the improved image. Figure 5.4(c)–(d) show the results of PCA and PCA with EOI. By visual comparison, PCA with EOI is closer to the original image in Figure 5.3(a) than PCA. For further quantitative comparison, the relative error of super-resolution reconstruction is used to evaluate the performance of PCA and PCA with EOI. Table 5.1 shows the reconstruction error of different types. It can be found that the reconstruction error of each class of PCA with EOI is lower than that of PCA. In addition, compared with PCA, the operating speed of PCA with EOI is increased by more than five times. This is because EOI is used to reduce the complexity of the algorithm and protect the interest classes. Therefore, to obtain a more ideal STHSM-PAN result, this paper chooses PCA with EOI as the pansharpening algorithm.

The subpixel mapping results of the six methods are shown in Figure 5.5. The intuitive comparison of the results shows that the proposed STHSM-PAN in Figure 5.5(f) provides the more ideal mapping result than the other five subpixel mapping methods in Figure 5.5(a)–(e). In Figure 5.5(a)–(d), there are many obvious burrs on the boundary between asphalt and brick. This is because the spatial-spectral information of the original image is not fully utilized. Although HNNF uses the panchromatic image as auxiliary information, HNNF sometimes has the problem of excessive constraints, which makes the class boundaries too smooth, and small target

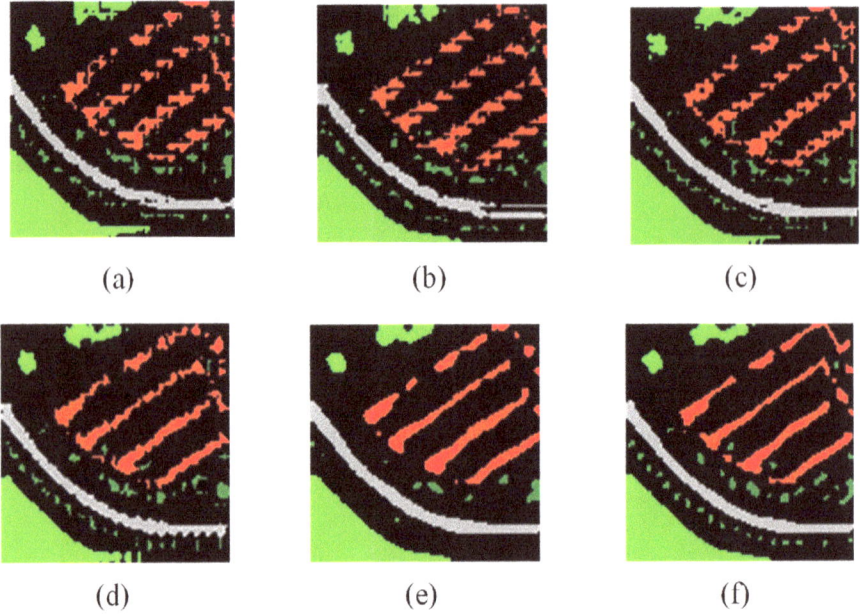

(a)   (b)   (c)

(d)   (e)   (f)

FIGURE 5.5   Subpixel mapping result: (a) EI, (b) SSBIC, (c) RBF, (d) HSAM, (e) HNNF, and (f) STHSM-PAN.

TABLE 5.2
Mapping Accuracy of the Six Methods in Experiment 1 ($S = 5$)

Land Cover Class	EI	SSBIC	RBF	HSAM	HNNF
Grass (%)	93.65	96.01	95.65	95.89	97.55
Asphalt (%)	87.31	84.45	97.65	95.01	98.53
Tree (%)	28.66	29.96	36.21	37.91	26.71
Brick (%)	58.03	61.12	69.50	76.03	81.09
PCC (%)	75.36	76.88	81.75	82.89	87.59
Kappa	0.6914	0.7074	0.7641	0.7760	0.8224

---

classes are ignored. With the help of pansharpening technology, this phenomenon is alleviated. Compared with the HNNF results in Figure 5.5(e), the STHSM-PAN results in Figure 5.5(f) show better continuity and smoother boundaries of land cover classes, and the subpixel mapping result is closer to the reference map.

The performance of the six methods is also quantitatively evaluated by the mapping accuracy (%) of each class, PCC (%) and Kappa. As shown in Table 5.2, the proposed STHSM-PAN achieves higher mapping accuracy than the other five methods. Compared with the HNNF algorithm, the accuracy of the grass, tree, and brick in the result of the STHSRM-PAN method are increased by 1.3%, 16.6%, and 4.2%, respectively. In addition, STHSRM-PAN can produce the highest PCC (%) and Kappa values.

5.2.3.2 Experiment 2

In order to get closer to the actual situation, the real multispectral image with a larger area and more complex distribution is selected as the second test data source. Figure 5.6(a) shows the Landsat 8 multispectral data collected in Rome with a size of 400 × 300 pixels. The test data have six bands including blue, green, red, near-infrared, short-wave infrared 1, and short-wave infrared 2. In the reference image shown in Figure 5.6(b), there are four classes (water, vegetation, building, and soil). As shown in Figure 5.7(a), the panchromatic image is generated by the method described in experiment 1. The original fine multispectral image is downsampled to produce a low-resolution image as shown in Figure 5.7(b). The result of pansharpening is shown in Figure 5.7(c).

The mapping results of EI, SSBIC, RBF, HSAM, HNNF, and STHSRM-PAN are shown in Figure 5.8. From Figure 5.8(a)–(d), it can be seen that there are obvious disconnected shapes in the water, and there are many burrs on the boundary of the building. In addition, since the constraints of HNNF are kept too large, the small target classes are hardly mapped, such as the water and soil in Figure 5.8(e). Since the proposed STHSRM-PAN uses the more spatial-spectral information, the final mapping result is closer to the reference image. As shown in Figure 5.8(f), the continuity of each land cover class is better, and the class boundary is smoother. Table 5.3 lists the mapping accuracy (%) of each class, PCC (%), and Kappa for the six methods. Similar to the conclusion drawn in experiment 1, it is found that the mapping accuracy of STHSRM-PAN is higher than that of the other five methods.

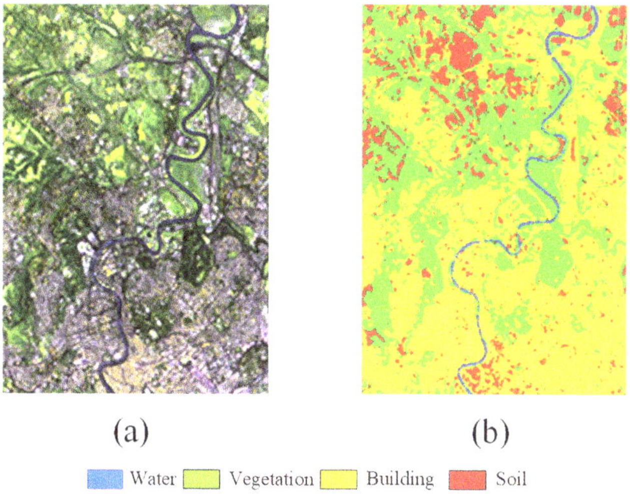

(a)   (b)

Water   Vegetation   Building   Soil

FIGURE 5.6 Rome data set: (a) false color composite image (RGB band: 6, 5, 2) and (b) reference image.

| (a) | (b) | (c) |

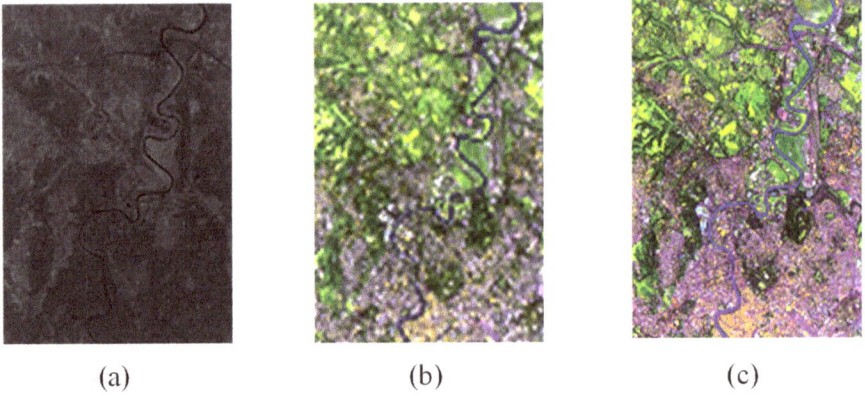

FIGURE 5.7   (a) Panchromatic image. (b) Low-resolution image ($S = 4$). (c) Pansharpening result.

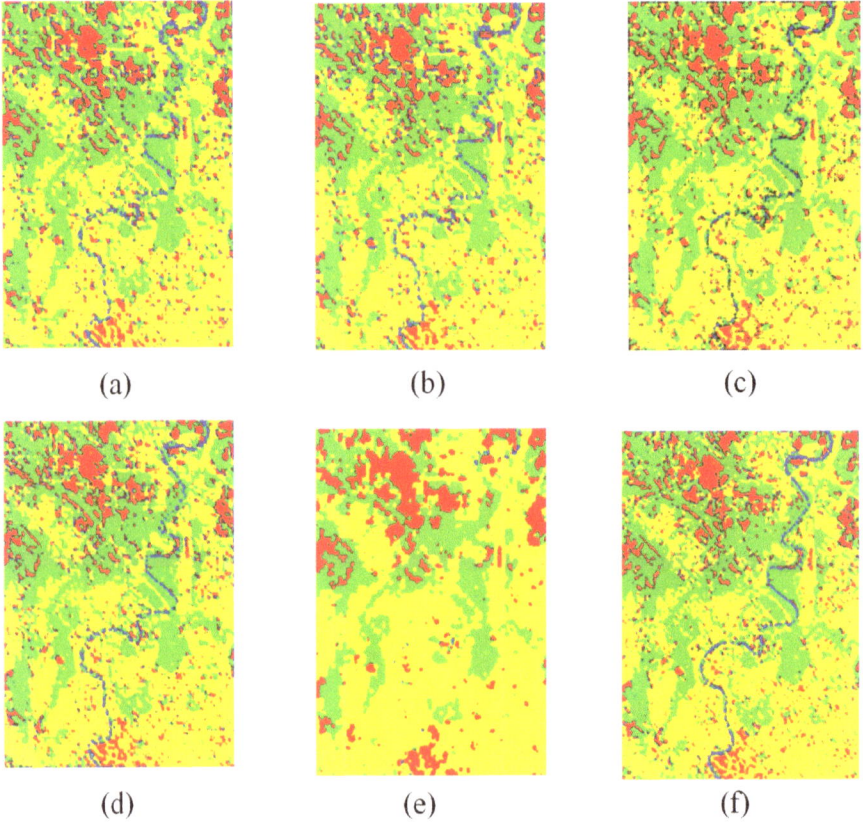

| (a) | (b) | (c) |

| (d) | (e) | (f) |

FIGURE 5.8   Subpixel mapping result: (a) EI, (b) SSBIC, (c) RBF, (d) HSAM, (e) HNNF, and (f) STHSM-PAN.

TABLE 5.3

Mapping Accuracy of the Six Methods in Experiment 2 ($S = 4$)

Land Cover Class	EI	SSBIC	RBF	HSAM	HNNF	STHSRM-PAN
Water (%)	46.95	47.19	47.47	52.66	11.17	66.25
Vegetation (%)	67.80	68.85	71.23	72.18	76.80	78.35
Architecture (%)	77.23	78.39	80.14	80.33	81.13	83.94
Soil (%)	56.06	56.04	62.74	60.14	65.71	68.50
PCC (%)	71.02	72.07	74.10	74.79	77.04	79.97
Kappa	0.5346	0.5445	0.5723	0.5888	0.6131	0.6703

To study the influence of the ratio scale S on the proposed method, the other two ratio scales $S = 8$ and $S = 10$ were tested. Figure 5.9 (a)–(b) show the PCC (%) and Kappa values of the six methods under all three scales. It can be found that with the increase of S, the PCC (%) and Kappa values of all six methods decrease. This is because the larger S value means a lower resolution of the coarse image is produced. But consistent with the results in Table 5.3, the proposed STHSRM-PAN still produces higher PCC (%) and Kappa values than the other five subpixel mapping methods.

5.2.3.3 Experiment 3

In experiment 3, the proposed STHSM-PAN was tested against the real hyperspectral image. As shown in Figure 5.10(a), the test data come from the Washington, DC, Shopping Center (250 ´ 300 area, 191 band, 0.4–2.4-μm spectral range and 3-m spatial resolution). The reference image is shown in Figure 5.10(b), which mainly includes seven classes: shadow, water, road, tree, grass, roof, and trail. The false color image is shown in Figure 5.11(a). In this experiment, the original hyperspectral image is downsampled with $S = 3$, 5, and 8 to obtain the simulated low-resolution image. The low-resolution image ($S = 5$) is shown in Figure 5.11(b). Figure 5.11(c) shows the improved image obtained by pansharpening technology.

The results of six subpixel mapping methods are shown in Figure 5.12. Due to the lack of sufficient spatial-spectral information, there are some broken hole patches in Figure 5.12(a)–(e). With the help of pansharpening technology, the subpixel mapping result generated by STHSRM-PAN is more ideal. Figure 5.13(a)–(b) is a schematic diagram of the PCC (%) and Kappa values of the six methods for three scales S, namely, 3, 5, and 8. Consistent with previous experimental results, the proposed STHSRM-PAN produces the higher PCC (%) and Kappa values than the other five subpixel mapping methods.

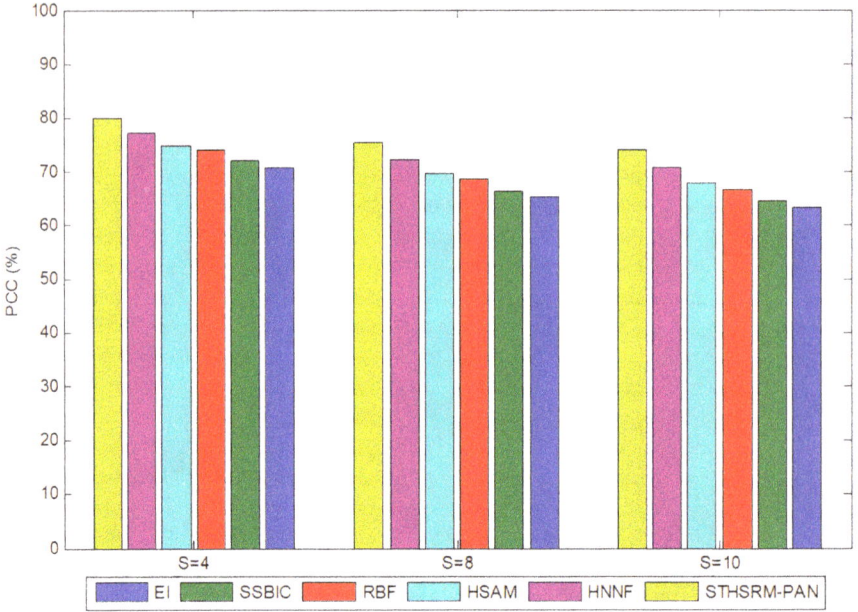

(a)

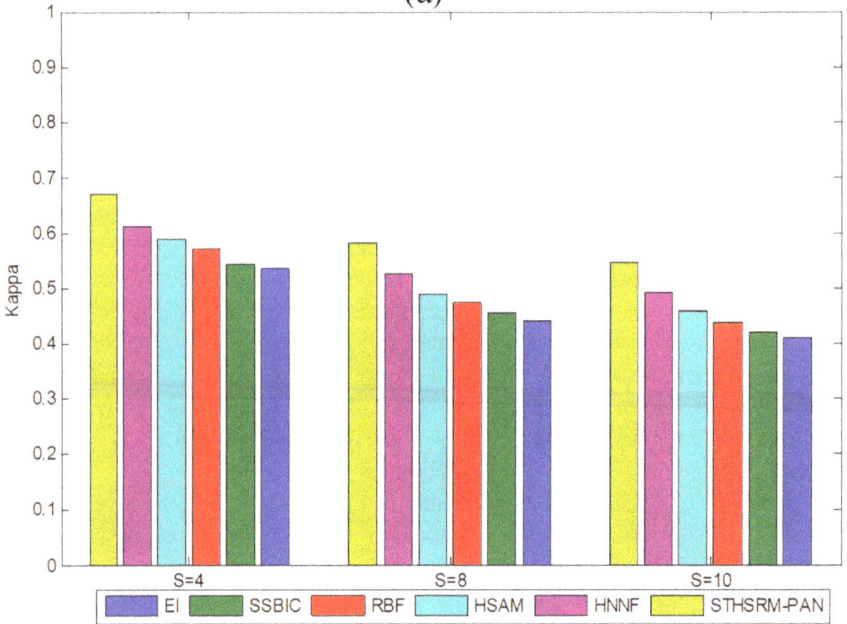

(b)

FIGURE 5.9   (a), (b) Relationship between PCC (%) and Kappa of the six methods and scale *S*.

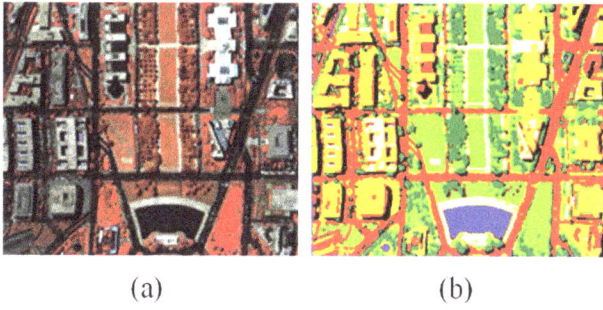

■ Shadow ■ Water ■ Road ■ Tree ■ Grass ■ Roof □ Trail

FIGURE 5.10   Washington, DC, data set: (a) false color composite image (RGB band: 65, 52, 36) and (b) reference image.

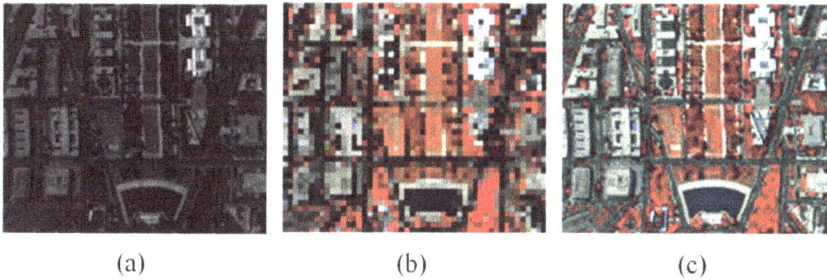

FIGURE 5.11   (a) Panchromatic image. (b) Low-resolution image ($S = 5$). (c) Pansharpening result.

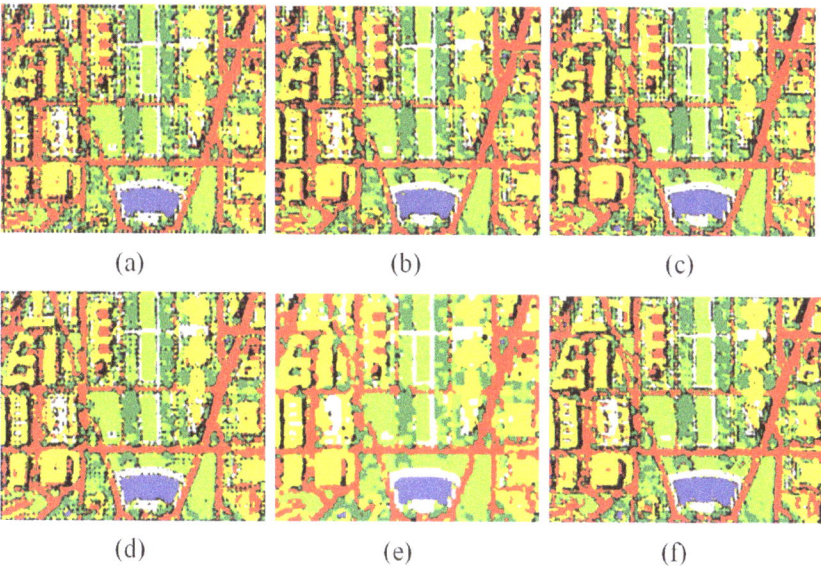

FIGURE 5.12   Subpixel mapping result: (a) EI, (b) SSBIC, (c) RBF, (d) HSAM, (e) HNNF, and (f) STHSM-PAN.

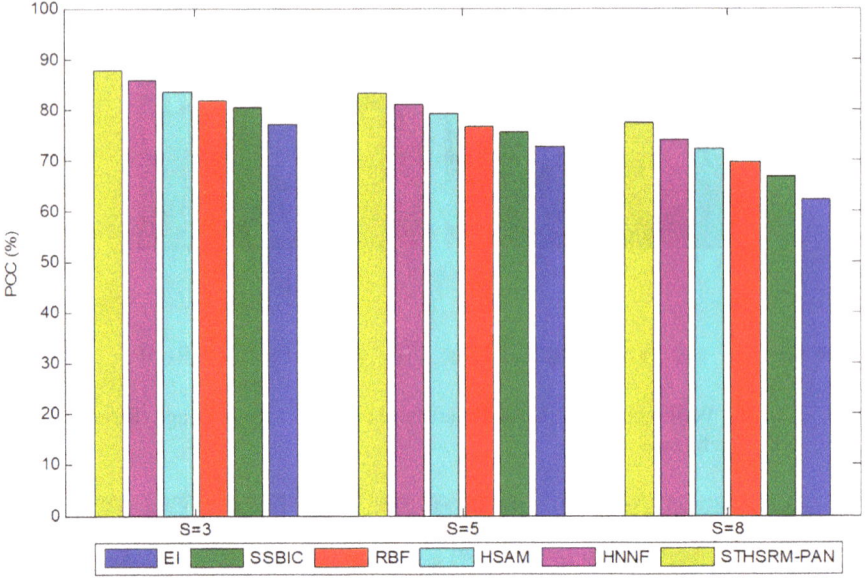

(a)

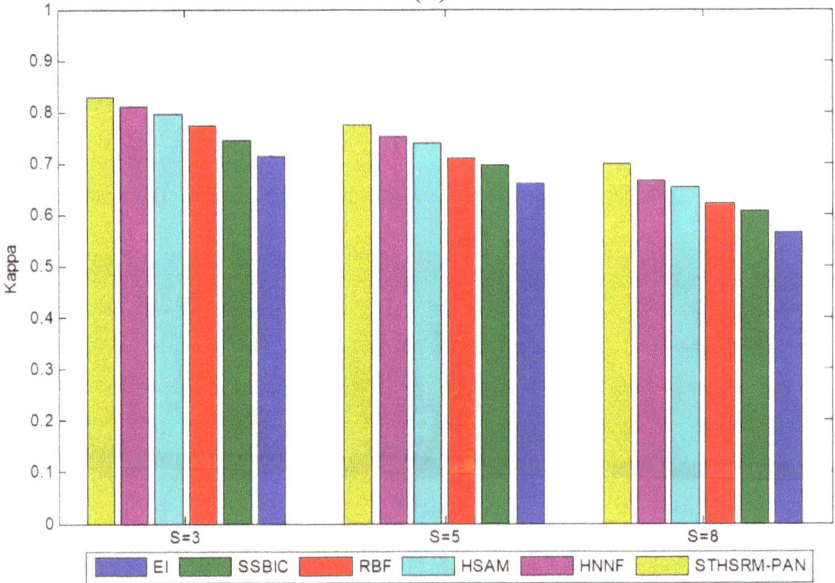

(b)

FIGURE 5.13   (a), (b) Relationship between PCC (%) and Kappa of the six methods and scale *S*.

5.3 SUBPIXEL LAND COVER MAPPING BASED ON PARALLEL PROCESSING PATH FOR HYPERSPECTRAL IMAGE

Through the introduction in the previous section, it is found that the resolution of the original remote sensing image can be improved with the help of pansharpening technology, and then the final subpixel mapping result can be improved. In view of this idea, for a hyperspectral remote sensing image, fusion technology and deep learning are introduced into hyperspectral image subpixel mapping, and subpixel land cover mapping based on dual processing paths (DPPs) for hyperspectral image is proposed. The algorithm consists of a fusion path and a deep learning path. In the fusion path, the original low spatial resolution hyperspectral image and the high spatial resolution multispectral image from the same area are fused through a PCA-based hyperspectral and multispectral fusion algorithm to produce an improved hyperspectral image with high spectral resolution and spatial resolution. Spectral unmixing is then performed on the improved hyperspectral image to obtain the high-resolution abundance images with rich spatial-spectral information. In the deep learning path, the low-resolution abundance images are obtained by spectral unmixing of the original low spatial resolution hyperspectral image. The subpixel sharpening of the constitutive model can obtain the high-resolution abundance images containing rich multiscale information; The two kinds of high-resolution abundance images with different information obtained in the two paths are integrated to obtain the finer abundance images with multi-scale spatial-spectral information; Finally, according to the accurate subpixel abundance values provided by the finer abundance images, the class labels are assigned to subpixels through the class allocation method to obtain the final mapping result. Three real hyperspectral remote sensing data are tested. The experimental results show that the subpixel result obtained by the proposed DPP method is better than the other subpixel mapping methods, which proves the effectiveness of using DPPs.

5.3.1 FUSION PATH

Due to the high quality of the fusion of spatial details and the simple and fast implementation method, the proposed algorithm uses the PCA-based hyperspectral and multispectral fusion algorithm [3] to combine the original low-resolution hyperspectral image with the high-resolution multispectral image, producing an improved hyperspectral image with both high spectral resolution and high spatial resolution. Figure 5.14 shows the principle diagram of the PCA-based hyperspectral and multispectral fusion algorithm.

First, the multispectral image with high spatial resolution is used to obtain a simulated low-resolution multispectral image through low-pass filtering. By performing spectral matching between the original coarse hyperspectral image and the simulated low-resolution multispectral image, the appropriate hyperspectral image bands are selected as the replacement component. PCA generates a set of scalar images called principal components by linearly transforming these replacement components. It is assumed that spatial information (shared by all bands) is concentrated on the first principal component scalar image, while spectral information

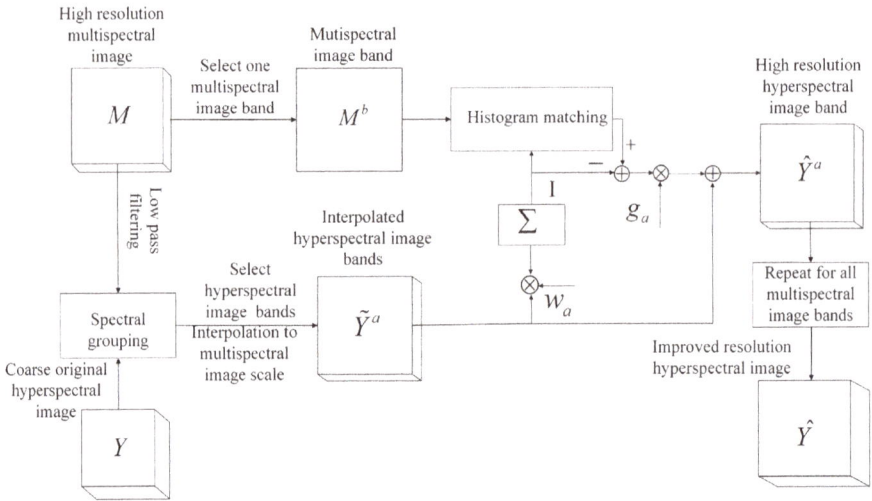

FIGURE 5.14 The PCA-based hyperspectral and multispectral fusion algorithm.

(specific to each frequency band) is on the other component scalar images. Then, these spatial details are replaced by the bands of a multispectral image containing the finer spatial details, which improves the spatial resolution of a hyperspectral image. The greater the correlation between the bands of the multispectral image and the replaced component means the smaller will be the spectral distortion introduced by the fusion method. Before the replacement occurs, the histogram matching of the bands of multispectral image with the replaced component is performed. The bands of multispectral image matched by the histogram will display the same mean and variance as the component to be replaced. This process is repeated for all bands of the multispectral image. Finally, the fused hyperspectral image is obtained through inverse transformation, and the PCA-based hyperspectral and multispectral fusion algorithm is completed. The formula of the PCA method is derived from the following formula:

$$\hat{Y}^a = \tilde{Y}^a + g_a\left(M^b - I\right) \tag{5.3}$$

where Y represents the original coarse hyperspectral image; $\hat{Y}$ is the hyperspectral image after fusion; $\hat{Y}^a$ ($a = 1, 2, \ldots, E$, E represents the number of selected hyperspectral image bands) is the ath band after fusion; $\tilde{Y}^a$ represents the size of the ath original coarse hyperspectral image band that is interpolated to the multispectral image; $g_a = [g_1, g_2, \ldots, g_E]$ is the gain vector; and M^b ($b = 1, 2, \ldots, B$, B is the number of the bands of multispectral image) is the bth band of the high spatial resolution multispectral image. At the same time, I is defined as

$$I = \sum_{a=1}^{E} w_a \tilde{Y}^a \tag{5.4}$$

where the weight vector $w_a = [w_1, w_2, \ldots, w_E]^T$ is used to measure the spectral overlap between the bands of multispectral image and the bands of hyperspectral image. When all the multispectral bands have been used for replacement, the final fusion result is obtained.

In particular, as with the previous pansharpening technology, since the hyperspectral image belongs to three-dimensional nautical data, the EOI of class of interest (COI) is used to reduce the complexity of fusion and improve the fusion result. The land cover classes with more pixels are selected as the COI. EOI is composed of spectral endmembers of COI. The EOI column vector is defined as an operator Φ that represents the low-dimensional data mapping to the original high-dimensional data. $\Phi_{inv} = \left(\Phi^T \Phi\right)^{-1} \Phi^T$ is the mapping operator that the original high-dimensional data is mapped to the low-dimensional data. The original high-dimensional hyperspectral data are first mapped to the low-dimensional transform space by Φ_{inv}. Then, the PCA-based hyperspectral and multispectral fusion algorithm is applied. Finally, this operator Φ is used to map the fusion result to the original dimensional space to obtain the fused hyperspectral image $\hat{Y}$. Therefore, formula (5.3) and formula (5.4) can be expressed as formula (5.5). As the dimensionality of the input data is reduced, the fusion process is simplified, and the fusion effect is improved.

$$\hat{Y}^a = \Phi\left[\Phi_{inv}\tilde{Y}^a + g_a\left(M^b - \sum_{a=1}^{E} w_a \Phi_{inv}\tilde{Y}^a\right)\right] \tag{5.5}$$

Next, the fused hyperspectral image $\hat{Y}$ is unmixed to obtain the high-resolution abundance images H_k^{sp} with the abundance values $H_k^{sp}(p_j)$ belonging to kth class ($k = 1, 2, \ldots, K$, K is the total number of land cover classes). Then P_j ($j = 1, 2, \ldots, M \times S^2$) represents subpixels and M is the number of pixels; when the scale is S, the number of subpixels is $M \times S^2$. Since the resolution of the original hyperspectral image is improved by fusion technology, the high-resolution abundance images H_k^{sp} can carry more spatial-spectral information.

5.3.2 DEEP LEARNING PATH

In the deep learning path, the super-resolution reconstruction model based on deep Laplace pyramid networks (DLPNs) [15, 16] is used to obtain high-resolution abundance images with multi-scale information. Figure 5.15 shows the schematic diagram of the super-resolution reconstruction model based on DLPNs.

First, the original coarse hyperspectral image Y is unmixed to obtain the low-resolution abundance images L_k as input. The low-resolution abundance images L_k contain pixels $P_J(J = 1, 2, \ldots, M$, M is the number of pixels) belonging to the kth class of abundance value $L_k(P_J)$. The super-resolution reconstruction model based on DLPN gradually predicts the residual images of the low-resolution abundance images at the $\log_2 S$th level networks. For example, the fine abundance images with $S = 8$ are derived by three level networks. The super-resolution reconstruction model based on DLPN is composed of two parts: feature extraction and image reconstruction. In feature extraction, each level of network extracts features from the low-resolution abundance images through the concatenated convolution layers, and the

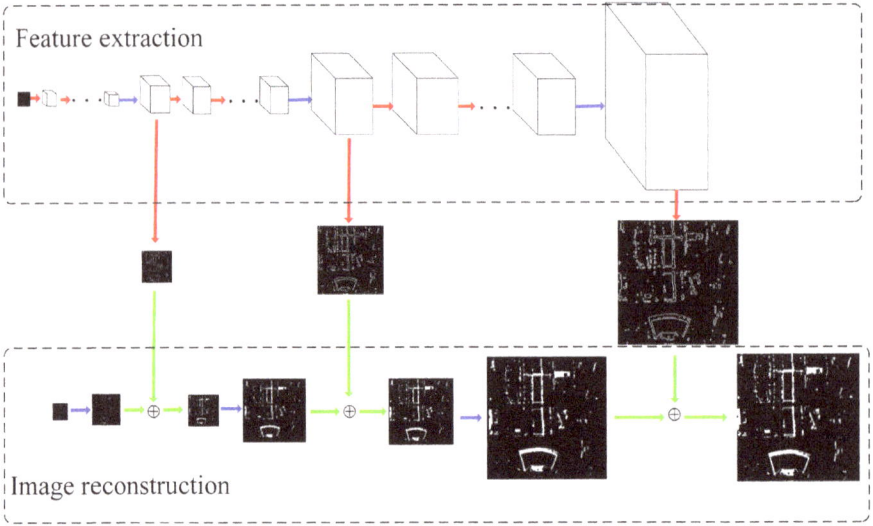

FIGURE 5.15   Deep Laplace pyramid network.

extracted features are upsampled at a scale of two by a deconvolution layer. There are two convolution layers behind the deconvolution layer. One convolution layer is used to continue extracting features, and the other convolution layer is utilized to predict the residual image at this level. In image reconstruction, the input coarse abundance images are upsampled at a scale of two through a deconvolution layer at each level, and the upsampled image is then added to the residual image at this level to obtain the reconstructing result at this level.

Let δ be the set of network parameters to be optimized. The goal of the super-resolution reconstruction model based on DLPN is to train a mapping function f to generate the high-resolution abundance images $H_k^{\mathrm{mp}}=f\left(L_k;\delta\right)$ that are close to real abundance images H_k. Assuming that the residual images of the low-resolution abundance images on the vth network are $r_k^{(v)}$, the upsampled low-resolution abundance images are $L_k^{(v)}$, and the corresponding fine abundance images are $H_k^{\mathrm{mp}(v)}=L_k^{(v)}+r_k^{(v)}$. The real abundance images H_k are downsampled by bicubic interpolation to adjust its size to be consistent with $H_k^{(v)}$ in each level of the network. The robust loss function used to handle outliers is defined as follows:

$$
\begin{aligned}
\phi\left(H_k,H_k^{\mathrm{mp}};\delta\right)&=\frac{1}{Z}\sum_{z=1}^{Z}\sum_{v=1}^{\log_2 S}y\left(H_{k(z)}^{(v)}-H_{k(z)}^{\mathrm{mp}(v)}\right)\\
&=\frac{1}{Z}\sum_{z=1}^{Z}\sum_{v=1}^{\log_2 S}y\left(H_{k(z)}^{(v)}-\left(L_{k(z)}^{(v)}+r_{k(z)}^{(v)}\right)\right)
\end{aligned}
\tag{5.6}
$$

where $y\left(x\right)=\sqrt{x^2+\varepsilon^2}$ is the Charbonnier penalty function, Z is the number of training samples in the two components, and ε is set to 0.001.

Finally, after the $\log_2 S$th level network training based on the loss function is completed, the high-resolution abundance images H_k^{mp} containing the subpixel predicted values $H_k^{mp}(p_j)$ are obtained. Since the multi-scale abundance images are trained in the super-resolution reconstruction model based on DLPN, the high-resolution abundance images H_k^{mp} with multi-scale information are generated in the deep learning path.

5.3.3 DUAL PROCESSING PATH

The fine abundance images H_k^{sp} with more spatial-spectral information from spatial-spectral path and the fine abundance images H_k^{sp} with multi-scale information from multi-scale path are integrated to produce the improved abundance images H_k^{sm} with multi-scale spatial-spectral information. A more accurate predicted value $H_k^{sm}(p_j)$ from the improved abundance images H_k^{sm} can be derived by integrating $H_k^{sp}(p_j)$ and $H_k^{mp}(p_j)$ according to the following:

$$H_k^{sm}(p_j) = wH_k^{sp}(p_j) + (1-w)H_k^{mp}(p_j) \qquad (5.7)$$

where $w\ (0 \le w \le 1)$ is a weight parameter that balances the abundance values $H_k^{sp}(p_j)$ and $H_k^{mp}(p_j)$.

The DPP flowchart we proposed is shown in Figure 5.16, including the following four steps:

Step 1. The PCA fusion algorithm is used to fuse the low spatial resolution hyperspectral image Y and the high spatial resolution multispectral image M on the fusion path to generate the high-resolution fused hyperspectral image $\hat{Y}$. At the same time, the original coarse hyperspectral image Y is unmixed on the deep learning path to generate the low-resolution abundance images L_k.

Step 2. The fused hyperspectral image $\hat{Y}$ is unmixed on the fusion path to obtain the high-resolution abundance images H_k^{sp} with more spatial-spectral information. On the deep learning path, a super-resolution reconstruction model based on DLPN works with the low-resolution abundance images L_k to produce the high-resolution abundance images H_k^{mp} with multi-scale information.

Step 3. The high-resolution abundance images H_k^{sp} containing the subpixel abundance values $H_k^{sp}(p_j)$ and the high-resolution abundance images H_k^{mp} containing the subpixel abundance values $H_k^{mp}(p_j)$ are integrated by using formula (5.7) to generate the higher resolution abundance images H_k^{sm} containing the more accurate subpixel abundance values $H_k^{sm}(p_j)$.

Step 4. According to the more accurate subpixel abundance values $H_k^{sm}(p_j)$ from the higher resolution abundance images H_k^{sm} generated in step 3, the class labels are assigned to subpixels through the class allocation method, obtaining the final subpixel mapping result.

As shown in Figure 5.16, it is noted that since the types of information in the abundance images are enriched by parallel processing paths, the proposed DPP can produce better subpixel mapping results.

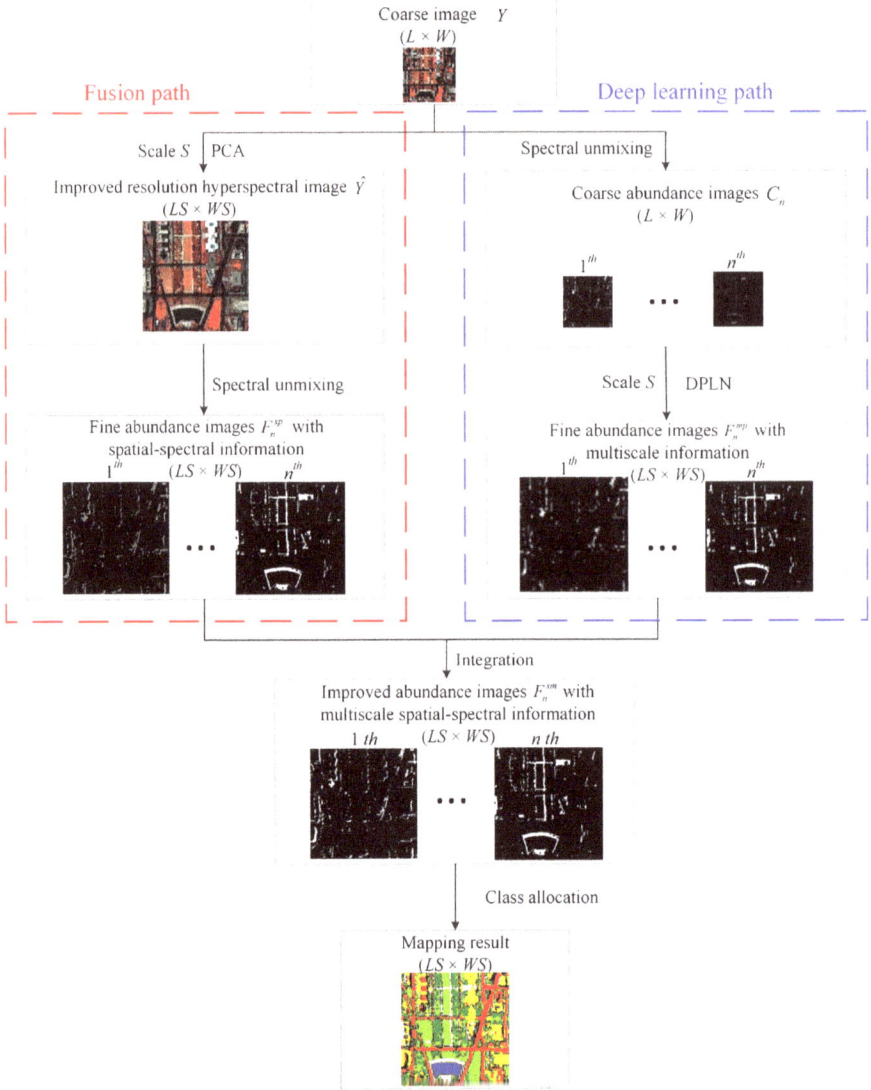

FIGURE 5.16 Dual processing path.

5.3.4 EXPERIMENTAL CONTENT AND RESULT ANALYSIS

Three real hyperspectral images are used to prove the effectiveness of the proposed DPP. To quantitatively evaluate the application of the proposed DPP in a coarse image, the fine image is downsampled in three experiments. In this case, the known subpixel-level land cover classes help to directly assess the impact of image registration errors. The least squares support vector machine (LSSVM) is used as the spectral unmixing method. UOC is selected as the class allocation method. In UOC, the order of class allocation is determined by calculating Moran's I from the fused hyperspectral image [17]. In addition, we use the spectral response of the IKONOS satellite to generate a suitable multispectral image to avoid the influence of errors caused by multispectral image acquisition. The satellite captured four multispectral bands (0.45–0.52 μm, 0.52–0.60 μm, 0.63–0.69 μm, and 0.76–0.90 μm) [6]. In summary, the purpose of using the simulated coarse hyperspectral image is to only consider the performance of the proposed method and avoid the influence of external errors on the final subpixel mapping result.

Six subpixel mapping methods are tested and compared: the edge-directed interpolation (EI) [10], hybrid interpolation by parallel paths (HIPP) [18], hybrid spatial attraction model (HSAM) [19], iterative interpolation deconvolution (IID) [20], HNN with panchromatic image (HNNP) [21], and the proposed DPP. The mapping accuracy of each class, the overall accuracy (PCC), and the Kappa coefficient (Kappa) are utilized as the quantitative evaluation indices.

5.3.4.1 Experiment 1

The hyperspectral image data set was generated during the flight activities of the University of Pavia in experiment 1. As shown in Figure 5.17(a), the size of the test area is 100×100 pixels and 103 spectral bands, and the spatial resolution is 1.3 m. The reference image shown in Figure 5.17(b) contains asphalt, grass, tree, and brick. The weight parameter w is set to 0.6.

As shown in Figure 5.18(a), the spectral response of the IKONOS satellite creates an appropriate multispectral image. Figure 5.18(b) shows the coarse hyperspectral image generated by downsampling the fine hyperspectral image with $S = 8$. It is difficult to obtain the spatial distribution information in a coarse hyperspectral image. In order to improve the resolution of the hyperspectral image and provide more spatial-spectral information, PCA-based hyperspectral and multispectral fusion algorithms are used

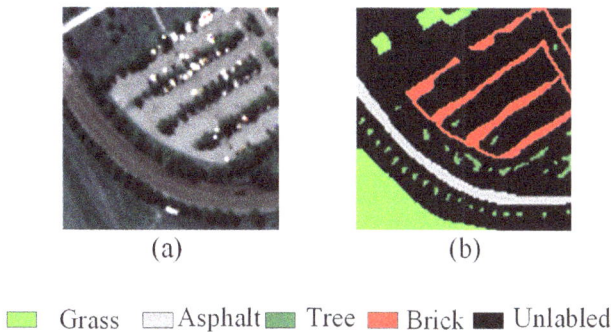

(a)   (b)

■ Grass ▫Asphalt■ Tree ■ Brick ■ Unlabled

FIGURE 5.17   Pavia University data set: (a) false color composite image (RGB band: 19, 30, 44) and (b) reference image.

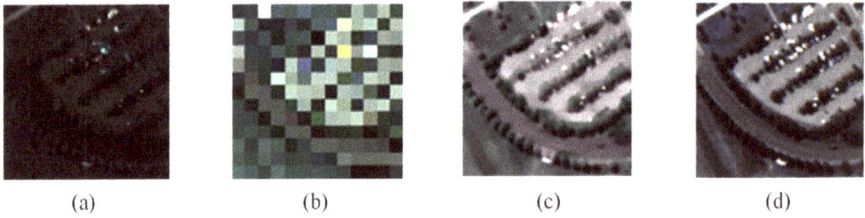

(a)   (b)   (c)   (d)

FIGURE 5.18   (a) Multispectral image (RGB band: 3, 2, 1). (b) Low-resolution image ($S = 8$).
(c) PCA. (d) PCA with EOI.

TABLE 5.4
Three Evaluation Indicators of PCA and PCA with EOI

Evaluation Indicator	PCA	PCA With EOI
ERGAS	3.632	1.949
SAM	9.194	5.425
UIQI	0.826	0.992

in the fusion path. Figure 5.18(c)–(d) shows the fused results of traditional PCA and
PCA with EOI. Through visual observation, the resolution of the coarse hyperspectral
image has been improved, and the high-resolution abundance images can obtain the
more spatial-spectral information from the improved hyperspectral image. Compared
with traditional PCA, PCA with EOI is closer to the original hyperspectral image
as shown in Figure 5.17(a). For quantitative comparison, three evaluation indicators
including *Erreur Relative Globale Adimensionnelle de Synthèse* (ERGAS), Spectral
Angle Mapper (SAM), and Universal Image Quality Index (UIQI) [9] are used. Lower
ERGAS, lower SAM, and higher UIQI mean better performance. Table 5.4 shows the
three evaluation indicators of PCA and PCA with EOI. PCA with EOI produces lower
ERGAS, lower SAM, and higher UIQI than PCA. Compared with PCA, the operating
speed of PCA with EOI is increased by more than three times. This is because the
use of EOI reduces the complexity of the algorithm and improves the fusion effect.
To obtain the better subpixel mapping result, this chapter uses PCA with EOI as the
hyperspectral and multispectral fusion algorithm.

The subpixel mapping results of the six methods are shown in Figure 5.19(a)–(f).
Visual comparison shows that the proposed DPP shown in Figure 5.19(f) is closer to
the reference image than the other five subpixel mapping methods shown in Figure
5.19(a)–(e). The brick and tree in Figure 5.19(a)–(b) have many disconnected places
and burrs on the border. Although HNNP uses panchromatic images as auxiliary
information, its constraints are sometimes excessive. Although HNNP uses the panchromatic image as the auxiliary information, the constraint of HNNP maintains in
excess. The small target trees are hardly mapped, and the boundaries of all classes

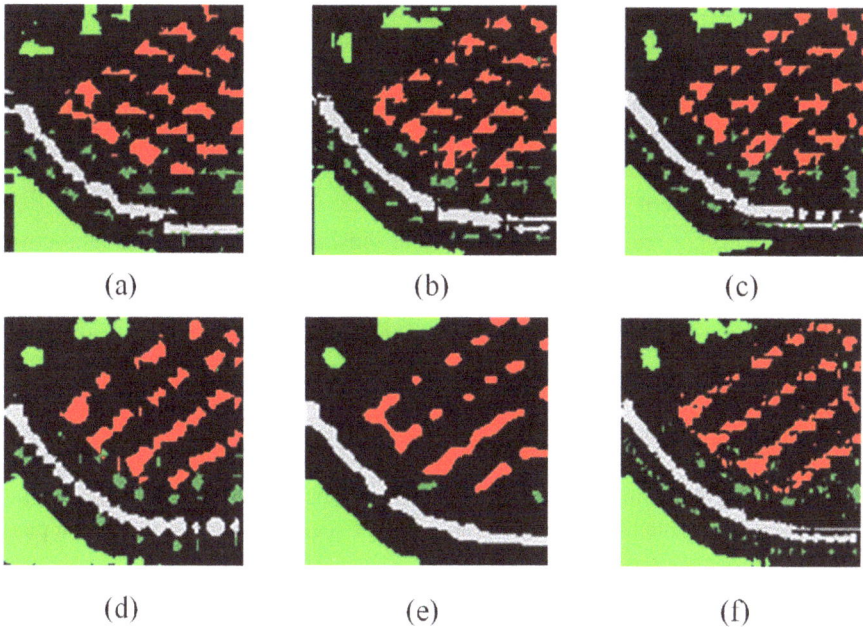

FIGURE 5.19 Subpixel mapping results: (a) EI, (b) HIPP, (c) HSAM, (d) IID, (e) HNNP, and (f) DPP.

TABLE 5.5

Mapping Accuracy of the Six Methods in Experiment 1 ($S = 8$)

Land Cover Class	EI	HIPP	HSAM	IID	HNNP	DPP
Grass (%)	87.79	90.59	91.06	92.38	94.99	95.65
Asphalt (%)	60.13	70.83	76.31	78.48	83.18	85.26
Trees (%)	20.91	22.90	23.47	28.88	10.30	30.10
Bricks (%)	40.01	45.51	52.55	53.42	57.93	60.11
PCC (%)	60.89	65.99	69.48	71.27	72.66	76.04
Kappa	0.5398	0.5915	0.6282	0.6457	0.6588	0.6985

are overly smoothed in Figure 5.19(e). Since the more spatial-spectral information and multi-scale information are obtained from the parallel processing paths, and the types of information from the abundance images are enriched, these problems are alleviated in the proposed DPP. Figure 5.19(f) shows that all land cover classes have better continuity and smoothness of boundaries.

The performance of six subpixel mapping methods is quantitatively evaluated by calculating three evaluation indicators of mapping accuracy (%) of each class, PCC (%), and Kappa. Table 5.5 lists the evaluation indicators for the six subpixel mapping methods. Compared with the other five subpixel mapping methods, the proposed

DPP has a higher evaluation index. Through the inspection of the mapping accuracy (%) of each class, when compared with the HNNP algorithm, the accuracy of the asphalt, tree, and brick of the DPP has been improved by about 2.1%, 19.8%, and 2.2%, respectively. In addition, when compared with the other five subpixel mapping methods, DPP can obtain the highest PCC (%) and Kappa.

5.3.4.2   Experiment 2

In experiment 2, a real hyperspectral remote sensing image with more land cover classes was selected as the experimental data. As shown in Figure 5.20(a), the test data come from Washington, DC, which has 240 × 240 pixels, 191 band, and 3-m spatial resolution. Figure 5.20(b) shows that the reference image covers seven land cover classes, including shadow, water, road, tree, grass, roof, and trail. The weight parameter is set to 0.5.

As shown in Figure 5.21(a), the multispectral image is generated by the method described in experiment 1. The original fine hyperspectral image is downsampled by the scale of $S = 8$ to generate the coarse hyperspectral image as shown in Figure 5.21(b). By fusing the high spatial resolution multispectral image in the fusion path,

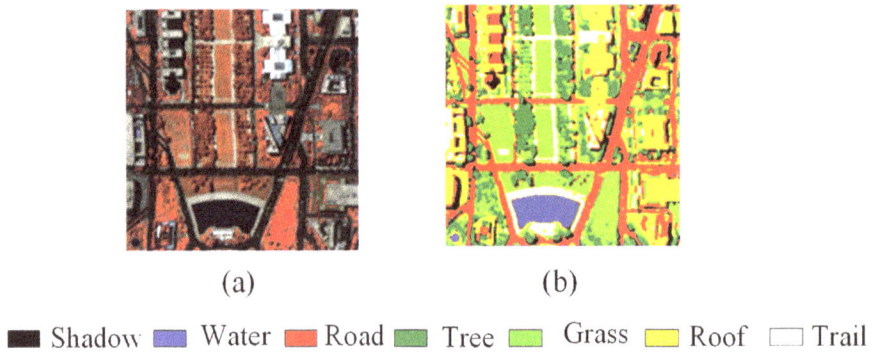

(a)   (b)

■ Shadow ■ Water ■ Road ■ Tree □ Grass □ Roof □ Trail

FIGURE 5.20   Washington, DC, data set: (a) false color composite image (RGB band: 65, 52, 36) and (b) reference image.

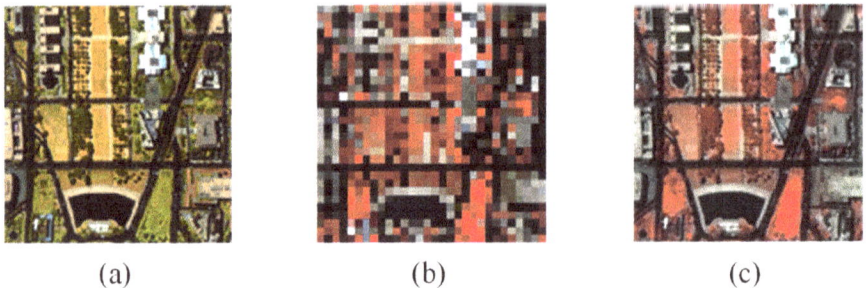

(a)   (b)   (c)

FIGURE 5.21   (a) Multispectral image (RGB band: 3, 2, 1). (b) Low-resolution image ($S = 8$). (c) Fused result.

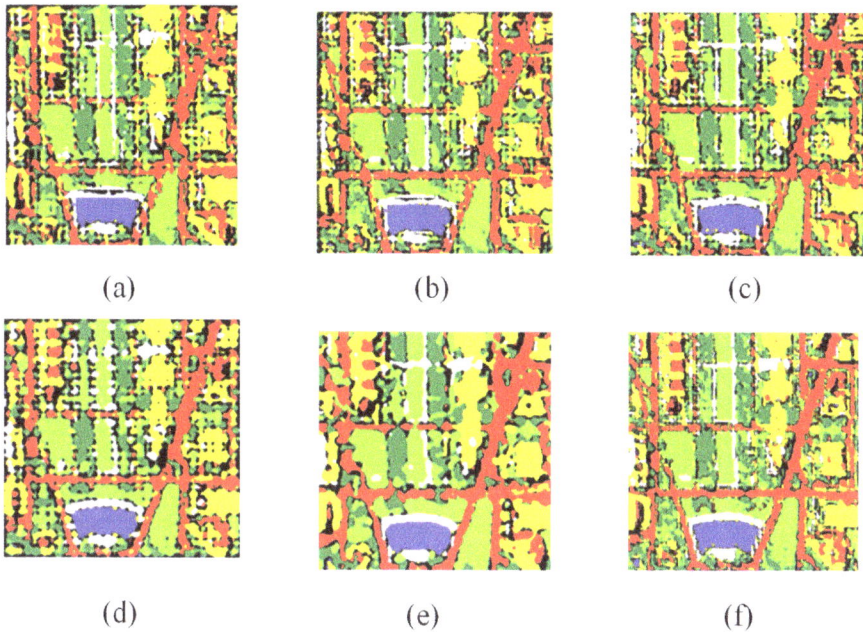

FIGURE 5.22 Subpixel mapping results: (a) EI, (b) HIPP, (c) HSAM, (d) IID, (e) HNNP, and (f) DPP.

the resolution of the coarse hyperspectral image is improved, and the fusion result shown in Figure 5.21(c) is obtained. It can be found that the fusion result is close to the original fine hyperspectral image as shown in Figure 5.20(a). Therefore, the fusion result is used as the input of subpixel mapping. Compared with the direct use of the coarse hyperspectral image, the more spatial-spectral information can be applied to improve the final mapping result.

The mapping results of EI, HIPP, HSAM, IID, HNNP, and DPP are shown in Figure 5.22. There is a lot of salt and pepper noise in the land cover classes in Figure 5.22(a)–(e). Since the abundance images in the proposed DPP have more information types, the result of DPP is more consistent with the reference image. As shown in Figure 5.22(f), land cover classes have better continuity and smoother boundaries.

Table 5.6 lists the mapping accuracy (%) of each class, PCC (%), and Kappa of the six subpixel mapping methods. Similar to the conclusion drawn in experiment 1, comparing the evaluation indicators, it can be found that the mapping accuracy (%) of each class in DPP is higher than that in the other five subpixel mapping methods. In addition, the proposed DPP can produce the highest PCC (%) of 71.79% and Kappa of 0.6347.

5.3.4.3 Experiment 3

To better reflect the effectiveness of the proposed DPP, a hyperspectral data set with a larger area and a more complex distribution was used in experiment 3. The data set was captured by NASA via AVIRIS over the Kennedy Space Center (KSC). The height of the hyperspectral data set is about 20 km, the spatial resolution is 18 m,

TABLE 5.6

Mapping Accuracy of the Six Methods in Experiment 2 ($S = 8$)

Land Cover Class	EI	HIPP	HSAM	IID	HNNP	DPP
Background (%)	48.60	50.02	50.87	52.57	40.37	57.14
Water (%)	80.80	83.53	83.65	86.29	88.54	89.30
Road (%)	52.02	56.71	57.99	58.47	62.30	65.88
Tree (%)	50.92	54.45	54.59	55.53	60.54	63.83
Grass (%)	61.78	65.04	65.54	66.05	69.77	72.22
Roof (%)	57.35	60.63	61.48	62.87	67.07	69.14
Trail (%)	45.20	49.66	50.88	54.66	58.97	61.96
PCC (%)	55.37	58.97	59.40	60.44	65.64	71.79
Kappa	0.4653	0.5053	0.5114	0.5260	0.5831	0.6347

and it has 176 spectral bands. As shown in Figure 5.23(a), the test area is composed of 512×608 pixels. The reference image is shown in Figure 5.23(b) and includes 13 land cover classes. The weight parameter w is selected as 0.5. The high spatial resolution multispectral image is shown in Figure 5.24(a). As shown in Figure 5.24(b), the coarse hyperspectral image is obtained by downsampling the fine hyperspectral image with $S = 8$. The proposed DPP will use the fused results shown in Figure 5.24(c) to generate a better subpixel mapping result.

Figure 5.25(a)–(f) shows the subpixel mapping results of the six methods. In Figure 5.25(a)–(e), there are many spot artifacts and broken holes in the land cover classes. The proposed DPP in Figure 5.25(f) can produce the more continuous and smoother land cover classes. This is because the information types of the abundance images in the proposed method are enriched through the parallel processing paths. Therefore, when compared with the other five subpixel mapping methods, the mapping result of DPP is closer to the reference image. Table 5.7 lists the PCC (%) and Kappa of the six subpixel mapping methods. Similar to the results of experiments 1 and 2, the mapping accuracy of DPP is better than the other five subpixel mapping methods.

5.3.4.4 Discussion

This section discusses some of the factors that influence the proposed DPP. First, the performance of DPP is affected by the scale S. Different scales S represent different upsampling scales. In three experiments, three scales S of six subpixel mapping methods were tested, namely $S = 2$, 4, and 8.

Figure 5.26(a)–(c) shows the PCC (%) of six subpixel mapping methods by three scales S. With the increase of S, the PCC (%) of the six subpixel mapping methods decreases. This is because a higher scale S means that the simulated coarse hyperspectral image has a lower resolution, and the lower resolution hyperspectral image brings more challenges to the subpixel mapping process. But consistent with the previous experimental results, the proposed DPP can still obtain the highest PCC (%). In addition, it is noted that the increase in PCC (%) obtained by the proposed DPP at $S = 8$ is higher than the increase in PCC (%) obtained by the proposed DPP at $S = 2$ and $S = 4$. This is because a higher S means that more level networks are needed in

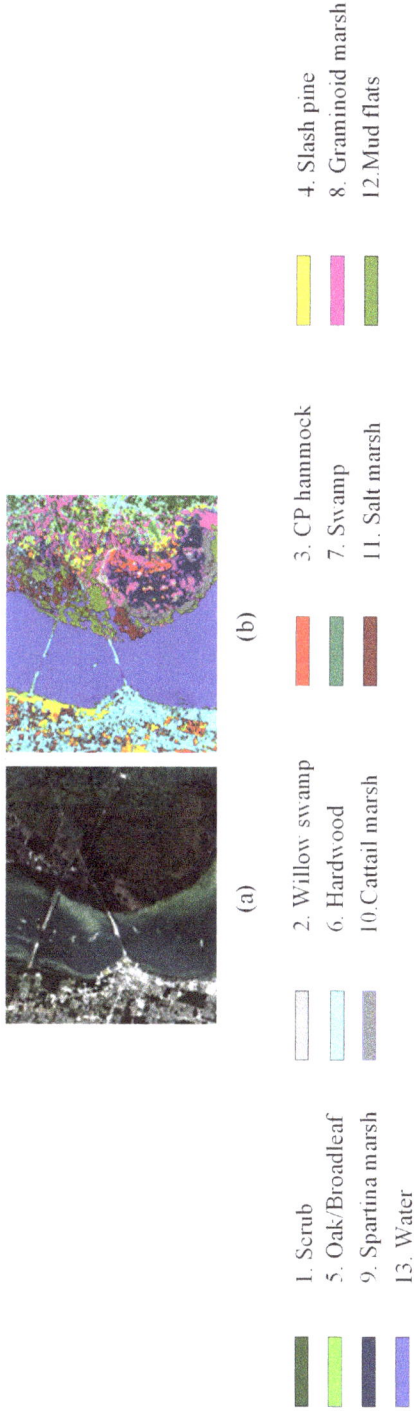

(a)   (b)

1. Scrub   2. Willow swamp   3. CP hammock   4. Slash pine
5. Oak/Broadleaf   6. Hardwood   7. Swamp   8. Graminoid marsh
9. Spartina marsh   10.Cattail marsh   11. Salt marsh   12.Mud flats
13. Water

FIGURE 5.23   KSC data set: (a) false color composite image (RGB band: 28, 19, 10) and (b) reference image.

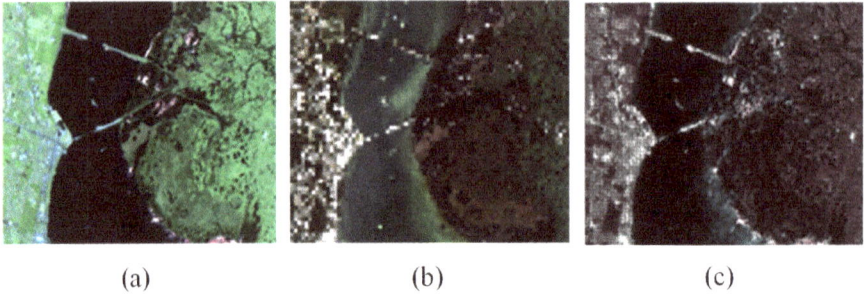

(a)   (b)   (c)

FIGURE 5.24   (a) Multispectral image (RGB band: 3, 2, 1). (b) Low-resolution image ($S = 5$).
(c) Fused result.

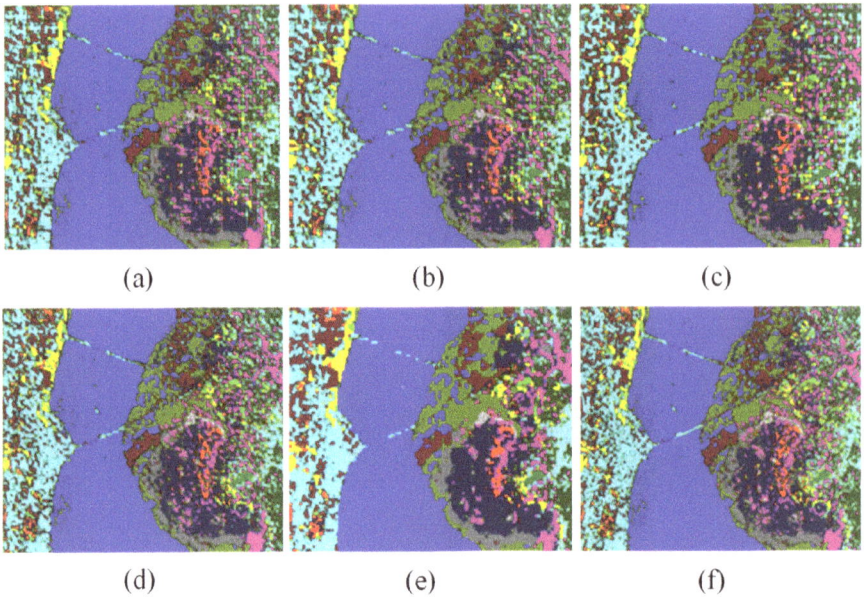

(a)   (b)   (c)

(d)   (e)   (f)

FIGURE 5.25   Subpixel mapping results: (a) EI, (b) HIPP, (c) HSAM, (d) IID, (e) HNNP,
and (f) DPP.

TABLE 5.7
Mapping Accuracy of the Six Methods in Experiment 3 ($S = 8$)

Evaluation Indicator	EI	HIPP	HSAM	IID	HNNP	DPP
PCC (%)	60.49	62.05	62.76	63.29	65.57	69.23
Kappa	0.5300	0.5507	0.5575	0.5639	0.5941	0.6349

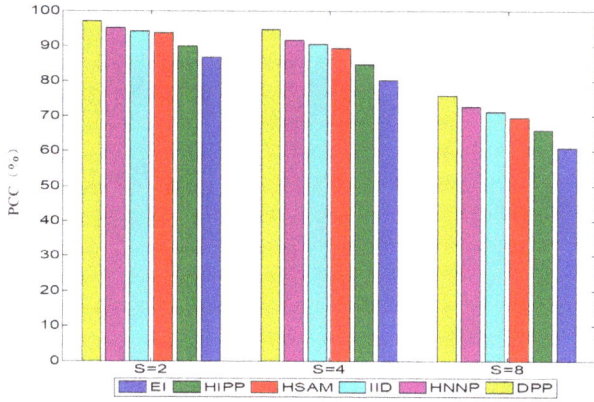

(a)

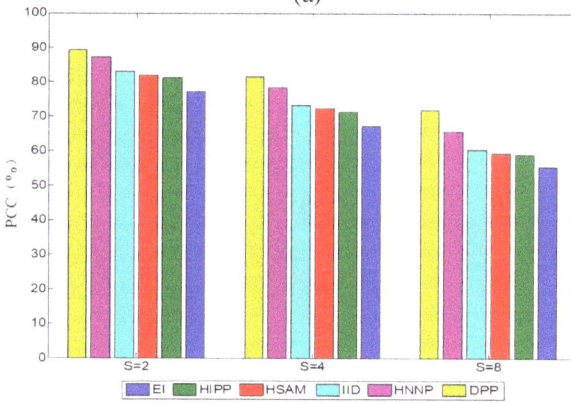

(b)

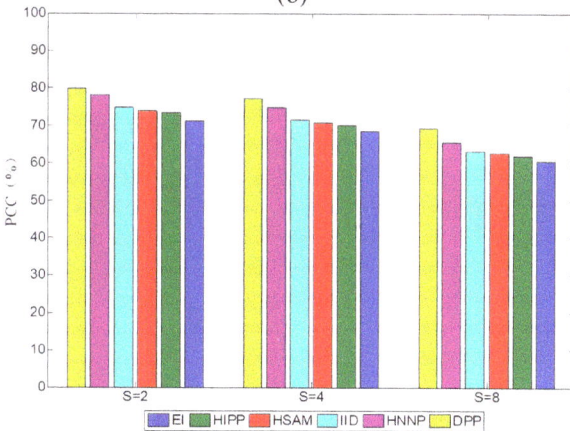

(c)

FIGURE 5.26 The relationship between PCC (%) and scale factor S of six methods: (a) experiment 1, (b) experiment 2, and (c) experiment 3.

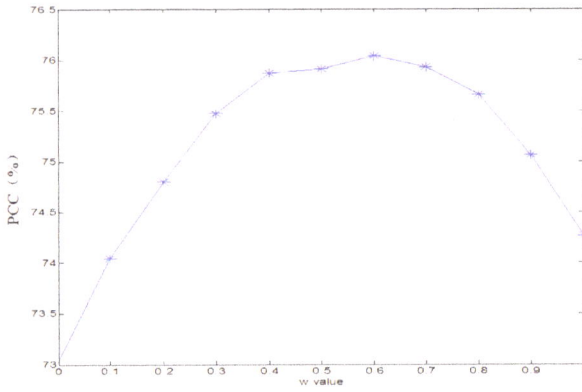

(a)

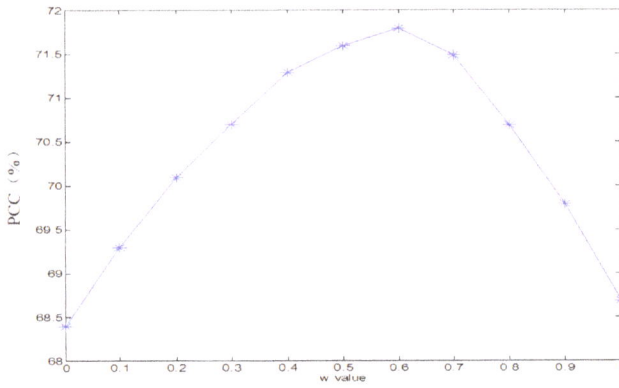

(b)

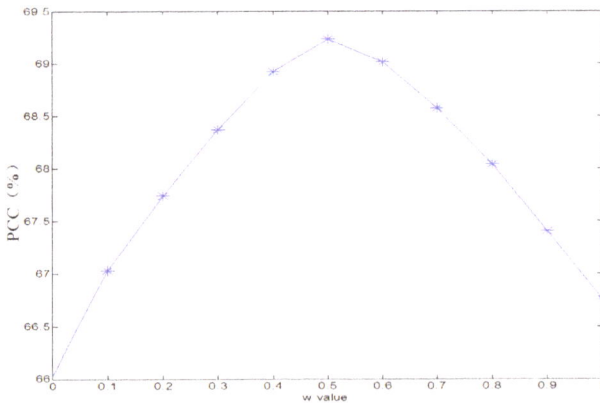

(c)

FIGURE 5.27   The relationship between PCC (%) and weight parameter θ in the DPP method: (a) experiment 1, (b) experiment 2, and (c) experiment 3.

the multi-scale path, and more multi-scale information can be supplied. Therefore, the DPP can perform better than the other five subpixel mapping methods under the condition of a coarser hyperspectral image. Moreover, because the number of level networks is $\log_2 S$, the scale factor S only is even in this paper.

Secondly, the weight parameter θ balances the influence of $F_n^{sp}(p_j)$ and $F_n^{mp}(p_j)$ in formula (5.7). To obtain the appropriate weight parameter θ, the PCC (%) values of DPP in three experimental data are tested under 11 combinations with an interval of 0.1 in the range of [0, 1]. In addition, the difference between the fusion path and the deep learning path can also be reflected by analyzing the changes in PCC (%). Figure 5.27(a)–(c) shows the three experimental results ($S = 8$). When $\theta = 0$ or $\theta = 1$, only the deep learning path or the fusion path works in the proposed DPP. Although the information types of the abundance images are not rich at this time, when compared with the other five subpixel mapping methods, PCC (%) in the proposed DPP is still improved. When $\theta = 0.6$ in experiment 1, $\theta = 0.6$ in experiment 2, and $\theta = 0.5$ in experiment 3, PCC (%) can get the highest value. This is because at this time, the abundance images generated by integrating $F_n^{sp}(p_j)$ and $F_n^{mp}(p_j)$ with appropriate weight parameters θ have the richest information types.

Third, since the three experiments have considered the processing process of spectral unmixing, we further analyze the influence of the choice of spectral unmixing on the performance of the proposed DPP. We tested the effects of two different spectral unmixing methods on DPP performance. The least squares linear mixture model (LSLMM) is selected as another spectral unmixing method and is compared with the LSSVM spectral unmixing method used in the three experiments ($S = 4$). Figure 5.28(a)–(c) shows the PCC (%) of the six subpixel mapping methods of LSSVM and LSLMM. Because LSSVM is more effective than LSLMM, when LSLMM is used for spectral unmixing, the PCC (%) of the six subpixel mapping methods are all reduced. But consistent with the previous experimental results, the proposed DPP can still obtain the highest PCC (%).

5.4 SUBPIXEL MAPPING BASED ON MULTI-SOURCE REMOTE SENSING FUSION DATA FOR LAND COVER CLASSES

With continuous development of remote sensing technology, another kind of high spatial resolution remote sensing data in the same region is selected as auxiliary data to improve the subpixel mapping result for hyperspectral images. Reference [22] introduces the elevation information from light detection and ranging (LiDAR) into the HNN model to improve the final mapping result. Reference [23] uses the panchromatic image with rich spatial information to fuse the original hyperspectral image by pansharpening technology, producing a better subpixel mapping result. However, through the analysis of the literature, we find that the type of auxiliary remote sensing data used in subpixel mapping is relatively single, which cannot provide multiple types of auxiliary information at the same time, resulting in hindering further improvement of the accuracy of the subpixel mapping result. To utilize multiple types of auxiliary information from multi-source remote sensing data, subpixel mapping based on multi-source remote sensing fusion data (MRSFD) for land-cover classes (SPM-MRSFD) is proposed.

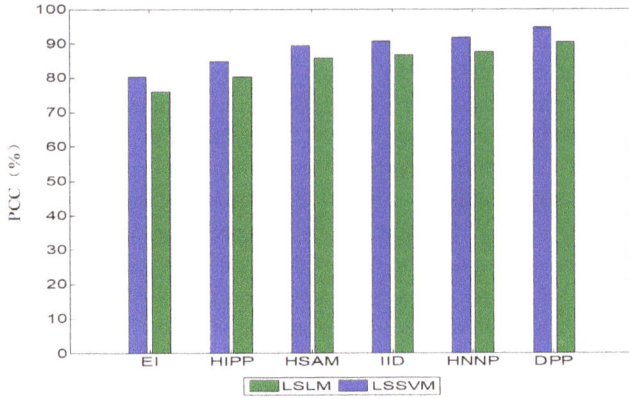

(a)

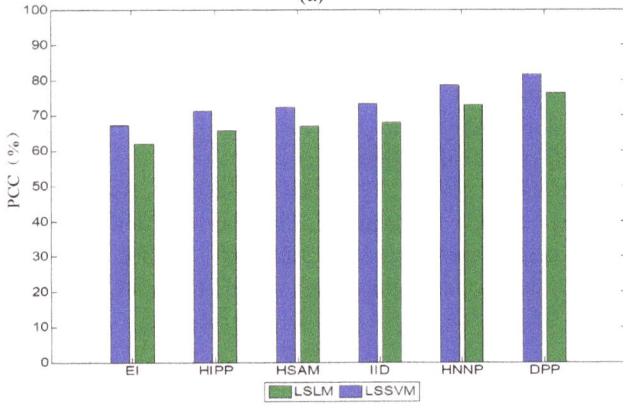

(b)

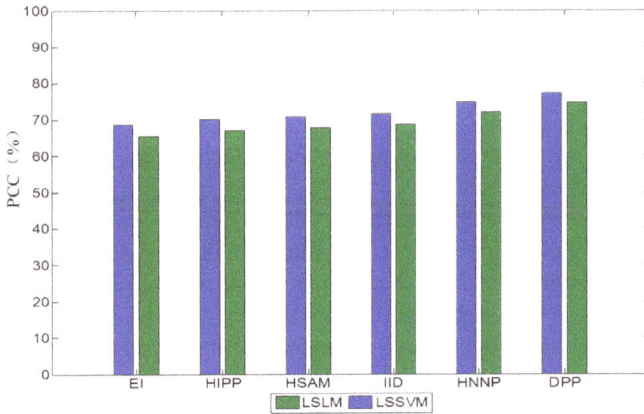

(c)

FIGURE 5.28 The relationship between PCC (%) of the six methods and the choice of spectral unmixing methods: (a) experiment 1, (b) experiment 2, and (c) experiment 3.

Since hyperspectral and panchromatic images are collected through a similar imaging method, their data properties are similar. Therefore, we choose the data-level fusion between the hyperspectral image and the panchromatic image through pansharpening technology. However, the method for collecting the LiDAR image is different from the imaging method for collecting the previous two kinds of images; the data properties are different between the LiDAR image and the previous two kinds of images. Because the images captured in the same region at close time intervals often have similar features, we chose the feature-level fusion between the LiDAR image and the previous two kinds of images. In addition, although the fusion method based on deep learning shows better performance than the fusion method based on traditional machine learning, deep learning usually needs a lot of fine training data to achieve the desired performance. In the absence of a large number of fine training data, we chose the fusion method based on traditional machine learning to achieve data- and feature-level fusion here.

The flowchart of SPM-MRSFD is shown in Figure 5.29. First, we utilize pansharpening technology to fuse the original hyperspectral image with the auxiliary

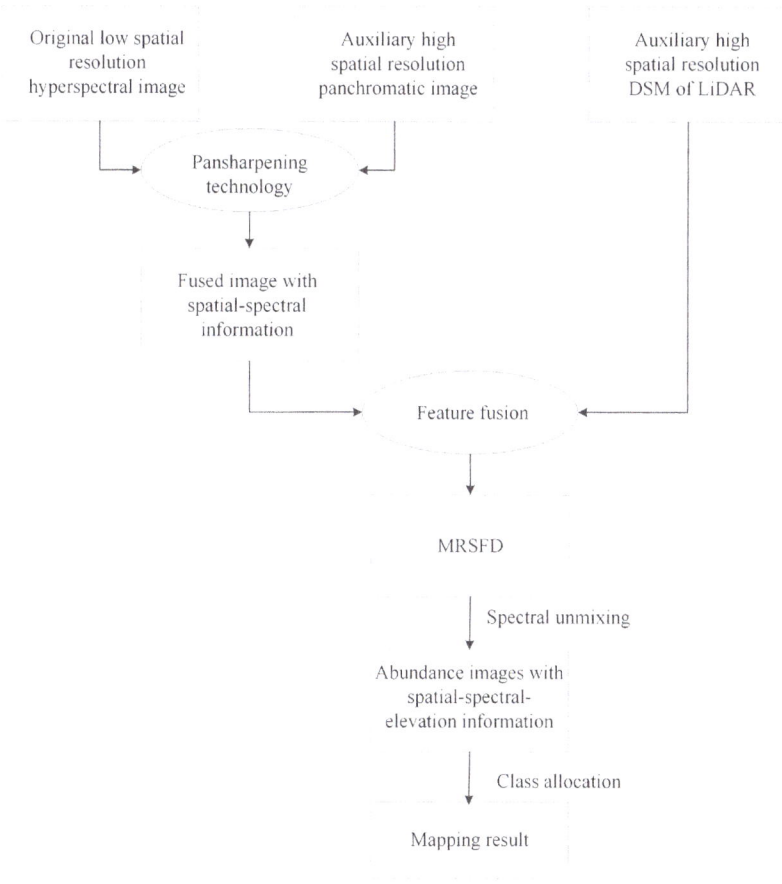

FIGURE 5.29 The flowchart of SPM-MRSFD.

panchromatic image to obtain the fused image including spatial-spectral information. Second, the fused image and the auxiliary digital surface model (DSM) of LiDAR are fused by feature fusion to produce the MRSFD including spatial-spectral-elevation information. Finally, the MRSFD is unmixed to obtain the abundance images with the proportions of subpixels belonging to land cover classes, and the class allocation method is used to assign the class labels to subpixels to obtain the subpixel mapping result according to information about these proportions. Three steps of the proposed SPM-MRSFD are described in detail as follows.

Let S be the ratio scale between the original hyperspectral image and the auxiliary panchromatic image or DSM of LiDAR, namely, a mixed pixel in hyperspectral image is considered to divide into S^2 subpixels in the auxiliary panchromatic image or DSM of LiDAR. The original hyperspectral image Y^H with M mixed pixels and B bands includes mixed pixels P_J ($J = 1, 2, \ldots, M$). The auxiliary panchromatic image Y^P and the auxiliary DSM of LiDAR Y^L are with MS^2 subpixels, respectively.

5.4.1   DATA-LEVEL FUSION

The hyperspectral image Y^H and the panchromatic image Y^P are fused to produce the fused image with MS^2 subpixels and B bands by pansharpening technique. Due to the high quality in rendering the spatial details, the pansharpening based on the PCA method is selected here.

In the PCA, the hyperspectral image is first converted into some principal components (PCs). According to reference, the hypothesis underlying PCA is that the first PC contains the spatial information, while the other PCs contain the spectral information. The first PC is then replaced by the rich spatial information of the panchromatic image according to formula (5.8). In addition, to reduce matching distortion, histogram matching of the panchromatic image and the first PC is implemented before replacement occurs. Finally, the pansharpening approach is completed by using the inverse transformation to produce the fused image with rich spatial-spectral information.

$$Y_b^F = \tilde{Y}_b^H + g_b \left(Y^P - \sum_{b=1}^{B} w_b \tilde{Y}_b^H \right) \tag{5.8}$$

where Y^F is the fused image after pansharpening, Y_b^F ($b = 1, 2, \ldots, B$) represents the bth band of the fused image, $\tilde{Y}_b^H$ denotes the bth band of the hyperspectral image Y^H interpolated at the ratio scale of the panchromatic image Y^P, g_b is the vector of the injection gains, w_b is the weight vector that calculates the spectral overlap between the hyperspectral image and the panchromatic image.

5.4.2   FEATURE FUSION

The fused image Y^F and the auxiliary DSM of LiDAR Y^L are fused by feature fusion to produce the MRSFD Y^M with MS^2 subpixels. Morphological opening or closing structural elements that are predefined size and shape are first used to extract spatial-spectral features $\mathbf{F}^S = \left\{ f_x^S \right\}_{x=1}^{h}$ (n is the number of features) with B bands from the fused image and elevation features $\mathbf{F}^{\mathrm{E}} = \{f_x^{\mathrm{E}}\}_{x=1}^{n}$ with B^* bands from the DSM of LiDAR.

Because the fused image has more bands, the band numbers of spatial-spectral features are more than these of elevation features (i.e., $B > B^*$). We then need to normalize the band numbers of spatial-spectral features to B^* by kernel principal component analysis (KPCA) [24] to facilitate feature fusion. We suppose $\mathbf{F}^{\mathrm{S\text{-}E}} = \left[\mathbf{F}^{\mathrm{S}}; \mathbf{F}^{\mathrm{E}}\right] = \{f_x^{\mathrm{S\text{-}E}}\}_{x=1}^{n}$ and $f_x^{\mathrm{S\text{-}E}} = \left[f_x^{\mathrm{S}}; f_x^{\mathrm{E}}\right]$ represent the vector stacked by the spatial-spectral and elevation features.

Finally, the MRSFD including the fusion features $\mathbf{F}^{\mathrm{M}} = \{f_x^{\mathrm{M}}\}_{i=1}^{n}$ is obtained by the feature fusion based on graph method (FFG). The goal of feature fusion based on graph is to find a transformation matrix $\mathbf{R}$ that can achieve a way of $f_x^{\mathrm{M}} = \mathbf{R}^T f_x^{\mathrm{S\text{-}E}}$. The matrix $\mathbf{R}$ is defined as

$$\arg\min_{R}\left(\sum_{x,y}^{n}\left\|\mathbf{R}^T f_x^{\mathrm{S\text{-}E}} - \mathbf{R}^T f_y^{\mathrm{S\text{-}E}}\right\|^2 z_{xy}\right) \tag{5.9}$$

where $z_{xy} \in \{0,1\}$ is from the matrix $\mathbf{z}$, which is defined as the edge of the graph. We utilize k-nearest neighbors based on Euclidean distance to find the features $f_y^{\mathrm{S\text{-}E}}$ that are close to the central feature $f_x^{\mathrm{S\text{-}E}}$. If $f_x^{\mathrm{S\text{-}E}}$ and $f_y^{\mathrm{S\text{-}E}}$ are close, $z_{xy} = 1$, and if $f_x^{\mathrm{S\text{-}E}}$ and $f_y^{\mathrm{S\text{-}E}}$ are far, $z_{xy} = 0$. In addition, we use the constraint in formula (5.10) to avoid degeneracy.

$$\mathbf{R}^T\left(\mathbf{F}^{\mathrm{S\text{-}E}}\right)\mathbf{D}\left(\mathbf{F}^{\mathrm{S\text{-}E}}\right)^T \mathbf{R} = \mathbf{I} \tag{5.10}$$

where $\mathbf{D}$ is the diagonal matrix with the value $D_{x,x}$ on the diagonal being $\sum_{y=1}^{n} z_{xy}$, and $\mathbf{I}$ is the identity matrix. We obtain the solution of $\mathbf{R} = (\mathbf{r}_1, \mathbf{r}_2, ..., \mathbf{r}_o)$ that is made up by o eigenvectors associated with the least o eigenvectors $\lambda_1 \leq \lambda_2 \leq \cdots \leq \lambda_o$ of the generalized problem in formula (5.11):

$$\left(\mathbf{F}^{\mathrm{S\text{-}E}}\right)\mathbf{W}\left(\mathbf{F}^{\mathrm{S\text{-}E}}\right)^T \mathbf{r} = \lambda\left(\mathbf{F}^{\mathrm{S\text{-}E}}\right)\mathbf{D}\left(\mathbf{F}^{\mathrm{S\text{-}E}}\right)^T \mathbf{r} \tag{5.11}$$

where $\mathbf{W} = \mathbf{D} - \mathbf{Z}$ is the Laplacian matrix.

5.4.3 Obtaining Mapping Result

After producing the MRSFD Y^{M} with MS^2 mixed pixels by feature fusion, we utilize the spectral unmixing based on the linear spectral mixture model (LSMM) to unmix the MRSFD to obtain the fractional images with the proportions $L(p_j)$ of subpixels p_j ($j = 1, 2, \ldots, MS^2$) belonging to land cover classes. The proportions $L(p_j)$ are defined as

$$\mathbf{D}_b = \mathbf{E}\mathbf{L} + \mathbf{n} \tag{5.12}$$

where $\mathbf{D}_b = [D_1, D_2, \ldots, D_{B^*}]^T$ is the vector of the spectral values of the MRSFD; $\mathbf{L} = [L(p_1), L(p_2), \ldots, L(p_N)]^T$ is the vector of the proportions; $\mathbf{E}$ is the matrix of the spectral endmembers; and **n** is the random noise. LSMM is used to obtain the optimal estimation under the minimum random noise condition.

Next, according to these proportions $L(p_j)$, class allocation method based on unit of class is utilized to allocate class labels to subpixels to obtain the final subpixel mapping result.

5.4.4 EXPERIMENTAL CONTENT AND RESULT ANALYSIS

Five subpixel mapping methods are tested and compared, including subpixel mapping based on intra- and inter-pixel dependence (NSAM) [25], subpixel mapping based on radial basis function interpolation by reducing point spread function effect (RBF-PSF) [26], subpixel mapping based on HNN with LiDAR image (HNN-LiDAR) [22], subpixel mapping based on pansharpening technology (SPM-PAN) [23], and the proposed SPM-MRSFD. We evaluate the performance of the five subpixel mapping methods by mapping accuracy of each class, overall accuracy (OA), and kappa coefficient (Kappa). The fractional images for five subpixel mapping methods are derived from the experiment data by LSMM. The experimental platform is the MATLAB R2018 software package in a Pentium dual-core processor (2.20 GHz).

5.4.4.1 Experiment 1

The data sets in experiment 1 are selected as 2013 IEEE GRSS Data Fusion Contest [27]. The data sets provided by the National Science Foundation–funded Center for Airborne Laser Mapping are collected over the University of Houston campus and neighboring area. The data sets include a fine hyperspectral image and a fine DSM of LiDAR. Since the two images have been registered well through preprocessing, the errors of image acquisition can be ignored. The hyperspectral image has 144 bands in 380 to 1050 nm spectral resolution, which was acquired on June 23, 2012. The DSM of LiDAR was acquired on June 22, 2012. As shown in Figure 5.30(a)–(b), the experimental area in the hyperspectral image and DSM of LiDAR are with 320 × 1800 pixels and 2.5-m spatial resolution.

To know the land cover classes at subpixel level, the simulated low-resolution hyperspectral image is derived by downsampling the fine hyperspectral image by $S \times S$ mean filter according to the general experimental process of subpixel mapping [3]. As shown in Figure 5.31(a), the fine hyperspectral image in Figure 5.30(a) is downsampled by an $S = 8$ mean filter to obtain the simulated low-resolution hyperspectral image as input experimental data. It is noted that the spatial distribution of land cover classes cannot be confirmed due to the large number of mixed pixels in Figure 5.31(a). In addition, we only consider pansharpening technique and avoid the influence of errors caused by the acquisition of the panchromatic image. The appropriate panchromatic image is created by applying the spectral response function of the IKONOS satellite to the fine hyperspectral image. The appropriate panchromatic image is shown in Figure 5.31(b). Fusing Figure 5.31(a) and Figure 5.31(b) by pansharpening, the fused image is shown in Figure 5.31(c). Because the spatial information of the panchromatic image is supplied to the low-resolution hyperspectral image, the fused image is clearer than Figure 5.31(a). To obtain high-quality

(a)

(b)

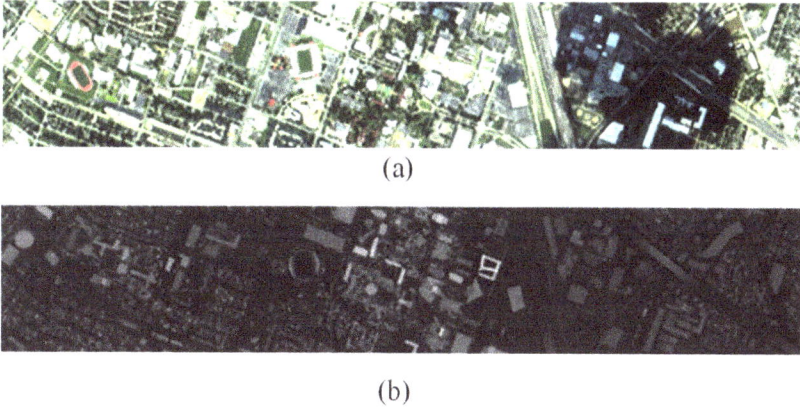

FIGURE 5.30   Data sets. (a) False color image of fine hyperspectral image. (b) DSM of LiDAR.

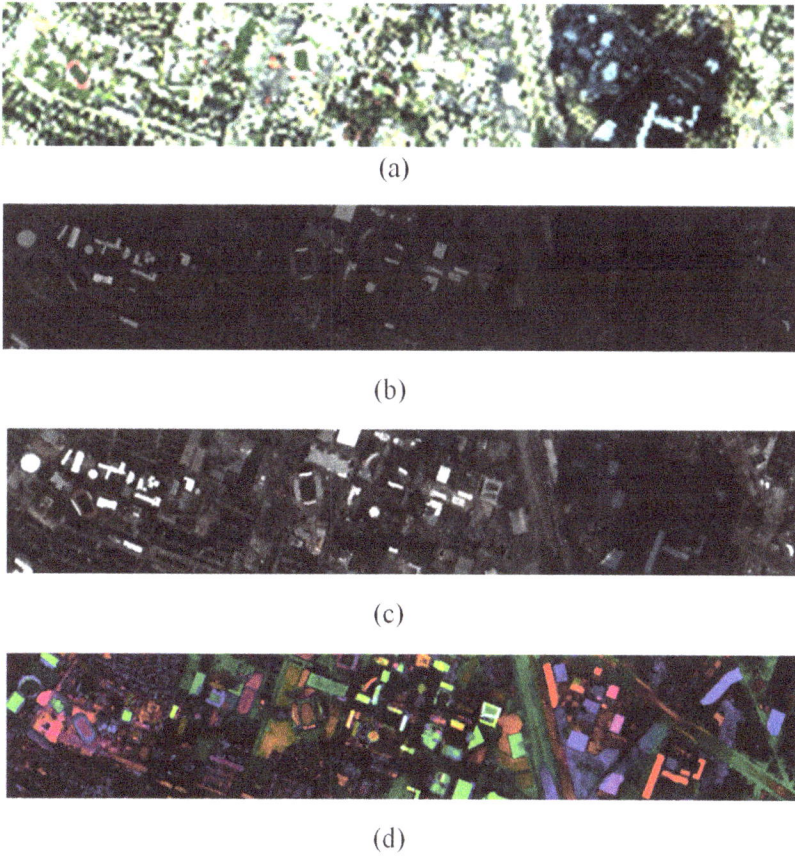

(a)

(b)

(c)

(d)

FIGURE 5.31   Experimental data. (a) Low-resolution hyperspectral image. (b) Panchromatic image. (c) Fused image. (d) Fused image MRSFD.

MRSFD, we then fuse Figure 5.31(c) with Figure 5.30(b) to obtain Figure 5.31(d) by feature fusion. By comparing Figure 5.31(c) with Figure 5.31(d), we could find that the boundaries and heights among the land cover classes are clearer in Figure 5.31(d) than those in Figure 5.31(c) due to the rich spatial-spectral-elevation information in Figure 5.31(d). The proposed SPM-MRSFD could produce better mapping results by utilizing Figure 5.31(d). In addition, for fair comparison, HNN-LiDAR uses eleva-tion information from Figure 5.30(b), and SPM-PAN utilizes the fused image of the pansharpening in Figure 5.31(c).

As shown in Figure 5.32(a), the ground truth including 15 classes is also from data sets selected as reference images. The mapping results of five subpixel mapping methods are shown in Figure 5.32(b)–(f). To facilitate observation, a sub-region with 100×100 pixels marked in Figure 5.32(a) of the mapping results is magnified and shown in Figure 5.33. By visual comparison, the mapping result of SPM-MRSFD is closer to the reference image than that of the other four methods. We could find that the subpixel mapping results of NSAM, RBF-PSF, HNN-LiDAR, and SPM-PAN are not ideal, for example, there are many boundaries of protruding burrs and areas of

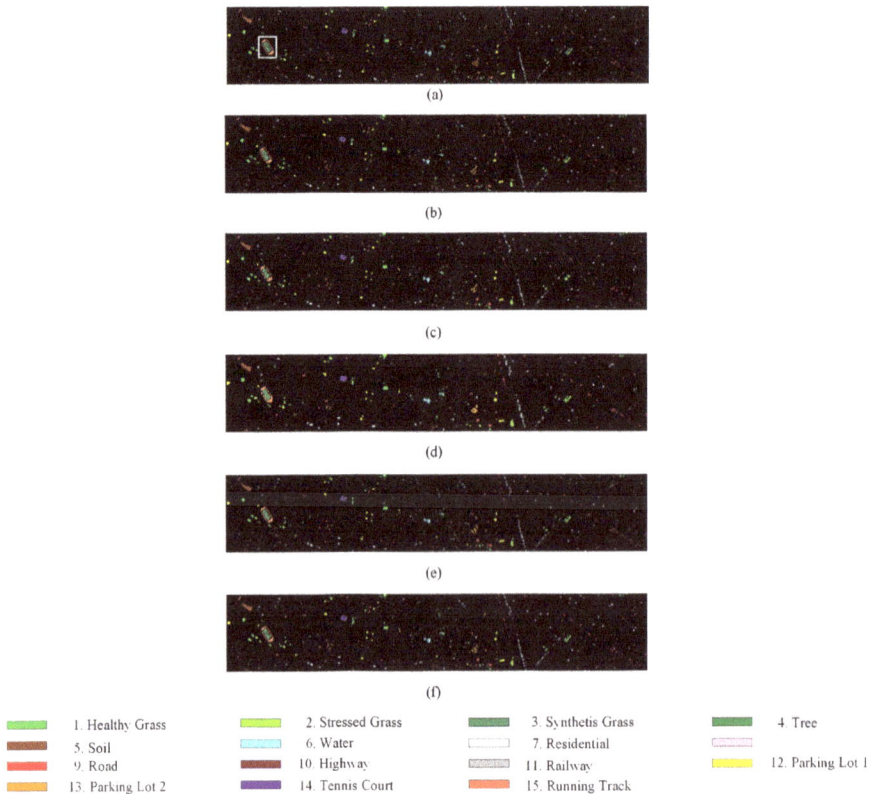

FIGURE 5.32 Subpixel mapping results: (a) reference image, (b) NSAM, (c) RBF-PSF, (d) HNN-LiDAR, (e) SPM-PAN, and (f) SPM-MRSFD.

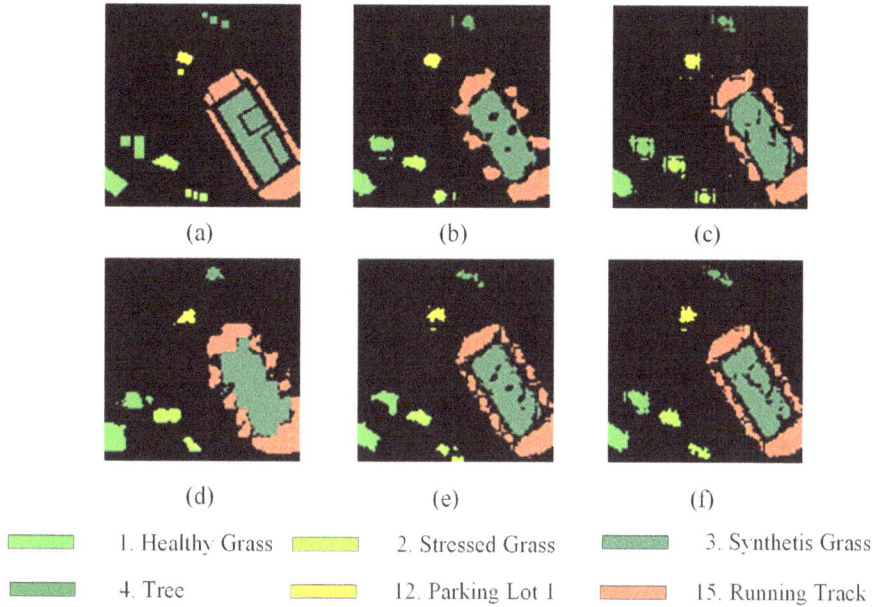

FIGURE 5.33 Sub-region of subpixel mapping results in experiment 1: (a) reference image, (b) NSAM, (c) RBF-PSF, (d) HNN-LiDAR, (e) SPM-PAN, and (f) SPM-MRSFD.

disconnected shapes in class 15 (Running Track) in Figure 5.33(b)–(e). This phenomenon is alleviated in the proposed SPM-MRSFD in Figure 5.33(f).

The precision of per class (%) and OA (%) for five subpixel mapping methods are listed in Table 5.8. According to quantitative evaluation about subpixel mapping [3], the precision of per class (%) is calculated by the ratio of the number of correct mappings per class obtained by comparing the reference image to the actual total number of this class. It is noted that the performance of the proposed SPM-MRSFD is superior to the other four subpixel mapping methods. Although the precision of class 4 (synthetic grass), class 13 (parking lot 2), and class 14 (tennis court) in HNN-LiDAR is higher than these in the proposed SPM-MRSFD, the precision of the other classes, especially the small target classes, such as class 4 (tree), class 7 (residential), and class 9 (road), is very low, resulting in the lowest OA (%) among the five subpixel mapping methods. When comparing with SPM-PAN, the OA (%) of the SPM-MRSFD increased by around 2.6%.

5.4.4.2 Experimental 2

The data sets in experiment 2 are selected as 2018 IEEE GRSS Data Fusion Contest [28]. The data sets are also acquired by the NCALM, covering the University of Houston campus and its surrounding urban areas, but the data sets in experiments 1 and 2 are collected from different sensors, areas, and time. In experiment 2, the hyperspectral image covers the 380 to 1050 nm spectral range with 48 bands at a 1-m spatial resolution, the DSM of LiDAR is at 0.5-m spatial resolution, and both

TABLE 5.8

Mapping Accuracy of the Six Methods in Experiment 1

Land Cover Class	NSAM	RBF-PSF	HNN-LiDAR	SPM-PAN	SPM-MRSFD
Class 1 (%)	78.28	76.33	81.74	81.57	**83.77**
Class 2 (%)	63.94	61.34	60.21	65.77	**68.81**
Class 3 (%)	87.23	87.66	**93.13**	89.38	90.10
Class 4 (%)	53.52	56.82	36.93	56.65	**61.32**
Class 5 (%)	72.91	72.81	74.69	72.81	**76.48**
Class 6 (%)	70.15	70.77	73.23	67.69	**78.46**
Class 7 (%)	53.56	59.10	38.49	64.44	**65.96**
Class 8 (%)	64.66	67.50	41.79	61.81	**70.31**
Class 9 (%)	52.72	53.58	35.53	49.76	**58.46**
Class 10 (%)	58.99	63.81	65.69	74.02	**77.37**
Class 11 (%)	57.40	59.37	54.44	63.32	**65.44**
Class 12 (%)	72.26	73.40	67.80	72.51	**75.83**
Class 13 (%)	51.91	53.03	**58.12**	48.76	54.16
Class 14 (%)	84.11	84.35	**88.33**	82.48	85.75
Class 15 (%)	69.85	77.12	81.06	79.85	**81.52**
OA (%)	65.12	66.80	61.29	68.28	**70.92**

Note: Boldface indicates highest value.

were collected on February 16, 2017, between 16:31 and 18:18. The panchromatic image is generated by the method described in experiment 1. The original hyperspectral image is degraded with $S = 10$ to produce the low-resolution image. As shown in Figure 5.34(a), the ground truth including 1200 × 1200 pixels and 20 classes is selected as the reference image. The mapping results of the five subpixel mapping methods are shown in Figure 5.34(b)–(f). The quantitative evaluation for five subpixel mapping methods is listed in Table 5.9. Similar to the conclusion drawn from experiment 1, the performance of SPM-MRSFD is still superior to the other four subpixel mapping methods.

5.4.4.3 Discussion

Several factors that affect the performance of the proposed SPM-MRSFD are discussed in this section. First, the selection of the pansharpening method and feature fusion method will affect the SPM-MRSFD. We utilize the pansharpening based on band-dependent spatial detail (BDSD) [29] to replace the PCA and keep the feature fusion based on graph unchanged in SPM-MRSFD. For experiments 1 and 2, the OA (%) in relation to two pansharpening methods is listed in Table 5.10. Since BDSD is more effective than PCA, the value of OA (%) in BDSD is higher than that in PCA. According to the same experimental method, the other feature fusion, named GGF [30], is selected to replace the existing FFG in the SPM-MRSFD. The experimental results are also listed in Table 5.10. Similar to the previous results, because GGF shows the better performance, GGF-based SPM-MRSFD obtains the higher OA (%).

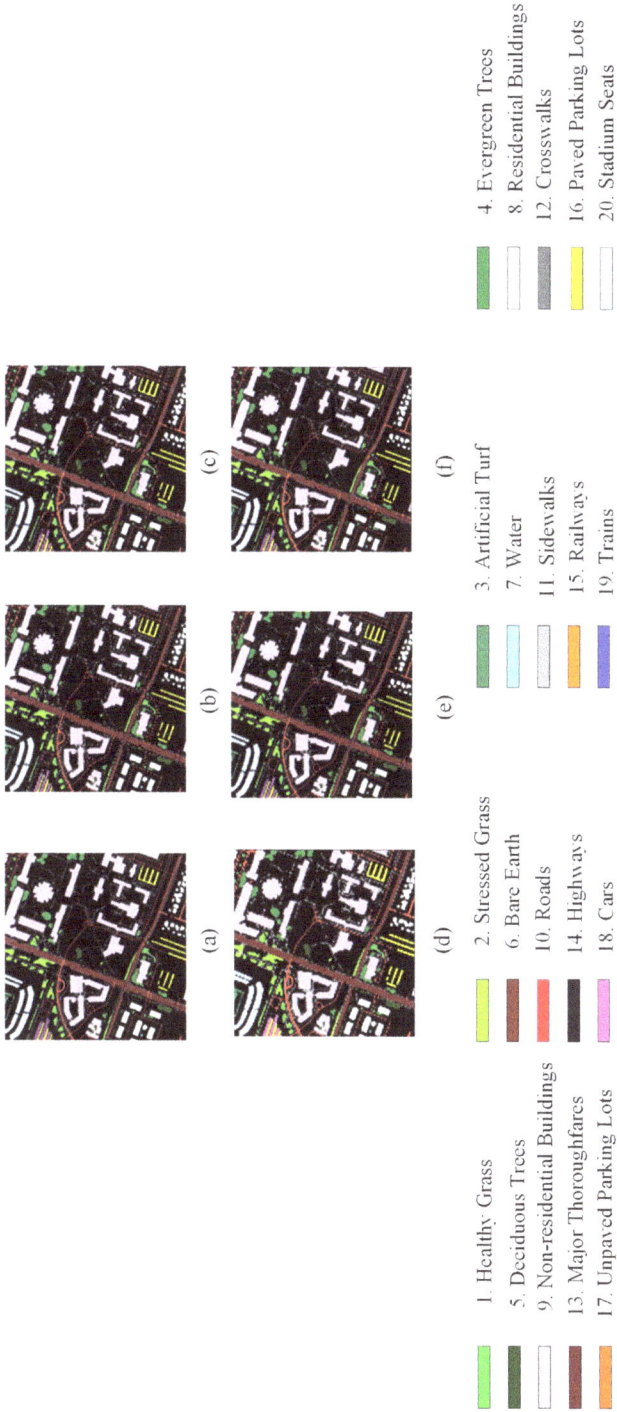

FIGURE 5.34 Subpixel mapping results in experiment 2: (a) reference image, (b) NSAM, (c) RBF-PSF, (d) HNN-LiDAR, (e) SPM-PAN, and (f) SPM-MRSFD.

1. Healthy Grass
2. Stressed Grass
3. Artificial Turf
4. Evergreen Trees
5. Deciduous Trees
6. Bare Earth
7. Water
8. Residential Buildings
9. Non-residential Buildings
10. Roads
11. Sidewalks
12. Crosswalks
13. Major Thoroughfares
14. Highways
15. Railways
16. Paved Parking Lots
17. Unpaved Parking Lots
18. Cars
19. Trains
20. Stadium Seats

TABLE 5.9

Mapping Accuracy of the Six Methods in Experiment 2

Evaluation Indicator	NSAM	RBF-PSF	HNN-LiDAR	SPM-PAN	SPM-MRSFD
OA (%)	86.90	87.81	84.81	88.74	90.78

TABLE 5.10

OA (%) in Relation to Pansharpening and Feature Fusion

	Pansharpening		Feature Fusion	
	PCA	BDSD	FFG	GGF
Experiment 1	70.92	73.38	70.92	74.07
Experiment 2	90.78	92.17	90.78	93.03

Hence, the more effective pansharpening method and feature fusion method can obtain the better mapping result.

Second, the performance of five subpixel mapping methods depends on the spectral unmixing method. The spectral unmixing method based on the least squares support vector machine (LSSVM) [31] is selected to replace LSMM in an experiment of five subpixel mapping methods. Figure 5.35(a) shows the OA (%) of the five subpixel mapping methods in relation to two spectral unmixing methods for experiment 1. Since LSSVM is more effective than LSMM, the OA (%) of the five subpixel mapping methods increases when using LSSVM. But the proposed SPM-MRSFD still has the highest OA (%).

Finally, the performance of five subpixel mapping methods is also affected by different S values. The five subpixel mapping methods are tested with three S values (i.e., 5, 8, and 10) for experiment 1. The values of OA (%) in relation to different S values are shown in Figure 5.35(b). Since the larger S means that the simulated hyperspectral image has lower resolution, OA (%) of the five subpixel mapping methods decreases. But the proposed SPM-MRSFD still obtains a higher OA (%) than the other four methods.

5.5 SUMMARY

In the STHSRM-PAN, the original low-resolution remote sensing image is fused with the high-resolution panchromatic image of the same area, and the remote sensing image with higher resolution is generated through pansharpening technology. By unmixing the improved image, the fine abundance images with more spatial-spectral information are obtained. Finally, according to the subpixel abundance values from the fine abundance images, the class allocation method based on UOC is used to generate the subpixel mapping result. The experimental results show that, when

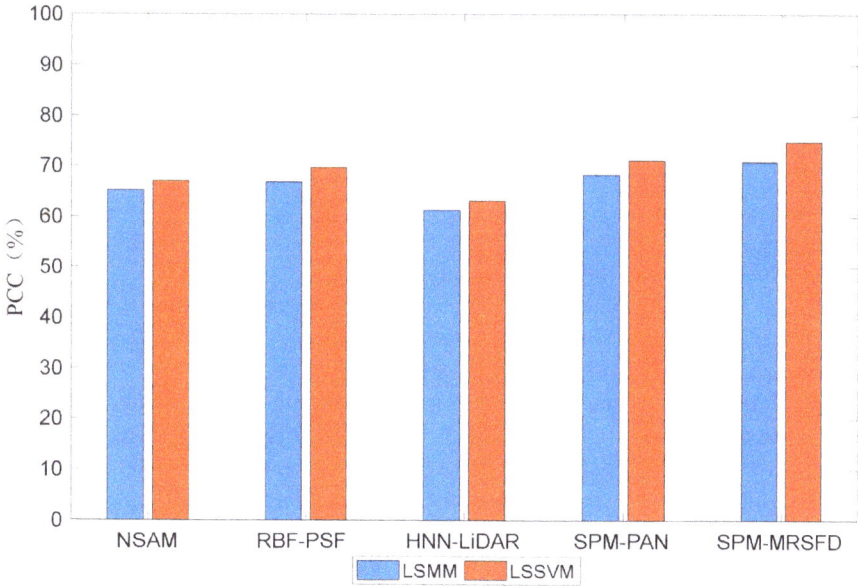

(a)

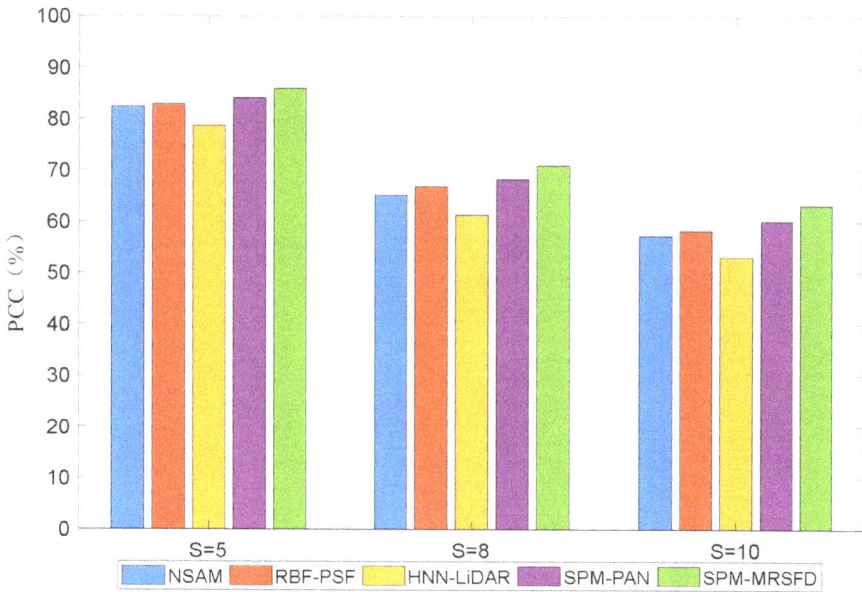

(b)

FIGURE 5.35   OA (%) of the five subpixel mapping methods in relation to (a) different spectral unmixing methods and (b) different S values.

compared with the existing subpixel mapping methods, the proposed STHSRM-PAN obtains a more ideal mapping result.

For a hyperspectral image, with the help of fusion technology and deep learning, DPP is proposed here. DPP includes a fusion path and a deep learning path. First, in the fusion path, the PCA-based hyperspectral and multispectral fusion algorithm is used to fuse the original coarse hyperspectral image and the high spatial resolution multispectral image to obtain an improved hyperspectral image. The improved hyperspectral image is unmixed to obtain the abundance images with richer spatial-spectral information. In the deep learning path, the coarse abundance images are handled by a super-resolution reconstruction algorithm based on DLPN, producing the high-resolution abundance images with multi-scale information. Then, the two different types of high-resolution abundance images are integrated to obtain the finer abundance images with multiple types of information. Finally, according to the subpixel abundance values provided by the finer abundance images, the class allocation method based on UOC is used to obtain the final subpixel mapping result. The visual and quantitative comparison with the existing subpixel mapping methods shows that the proposed DPP can obtain the more ideal subpixel mapping result.

To improve the subpixel mapping result by utilizing multiple types of auxiliary information from MRSFD, SPM-MRSFD is proposed in this chapter. In SPM-MRSFD, hyperspectral image, panchromatic image, and DSM of LiDAR are fused by pansharpening and feature fusion in turn to produce MRSFD with spatial-spectral-elevation information. The MRSFD is then unmixed to obtain the fractional images. According to the proportions information from the fractional images of MRSFD, the class allocation method is utilized to obtain the subpixel mapping result. The experimental results show that the SPM-MRSFD achieves better mapping results than state-of-the-art subpixel mapping methods.

REFERENCES

[1] Xu X. Research on the Subpixel Mapping Theory and Method of Remote Sensing Image Taking Into Account the Spatial Characteristics of Ground Objects[D]. Wuhan: Wuhan University, 2012.

[2] Vivone G, Alparone L, Chanussot J, Mura M D, Garzelli A, Licciardi G, Restaino R, Wald L. A critical comparison among pansharpening algorithms [J]. IEEE Transactions on Geoscience and Remote Sensing, 2015, 53(5): 2565–2585.

[3] Loncan L, Almeida L, et al. Hyperspectral pansharpening: A review[J]. IEEE Geoscience and Remote Sensing Magazine, 2015, 3(3): 27–46.

[4] Thomas C, Ranchin T, Wald L, Chanussot J. Synthesis of multispectral images to high spatial resolution: A critical review of fusion methods based on remote sensing physics[J]. IEEE Transactions on Geoscience and Remote Sensing, 2008, 46(5): 1301–1312.

[5] Chavez Jr P S, Sides S C, Anderson J A. Comparison of three different methods to merge multiresolution and multispectral data: Landsat TM and SPOT panchromatic[J]. Photogrammetric Engineering and Remote Sensing, 1991, 57(3): 295–303.

[6] Aiazzi B, Baronti S, Selva M. Improving component substitution pansharpening through multivariate regression of MS+Pan data [J]. IEEE Transactions on Geoscience and Remote Sensing, 2007, 45(10): 3230–3239.

[7] Jia S, Qian Y. Spectral and spatial complexity-based hyperspectral unmixing[J]. IEEE Transactions on Geoscience and Remote Sensing, 2006, 45(12): 3867–3879.

[8] Tu T M, Huang P S, Hung C L, Chang C P. A fast intensity-hue-saturation fusion technique with spectral adjustment for IKONOS imagery [J]. IEEE Geoscience and Remote Sensing Letters, 2004, 1(4): 309–312.

[9] Simões M, Bioucas-Dias J, Almeida L B, Chanussot J. A convex formulation for hyperspectral image superresolution via subspace-based regularization [J]. IEEE Transactions on Geoscience and Remote Sensing, 2015, 53(6): 3373–3388.

[10] Wang L, Wang Z, Dou Z, Wang Y. Edge-directed interpolation-based sub-pixel mapping[J]. Remote Sensing Letters, 2013, 12(4): 1195–1203.

[11] Wang P, Wang L, Chanussot J. Soft-then-hard subpixel land cover mapping based on spatial-spectral interpolation[J]. IEEE Geoscience and Remote Sensing Letters, 2016, 13(12): 1851–1854.

[12] Wang Q, Shi W, Atkinson P M. Sub-pixel mapping of remote sensing images based on radial basis function interpolation[J]. ISPRS Journal of Photogrammetry and Remote Sensing, 2014, 92(1): 1–15.

[13] Chen Y, Ge Y, Song D. Superresolution land-cover mapping based on high-accuracy surface modeling[J]. IEEE Geoscience and Remote Sensing Letters, 2015, 12(12): 2516–2520.

[14] Nguyen M Q, Atkinson P M, Lewis H G. Superresolution mapping using a Hopfield neural network with fused images[J]. IEEE Transactions on Geoscience and Remote Sensing, 2006, 44(3): 736–749.

[15] Lai W, Huang J, Ahuja N, Yang M. Deep Laplacian pyramid networks for fast and accurate super-resolution[C]. IEEE Computer Vision and Pattern Recognition (CVPR), 2017: 5835–5843.

[16] Lai W, Huang J, Ahuja N, Yang M. Fast and accurate image super-resolution with deep Laplacian pyramid networks[J]. IEEE Transactions on Pattern Analysis and Machine Intelligence, 2019, 41(11): 2599–2613.

[17] Wang Q, Shi W, Wang L. Allocating classes for soft-then-hard sub-pixel mapping algorithms in units of class[J]. IEEE Transactions on Geoscience and Remote Sensing, 2014, 5(5): 2940–2959.

[18] Wang P, Zhang G, Kong Y, Leung H. Superresolution mapping based on hybrid interpolation by parallel paths[J]. Remote Sensing Letters, 2019, 10(2): 149–157.

[19] Chen Y, Ge Y, Wang Q, Jiang Y. A subpixel mapping algorithm combining pixel-level and subpixel-level spatial dependences with binary integer programming[J]. Remote Sensing Letters, 2014, 5(10): 902–911.

[20] Ling F, Foody G M, Ge Y, Li X, Du Y. An iterative interpolation deconvolution algorithm for superresolution land cover mapping[J]. IEEE Transactions on Geoscience and Remote Sensing, 2016, 54(12): 7210–7222.

[21] Nguyen M Q, Atkinson P M, Lewis H G. Super-resolution mapping using Hopfield neural network with panchromatic image[J]. International Journal of Remote Sensing, 2011, 32(21): 6149–6176.

[22] Nguyen M Q, Atkinson P M, Lewis H G. Superresolution mapping using Hopfield neural network with LIDAR data[J]. IEEE Geoscience and Remote Sensing Letters, 2005, 2(3): 366–370.

[23] Wang P, Mura M D, Chanussot J, Zhang G. Soft-then-hard super-resolution mapping based on pansharpening technique for remote sensing image[J]. IEEE Journal of Selected Topics in Applied Earth Observations and Remote Sensing, 2019, 12(1): 334–344.

[24] Scholkopf B, Smola A J, Muller K R. Nonlinear component analysis as a kernel eigenvalue problem[J]. Neural Computation, 1998, 10(5): 1299–1319.

[25] Ling F, Li X, Du Y, Xiao F. Sub-pixel mapping of remotely sensed imagery with hybrid intra- and inter-pixel dependence[J]. International Journal of Remote Sensing, 2013, 34(1): 341–357.

[26] Wang Q, Zhang C, Tong X, Atkinson P M. General solution to reduce the point spread function effect in subpixel mapping[J]. Remote Sensing of Environment, 2020, 251: 112054.

[27] Wang Q, Huang W, Xiong Z, Li X. Looking closer at the scene: Multiscale representation learning for remote sensing image scene classification. IEEE Transactions on Neural Networks and Learning Systems, online, 2020.

[28] Xu Y, et al. Advanced multi-sensor optical remote sensing for urban land use and land cover classification: Outcome of the 2018 IEEE GRSS Data Fusion Contest[J]. IEEE Journal of Selected Topics in Applied Earth Observations and Remote Sensing, 2019, 12(6): 1709–1724.

[29] Loncan L, Almeida L, et al. Hyperspectral pansharpening: A review[J]. IEEE Geoscience and Remote Sensing Magazine, 2015, 3(3): 27–46.

[30] Liao W, Bellens R, Pizurica A, Gautama S, Philips W. Generalized graph-based fusion of hyperspectral and LiDAR data using morphological features[J]. IEEE Geoscience and Remote Sensing Letters, 2015, 12(3): 552–556.

[31] Wang L, Liu D, Wang Q. Spectral unmixing model based on least squares support vector machine with unmixing residue constraints[J]. IEEE Geoscience and Remote Sensing Letters, 2013, 10(6): 1592–1596.

6 Remote Sensing Image Subpixel Mapping Based on Classification Then Reconstruction

6.1 INTRODUCTION

The final result of subpixel mapping is to obtain thematic mapping with accurate spatial distribution information of each class. Due to the imperfections of the existing spectral unmixing technology, a large number of unmixing errors will be generated during the process of spectral unmixing. However, as a subsequent processing step of the spectral unmixing, the subpixel mapping results will inevitably be affected by these unmixing errors. Therefore, how to reduce the influence of unmixing errors on the subpixel mapping result has become a key issue for improving the subpixel mapping method.

This chapter first proposes the subpixel mapping based on maximum a posteriori (MAP) super-resolution recovery then classification (MTC). The proposed method not only makes full use of the supervision information of the original image but also uses the classification algorithm to effectively avoid the spectral unmixing step, thereby avoiding the influence of the spectral unmixing error on the subpixel mapping result. Further, through the related research on pansharpening, it can be known that the pansharpening algorithm uses the panchromatic as auxiliary data to supplement the more spatial-spectral information. Therefore, when compared with the super-resolution reconstruction algorithm, the pansharpening algorithm better improved the resolution of the original remote sensing image. Based on previous research, this chapter proposes subpixel mapping based on pansharpening technique recovery, which uses the previously proposed principal component analysis (PCA) pansharpening technology to replace the MAP super-resolution reconstruction algorithm in pansharpening then classification (PTC). The pansharpening results are then directly classified to obtain the final subpixel mapping result. The following briefly introduces the super-resolution technology and fully supervised information classification technology used in this chapter.

6.2 THEORETICAL BASIS

6.2.1 SUPER-RESOLUTION ALGORITHM

Due to the limitations of imaging conditions and imaging methods, the quality of the acquired image will often decrease. Considering the hardware level and cost issues, how to improve the quality of low-resolution image through image processing

DOI: 10.1201/9781003279082-6

technology has become a research hot topic in recent years. The super-resolution algorithm can solve these problems. It estimates a high-resolution image from one or more low-resolution images and eliminates noise and blurring caused by hardware components. Because super-resolution technology has the advantages of not involving hardware, low cost, and can be realized by existing imaging systems, the technology has been widely used in video, imaging, remote sensing, medicine, surveillance, and military fields. According to its technical characteristics, the existing super-resolution methods can be divided into three main categories [1]: super-resolution based on interpolation, super-resolution based on reconstruction, and super-resolution based on learning.

6.2.1.1   Super-Resolution Based on Interpolation

Super-resolution based on interpolation is to use the gray value of the sampling point of the known low-resolution image to estimate the gray value of the unknown sampling point of the high-resolution image, which is to recover the original continuous image signal from the limited discrete sampling data. At the same time, the effective restoration techniques are used to eliminate blur and reduce image noise. Because the interpolation algorithm has the advantages of simple operation, small calculation amount, and no prior information, the super-resolution based on the interpolation algorithm is the most commonly used super-resolution method in the engineering field at present. Classical interpolation methods include nearest neighbor interpolation, bilinear interpolation, bicubic interpolation, polynomial interpolation, and spline interpolation [2].

6.2.1.2   Super-Resolution Based on Reconstruction

Super-resolution based on reconstruction is currently the most widely used super-resolution algorithm with the most ideal results. Its principle is the process of gradually recovering high-resolution images from low-resolution images through an imaging model. Therefore, the key to super-resolution based on reconstruction is to establish an imaging model that links the actual low-resolution image with the high-resolution image. At present, the popular super-resolution based on reconstruction algorithms [3] includes an iterative backprojection algorithm, convex set projection (POCS) algorithm, MAP algorithm, and maximum posterior probability and convex set projection hybrid (MAP/POCS) algorithm [4]. The super-resolution based on reconstruction algorithm is closer to the actual application situation, but with the increase of the resolution magnification factor, it is more and more difficult to improve the reconstruction effect, and the algorithm is highly complex and computationally intensive.

6.2.1.3   Super-Resolution Based on Learning

The super-resolution based on learning has now become a cutting-edge super-resolution research direction [5–6], which is different from the super-resolution based on reconstruction that relies too much on the image information generated from the input images. The high-quality and high-resolution images that contain the same information with the input images are used as training samples in the super-resolution based on learning, and the prior information used as the basis for super-resolution is obtained from a large number of training sample sets. However, in the implementation of the super-resolution based on learning, because it is necessary to know enough high-quality and high-resolution images as training samples to establish a learning model to provide more complete prior knowledge, the computational burden is bound to increase.

The differences between super-resolution and subpixel mapping are obvious. The super-resolution reconstructs a high-resolution image by inputting one or more images with continuous gray values, and the output is an image with continuous gray values. However, the subpixel mapping is mainly used in mixed pixels to reduce the influence of the mixed pixels. The high-resolution image obtained by subpixel mapping has the land cover labels, and the output is with a discrete gray values image.

6.2.2 FULLY SUPERVISED INFORMATION CLASSIFICATION ALGORITHM

According to whether the traditional classifier uses prior information in the classification process, the classification algorithms can be roughly divided into three kinds: unsupervised information classification algorithm, semi-supervised information classification algorithm, and fully supervised information classification algorithm. The classification algorithm used in the method proposed in this chapter is the fully supervised information classification algorithm, so the following is a detailed introduction to the fully supervised information classification algorithm. The fully supervised information classification algorithm refers to the learning process of the established classifier model under the condition of sufficient supervised information of the land cover classes of the remote sensing image and obtains the corresponding model parameters and the characteristics of each class. According to the classification decision function, the unknown samples are then classified. Compared with the unsupervised information classification algorithm and the semi-supervised information classification algorithm, the fully supervised information classification algorithm has the most significant feature of high mapping accuracy. However, the fully supervised information classification algorithm is based on certain prior-supervised information. Therefore, when the prior supervision information is not sufficient, it also causes certain limitations of the algorithm. Typical fully supervised classification methods include maximum likelihood classification, spectral angle matching classification, and support vector machine classification [7].

6.2.2.1 Maximum Likelihood Classification

The maximum likelihood (ML) classification algorithm is a classic statistical-based pattern recognition method, also known as Bayes criterion. The ML classification algorithm has been widely used in the classification of remote sensing images due to assigning test samples to their classes quickly. The core of the algorithm is based on statistical variables such as the mean, variance, and covariance of the training sample, and the Bayes criterion is used to estimate the category of the test sample.

Suppose there are K classes in a remote sensing image, expressed by ω_k ($k = 1, 2, \ldots, K$). Assume that the prior probability of occurrence of each class is $p(\omega_k)$ and the prior probability of each class is equal, namely, $p(\omega_k) = 1/k$. According to the

Bayesian criterion, the probability that a sample $\mathbf{x}$ belongs to a certain class can be expressed as

$$p(\omega_k/\mathbf{x}) = \frac{p(\mathbf{x}/\omega_k) \cdot p(\omega_k)}{p(\mathbf{x})} \tag{6.1}$$

$$p(\mathbf{x}) = \sum_{k=1}^{K} p(\mathbf{x}/\omega_k) \cdot p(\omega_k) \tag{6.2}$$

Since $p(\mathbf{x})$ does not affect the discrimination effect and the prior probability $p(\mathbf{x}/\omega_k)$ of each class are equal, the classification problem can be transformed into a problem of comparing the probability of $p(\omega_k)$. The discriminant function of the ML classification algorithm is as follows:

$$p(\mathbf{x}/\omega_k) = -\ln\left|\sum_k\right| - (\mathbf{x}\text{-}\mathbf{m}_k)^{\mathbf{T}} \sum_k^{-1} (\mathbf{x}\text{-}\mathbf{m}_k) \tag{6.3}$$

where $\mathbf{x}$ represents the spectral vector of the test sample, $\mathbf{m}_k$ represents the kth class mean vector, and $\sum_k$ represents the covariance matrix of the kth class. When $j \in \{1,2,...,K\}$ and $j \neq k$, if $p(\mathbf{x}/\omega_k) > p(\mathbf{x}/\omega_j)$, the sample $\mathbf{x}$ belongs to the kth class, namely,

$$p(\omega_k/\mathbf{x}) = \max_{j=1}^{K} p(\omega_k/\mathbf{x}) \tag{6.4}$$

The advantage of the ML classification algorithm is that it is easy to code, but due to the large number of feature dimensions and the total number of classes of remote sensing data, the calculation speed of the ML classification algorithm will be significantly affected.

6.2.2.2   Spectral Angle Matching Classification

The spectral angle mapping (SAM) classification algorithm mainly uses the conclusion that the radiation curves or reflection curves from the same type of land cover classes have a certain similarity to classify samples. This method regards each pixel as a spectral vector and calculates the generalized angle between this vector and other vectors. The smaller the angle means the two pixels are more similar and have a greater probability of belonging to the same class.

First, the two spectral vectors are defined as $\mathbf{X}(x_1, x_2, ..., x_n)$ and $\mathbf{Y}(y_1, y_2, ..., y_n)$, respectively, where n represents the number of spectral bands and θ represents the generalized angle of two vectors. The generalized angle of the two vectors can be expressed by

$$\theta = \cos^{-1}\left[\frac{\mathbf{X} \cdot \mathbf{Y}}{\|\mathbf{X}\| \cdot \|\mathbf{Y}\|}\right] = \cos^{-1}\left[\frac{\sum_{i=1}^{n}(x_i \cdot y_i)}{\sqrt{\sum_{i=1}^{n} x_i^2}\sqrt{\sum_{i=1}^{n} y_i^2}}\right] \tag{6.5}$$

The generalized angle θ represents the matching degree of the two pixels, $\theta \in [0, \pi/2]$. When $\theta = 0$, it means that the two spectral curves are completely consistent. When $\theta = \pi/2$, the two spectral curves have the greatest degree of difference. The smaller the θ means the greater the similarity between the two spectral curves and the greater the possibility of them belonging to the same class, and vice versa. Under normal circumstances, the geometric mean vector of the training samples of each class is used as the reference spectral vector of the class.

The advantage of the SAM classification algorithm is that it is not affected by the vector modulus and is only related to the direction of the vector. However, this method only uses the mean value information of the pixel and ignores the other information of the pixel, which results in serious defects in the learning ability and classification result.

6.2.2.3 Support Vector Machine Classification

When classifying the remote sensing image with high-dimensional data structures, the training samples include limited prior knowledge obtained based on historical experience. In this case, the two classification methods introduced earlier are often difficult to satisfy in terms of a reliability requirement. A support vector machine (SVM) can just solve this contradiction. It is developed from statistical learning theory, which measures the best compromise between experience risk and confidence risk and has high generalization and learning ability. The kernel function that is an important part of the method cleverly solves the problem of data classification in the case of linear inseparability. Therefore, SVM is used in the classification of high-dimensional remote sensing images and has become a hot topic in recent research. The selected classification algorithm in the proposed method in this chapter is the least squares support vector machine (LSSVM). Therefore, the following introduces the SVM classification algorithm in detail by the linear classification problem and the non-linear classification problem.

6.2.2.3.1 Linear Classification Problem

SVM theory starts from the perspective of two types of classification problems. The basic principle is to find a classification hyperplane. This hyperplane needs to ensure that samples of different types are distributed on different sides of the hyperplane, and the distance between the closest sample point to the hyperplane and the hyperplane should be maximized. SVM will comprehensively consider the empirical risk and the confidence risk to minimize the expected risk. Figure 6.1 shows the binary classification in the linear classification problem. The solid point and the hollow point represent two types of samples, H is the classification surface, and the distance between H_1 and H_2 is the classification interval.

For linear binary classification problems, suppose n sample training sets is $\mathbf{x}_i \in R^d$, $y_i \in \{+1, -1\}$ are the corresponding class labels, and $i = 1, \ldots, n$. A linear function is constructed in d dimensional space to represent the optimal classification plane:

$$g(\mathbf{x}) = \langle \mathbf{w}, \mathbf{x} \rangle + b \tag{6.6}$$

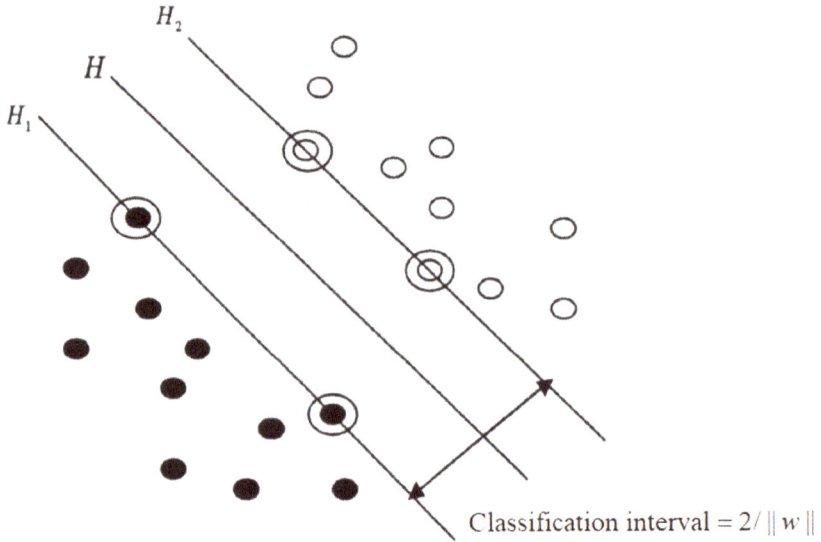

FIGURE 6.1   Two-class linear classification.

where $\mathbf{x}$ is the d dimensional feature vector; $\mathbf{w}$ is the weight vector and is expressed as $\mathbf{w} = \left[w_1, w_2, ..., w_d \right]^{\mathrm{T}}$; and b is called a constant, which is the threshold weight.

The decision rule for the second-class linear classification is

$$\begin{cases} g\left(\mathbf{x}\right) = \langle \mathbf{w}, \mathbf{x} \rangle + b > 0 \Rightarrow \mathbf{x} \in w_1 \\ g\left(\mathbf{x}\right) = \langle \mathbf{w}, \mathbf{x} \rangle + b < 0 \Rightarrow \mathbf{x} \in w_2 \\ g\left(\mathbf{x}\right) = \langle \mathbf{w}, \mathbf{x} \rangle + b = 0 \Rightarrow \mathbf{x} \in w_1 \ or \ w_2 \end{cases} \tag{6.7}$$

The corresponding classification decision plane is $g(\mathbf{x}) = 0$, which separates the points belonging to different classes, and this decision plane is recorded as H. Then $g(\mathbf{x})$ is regarded as an algebraic measure of the distance from a certain point $\mathbf{x}$ in the space to the decision plane H, and $\mathbf{x}$ can be expressed as

$$\mathbf{x} = \mathbf{x}_p + r \frac{\mathbf{w}}{\|\mathbf{w}\|} \tag{6.8}$$

where $\mathbf{x}_p$ is the projection vector of $\mathbf{x}$ on H; r is the vertical distance from $\mathbf{x}$ to H; and $\frac{\mathbf{w}}{\|\mathbf{w}\|}$ is the unit vector in the $\mathbf{w}$ direction.

Combining formulas (6.7) and (6.8), formula (6.9) can be obtained:

$$g\left(\mathbf{x}\right) = \left\langle \mathbf{w}, \left[\mathbf{x}_p + r \frac{\mathbf{w}}{\|\mathbf{w}\|} \right] \right\rangle + b = \left\langle \mathbf{w}, \mathbf{x}_p \right\rangle + b + r \frac{\mathbf{w}}{\|\mathbf{w}\|} = r \|\mathbf{w}\| \tag{6.9}$$

To separate the samples to be classified as much as possible, the classification interval should be the largest, and |w| should be the smallest. Therefore, formula (6.9) can be transformed into the following optimization problem:

$$\min \frac{1}{2}\|\mathbf{w}\|^2$$

$$s.t. \quad y_i \left[\langle \mathbf{w}, \mathbf{x}_i \rangle + b \right] - 1 \geq 0 \quad i = 1,2,...,n$$

(6.10)

Formula 6.10 is used to construct the Lagrange function, with α_i as the Lagrange operator, and then the dual problem of the original problem can be obtained, that is the following objective function of the maximum optimization:

$$L(\mathbf{w},b,\alpha) = \sum_{i=1}^{n} \alpha_i - \frac{1}{2} \sum_{i,j=1}^{n} \alpha_i \alpha_j y_i y_j \langle \mathbf{x}_i, \mathbf{x}_j \rangle$$

$$s.t. \quad \sum_{i=1}^{n} \alpha_i y_i = 0, \alpha_i \geq 0 \quad i = 1,2,...,n$$

(6.11)

This dual problem is usually easier to solve than the original problem. According to the Kuhn-Toucher theorem, the optimal solution satisfies:

$$\alpha_i \left[y_i \left(\langle \mathbf{w}, \mathbf{x}_i \rangle + b \right) - 1 \right] = 0$$

(6.12)

Let $(\boldsymbol{\alpha}^*, b^*)$ be the optimal solution of the maximization formula (6.10), then the corresponding discriminant function formula is

$$f(\mathbf{x}) = \text{sgn}\left\{ \langle \mathbf{w}^*, \mathbf{x}_i \rangle + b^* \right\} = \text{sgn}\left\{ \sum_{i=1}^{n} \alpha_i^* y_i (\mathbf{x}_i, \mathbf{x}) + b^* \right\}$$

(6.13)

where the vector $\mathbf{w}^* = \sum_{i=1}^{n} \alpha_i^* y_i \mathbf{x}_i$; sgn(·) is symbolic function. The form of b^* is not unique, it can be deduced from the Kuhn-Toucher theorem. The SVM classification algorithm can predict the test sample class labels by calculating the previously mentioned discriminant function $f(\mathbf{x})$.

6.2.2.3.2 Non-Linear Classification Problem

When dealing with a non-linear classification problem, SVM introduces a non-linear mapping φ, which maps data points that cannot be linearly separated from the original data to linearly separable points in the transformation space, as shown in Figure 6.2. Under this condition, the x_i in each optimization expression of the linear problem introduced earlier needs to be replaced with $\varphi(\mathbf{x}_i)$, and the inner product $\langle \mathbf{x}_i, \mathbf{x}_j \rangle$ will be replaced by the following formula:

$$\mathbf{K}(i,j) = K(\mathbf{x}_i, \mathbf{x}_j) = \langle \varphi(\mathbf{x}_i), \varphi(\mathbf{x}_j) \rangle$$

(6.14)

where $K(\mathbf{x}_i, \mathbf{x}_j)$ represents the inner product operation of two vectors in a high-dimensional space, that is the kernel function operator, and $\mathbf{K}(i, j)$ is the kernel

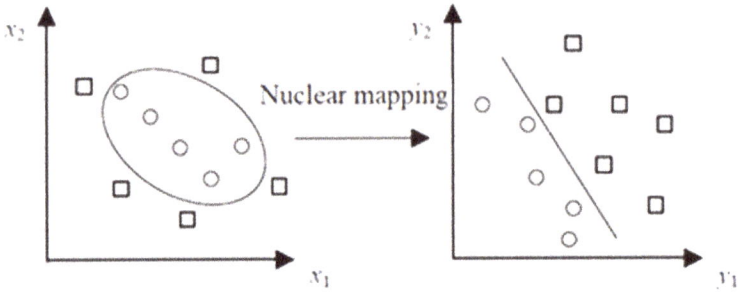

FIGURE 6.2   Kernel function mapping.

function matrix composed of the kernel function operator. According to the Hilbert-Schmidt principle in the statistical theory, any symmetric function $K\left(\mathbf{x}_i,\mathbf{x}_j\right)$ as long as it satisfies for the necessary and sufficient condition, which is any $\varphi\left(\mathbf{x}_i\right)\neq 0$ and $\int \varphi^2\left(\mathbf{x}_i\right)d\mathbf{x} < \infty$:

$$\iint K\left(\mathbf{x}_i,\mathbf{x}_j\right)\varphi\left(\mathbf{x}_i\right)\varphi\left(\mathbf{x}_j\right)d\mathbf{x}_i d\mathbf{x}_j > 0 \tag{6.15}$$

Such a symmetric function can be used as a kernel function.

The selection of the kernel function is closely related to the performance of the SVM classification algorithm. Selection of the kernel function and kernel function parameters can still only be determined by empirical methods, and there is no unified theoretical guidance. Different kernel functions and related parameters will lead to different classification results. In this chapter, the kernel function of the SVM classification algorithm selects the most widely used Gaussian radial basis kernel function, namely, $K\left(\mathbf{x}_i,\mathbf{x}_j\right)=\exp\left[-\left\|\mathbf{x}_i-\mathbf{x}_j\right\|^2/2\sigma^2\right]$. The following uses the principle of kernel function to map the original space to a newly transformed space. At this time, the optimization problem can be transformed into the following formula:

$$\min \frac{1}{2}\|\mathbf{w}\|^2$$
$$s.t. \quad y_i\left[\left\langle \mathbf{w},\varphi\left(\mathbf{x}_i\right)\right\rangle + b\right] \; 1\geq 0 \quad i-1,2,\ldots,n \tag{6.16}$$

The corresponding dual problem obtained by the Lagrange function is

$$L\left(\mathbf{w},b,\alpha\right)=\sum_{i=1}^{n}\alpha_i - \frac{1}{2}\sum_{i,j=1}^{n}\alpha_i\alpha_j y_i y_j K\left(\mathbf{x}_i,\mathbf{x}_j\right)$$
$$s.t. \quad \sum_{i=1}^{n}\alpha_i y_i = 0, \; \alpha_i \geq 0 \quad i=1,2,\ldots,n \tag{6.17}$$

Then the corresponding discriminant function is as follows:

$$f\left(\mathbf{x}\right)=\text{sgn}\left\{\left\langle \mathbf{w}^*,\varphi\left(\mathbf{x}_i\right)\right\rangle + b^*\right\}=\text{sgn}\left\{\sum_{i=1}^{n}\alpha_i^* y_i K\left(\mathbf{x}_i,\mathbf{x}\right)+b^*\right\} \tag{6.18}$$

6.3 SUBPIXEL MAPPING BASED ON MAP SUPER-RESOLUTION RECONSTRUCTION THEN CLASSIFICATION

Figure 6.3 shows the flowchart of the proposed subpixel mapping based on the MAP super-resolution recovery then classification (MTC) model. First, a super-resolution reconstruction model based on MAP is proposed. Especially, the end-member of interest (EOI) is applied to the reconstruction model, and the original high-dimensional image is transformed into the low-dimensional. We named this reconstruction model as a MAP super-resolution reconstruction algorithm based on the transformed space (T-MAP-SR). This algorithm reduces the amount of calculation of the super-resolution reconstruction algorithm and speeds up the running speed of the algorithm. Then, the original coarse remote sensing image is transformed into a high-resolution image using the imaging model according to the scale S. Finally, a classification algorithm based on LSSVM is used to obtain the final subpixel mapping result from the high-resolution image. The following section introduces the specific implementation process of the model framework from two aspects of T-MAP-SR and LSSVM classification algorithms.

6.3.1 TRANSFORMED MAP-BASED SUPER-RESOLUTION RECONSTRUCTION

The original high-dimensional remote sensing data are mapped to the low-dimensional through the EOI, and the data are then subjected to MAP-based super-resolution reconstruction processing (T-MAP-SR) in the transform domain. The complexity of the algorithm is greatly reduced, and the class of interest (COI) can be protected. The original high-dimensional remote sensing image g^{H} and f^{H} can be expressed by the low-dimensional transform domain remote sensing image g^{L} and f^{L} by the function $\mathbf{\Phi}$:

$$g^{\mathrm{H}} = \mathbf{\Phi} g^{\mathrm{L}} \text{ and } f^{\mathrm{H}} = \mathbf{\Phi} f^{\mathrm{L}} \tag{6.19}$$

where $\mathbf{\Phi}$ is a column vector composed of EOI, which represents the operator that maps the original high-dimensional data to the low-dimensional data.

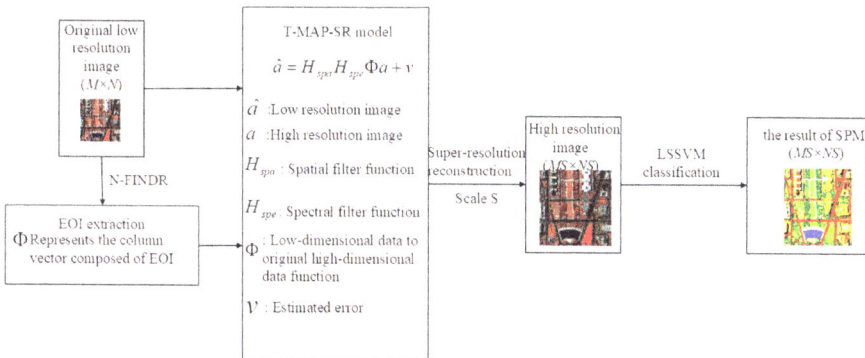

FIGURE 6.3 Flowchart of MTC model.

Suppose a is the input data from the high-resolution target image f^{L} after the dimensionality reduction; the data of the high-resolution target image f^{H} can be represented by Φa. Then $\hat{a}$ estimates the output data from the low-resolution observation image g^{H}, v is the corresponding estimation error, H_{spa} is the spatial domain filter function, and H_{spe} is the spectral domain filter function. Then the super-resolution reconstruction model can be expressed as

$$\hat{a} = H_{\text{spa}} H_{\text{spe}} \Phi a + v \tag{6.20}$$

When $\hat{a}$ is known, the probability density of a can be expressed as $P(a/\hat{a})$:

$$P(\hat{a}/a) = \frac{P(\hat{a}/a) \cdot P(a)}{P(\hat{a})} \tag{6.21}$$

where $P(\hat{a}/a)$ is the posterior probability, $P(\hat{a}/a)$ is the conditional probability, $P(a)$ is the prior probability of the high-resolution target image, and $P(\hat{a})$ is the prior probability of the low-resolution observation image.

MAP aims to seek a_{MAP} to estimate a, which is to maximize the posterior probability $P(\hat{a}/a)$:

$$a_{\text{MAP}} = \arg\max[P(a/\hat{a})] = \arg\max\left[\frac{P(\hat{a}/a) \cdot P(a)}{P(\hat{a})}\right] \tag{6.22}$$

Since $P(\hat{a})$ is a constant and has no effect on the result, $P(\hat{a})$ can be omitted. Formula (6.22) can be further simplified as

$$a_{\text{MAP}} = \arg\max[P(\hat{a}/a) \cdot P(a)] \tag{6.23}$$

The monotonic logarithmic function is added to the formula (6.23), and the formula (6.23) can be re-expressed as

$$a_{\text{MAP}} = \arg\max[\log P(\hat{a}/a) + \log P(a)] \tag{6.24}$$

The maximization problem is transformed into the following minimization problem:

$$a_{\text{MAP}} = \arg\min[-\log P(\hat{a}/a) - \log P(a)] \tag{6.25}$$

According to the previously mentioned super-resolution reconstruction model, the following formula can be obtained:

$$P(\hat{a} - H_{\text{spa}} H_{\text{spe}} \Phi a) = P(v) \tag{6.26}$$

The conditional probability $P(\hat{a}/a)$ and the prior probability $P(a)$ depend on the assumed statistical model of the image. The multivariate Gaussian model is an important model for super-resolution reconstruction. According to the multivariate Gaussian model, the conditional probability $P(\hat{a}/a)$ can be expressed as

$$
\begin{aligned}
P(\hat{a}/a) &= P(\hat{a} - H_{\text{spa}} H_{\text{spe}} \Phi a) \\
&= \frac{1}{Z} \exp(-v^T \mathbf{K}^{-1} v) \\
&= \frac{1}{Z} \exp(-(\hat{a} - H_{\text{spa}} H_{\text{spe}} \Phi a)^T \mathbf{K}^{-1} (\hat{a} - H_{\text{spa}} H_{\text{spe}} \Phi a))
\end{aligned}
\tag{6.27}
$$

where Z is a constant, and $\mathbf{K}^{-1}$ is the covariance matrix of the estimated error v.

The prior probability $P(a)$ can also be expressed as formula (6.28) by a multivariate Gaussian model:

$$
P(a) = \frac{1}{A} \exp(-a^T \Lambda^{-1} a)
\tag{6.28}
$$

where A is a constant, and Λ^{-1} is the covariance matrix of input data a.

Because the correlation between the pixel and the neighboring pixel is relatively large, the correlation decreases with an increase of the distance. Therefore, in the actual super-resolution reconstruction model, local analysis is used instead of global analysis to perform spatial transformation. This method avoids large-scale matrix operations and reduces the complexity of the algorithm. Suppose that $a^{(i)}$ $(i = 1, 2, ..., M)$ is a local pixel in the high-resolution target image. At this time, $\hat{a}$ represents the local pixel in all corresponding low-resolution observation images; H_{spa} is described as the corresponding weight α_i $(i = 1, 2, ..., M)$ of the input vector, and then formula (6.20) can be transformed into the following formula:

$$
\hat{a} = \sum_{i=1}^{M} \alpha_i H_{\text{spe}} \Phi a^{(i)} + v
\tag{6.29}
$$

This relationship considers the high-resolution target image and related input as the basic analysis unit. It can be seen that the dimensionality of the input data is reduced by the EOI, and the complexity of the reconstruction algorithm is reduced. Therefore, the use of this super-resolution method can increase the speed of the algorithm.

According to the previously mentioned probability density function and the relationship between input and output types, in the case of spatial domain transform local analysis, the optimal estimation based on T-MAP-SR can be expressed as

$$
\begin{aligned}
a_{\text{MAP}} &= \arg\min[-\log P(\hat{a}/a) - \log P(a)] \\
&= \arg\min[(\hat{a} - H_{\text{spa}} H_{\text{spe}} \Phi a)^T \mathbf{K}^{-1}(\hat{a} - H_{\text{spa}} H_{\text{spe}} \Phi a) + (-a^T \Phi^{-1} a)] \\
&= \arg\min[(\hat{a} - \sum_{i=1}^{M} \alpha_i H_{\text{spe}} \Phi a^{(i)})^T \mathbf{K}^{-1}(\hat{a} - \sum_{i=1}^{M} \alpha_i H_{\text{spe}} \Phi a^{(i)}) + \\
&\qquad \sum_{i=1}^{M} \alpha_i (a^{(i)})^T \Lambda^{-1}(a^{(i)})]
\end{aligned}
\tag{6.30}
$$

Suppose the spectral domain filter function does not change with the spatial position change, then this optimization process can be completed through a spatial iterative model. In the spatial iterative model, the image degradation process can be regarded as a linear process $g = Tf + v$. The existing image g is derived from the product of the ideal image f and the linear transformation coefficient T, and random noise v is added at the same time. When the image g is known, the spatial iterative algorithm is used to solve the inverse transformation of the linear formulas, obtaining the ideal image f. The process includes multiple super-resolution reconstruction, analog sampling, and spatial domain contrast correction, and then gradually approximates the ideal image f. First, an ideal initialization f_0 needs to be estimated, and then $g_0 = Tf_0 + v$ will be obtained. Then g_0 is compared with the obtained result g, and the difference $g - g_0$ can be obtained. Finally, we use this difference and correct the initialization f_0 through the appropriate transformation parameter θ, namely, $f_1 = f_0 + \theta(g - g_0)$. This process $f_n = f_{n-1} + \theta(g - g_{n-1})$ will be repeated continuously until the desired ideal image is obtained.

Suppose $E(a_n)$ is the cost function in the optimization process:

$$
\begin{aligned}
E(a_n) = &\frac{1-\lambda}{2}\left(\hat{a} - \sum_{i=1}^{M}\alpha_i H_{spe}\Phi a_{n-1}^{(i)}\right)^T \mathbf{K}^{-1}\left(\hat{a} - \sum_{i=1}^{M}\alpha_i H_{spe}\Phi a_{n-1}^{(i)}\right) \\
&+ \frac{\lambda}{2}\sum_{i=1}^{M}\alpha_i (a_{n-1}^{(i)})^T \Lambda^{-1}(a_{n-1}^{(i)})
\end{aligned}
\tag{6.31}
$$

where n is the number of iterations, $\mathbf{K}$ is the covariance matrix of the original domain data Φg^L, and Λ is the covariance matrix of transform domain data g^L. The initial estimate $a_0^{(i)}$ ($i = 1, 2, \ldots, M$) is from the initial value Φf_0^L of Φf^L, and Φf_0^L can be obtained by linear interpolation Φg^L.

Suppose the step parameter in the iterative model is u_n, and the gradient estimate is $\nabla E(a_n)$. The iterative model from the $n - 1$ th time optimal estimation to the nth time optimal estimation is

$$
a_n = a_{n-1} - u_n \nabla E(a_n)
\tag{6.32}
$$

The gradient estimation $\nabla E(a_n)$ and the step size parameter u_n can be expressed as

$$
\nabla E(a_n) = (\lambda - 1)\left(\sum_{i=1}^{M}\alpha_i H_{spe}\Phi a_{n-1}^{(i)}\right)^T \mathbf{K}^{-1}\left(\hat{a} - \sum_{i=1}^{M}\alpha_i H_{spe}\Phi a_{n-1}^{(i)}\right) + \lambda\alpha_i \Lambda^{-1}a_{n-1}^{(i)}
\tag{6.33}
$$

$$
u_n = \frac{(\nabla E(a_{n-1}))^T (\nabla E(a_{n-1}))}{(\nabla E(a_{n-1}))^T \Theta(\nabla E(a_{n-1}))}
\tag{6.34}
$$

where $\Theta = \lambda\alpha_i \Lambda^{-1}a_{n-1}^{(i)} + (1-\lambda)(H_{spe}\Phi)^T \mathbf{K}^{-1}H_{spe}\Phi$.

In this chapter, the T-MAP-SR algorithm is applied in the MTC. When Φ is set to the identity matrix, the existing MAP super-resolution reconstruction algorithm (MAP-SR) of the original domain can be obtained.

6.3.2 LSSVM CLASSIFICATION ALGORITHM

The original coarse remote sensing image is processed by the previously mentioned T-MAP-SR to obtain a high-resolution image. Then, the high-resolution image can obtain the thematic mapping with specific spatial distribution information of each class through a classification algorithm based on supervision information. The LSSVM classification algorithm replaces the inequality constraints in the standard SVM with equality constraints, which greatly simplifies the solution process. Therefore, the LSSVM classification algorithm is selected as the classification algorithm in the MTC model in this chapter.

The expression of the optimization problem of LSSVM is

$$\min_{\mathbf{w},b,\mathbf{e}} J(\mathbf{w},\mathbf{e}) = \frac{1}{2}\|\mathbf{w}\|^2 + \frac{r}{2}\sum_{i=1}^{n} e_i^2 \tag{6.35}$$

$$s.t. \quad y_i = \langle \mathbf{w}, \varphi(\mathbf{x}_i) \rangle + b + e_i \qquad i = 1,2,...,n \quad r > 0$$

where $\mathbf{x}_i \in R^d$ is the sample data set; $y_i \in \{+1, -1\}$ is the corresponding class label; $i = 1,\cdots,n$; and e_i is the discrimination error.

The dual problem is

$$\min_{\mathbf{w},b,\mathbf{e},\pm} \mathbf{L}(\mathbf{w},b,\mathbf{e},\pm) = J(\mathbf{w},\mathbf{e}) - \sum_{i=1}^{n} \alpha_i \left[\langle \mathbf{w}, \varphi(\mathbf{x}_i) \rangle + b + e_i - y_i \right] \tag{6.36}$$

The optimal KKT conditions are

$$\begin{cases} \dfrac{\partial \mathbf{L}}{\partial \mathbf{w}} = 0 \rightarrow \mathbf{w} = \sum_{i=1}^{n} \alpha_i \varphi(\mathbf{x}_i) \\[2mm] \dfrac{\partial \mathbf{L}}{\partial b} = 0 \rightarrow \sum_{i=1}^{n} \alpha_i = 0 \\[2mm] \dfrac{\partial \mathbf{L}}{\partial e_i} = 0 \rightarrow \alpha_i = re_i \\[2mm] \dfrac{\partial \mathbf{L}}{\partial \alpha_i} = 0 \rightarrow \langle \mathbf{w}, \varphi(\mathbf{x}_i) \rangle + b + e_k - y_i = 0 \end{cases} \qquad i = 1,2,...,n \tag{6.37}$$

The elimination method is used to eliminate $\mathbf{w}$ and $\mathbf{e}$, and formula (6.37) can be expressed as

$$\begin{bmatrix} 0 & \mathbf{1}_v^{\mathrm{T}} \\ \mathbf{1}_v & \mathbf{K}+\mathbf{I}/r \end{bmatrix} \begin{bmatrix} b \\ \alpha \end{bmatrix} = \begin{bmatrix} 0 \\ \mathbf{y} \end{bmatrix} \tag{6.38}$$

where n represents the number of training samples; $\mathbf{y} = \left[y_1, y_2, ..., y_n \right]^{\mathrm{T}}$; $\mathbf{1}_v = [1,1,\ldots 1]^{\mathrm{T}}$; $\alpha = [\alpha_1, \alpha_2, \ldots, \alpha_n]^{\mathrm{T}}$; $\mathbf{K}$ is the training sample kernel matrix of $n \times n$; and $\mathbf{I}$ is the identity matrix with $n \times n$. According to this, the optimal (α^*, b^*) can be solved, and then the discriminant function of LSSVM is

$$f(\mathbf{x}) = \sum_{i=1}^{n} \alpha_i^* \mathbf{K}(\mathbf{x}_i, \mathbf{x}) + b^* \tag{6.39}$$

It can be seen from this formula that if n is used to represent the number of training samples, the coefficient matrix of the linear formula system is $(n + 1) \times (n + 1)$. When using the ordinary SVM classification algorithm, the size of the coefficient matrix is $n^2 \times n^2$. Therefore, LSSVM not only reduces the computational complexity but also greatly reduces the required storage space. LSSVM classification can predict the test sample class labels by calculating the previously mentioned discriminant function $f(\mathbf{x})$. For multi-class problems, the multi-classification can be decomposed into a set of several binary classification problems. The classic multi-classifier structure has two types: one-against-rest (1-a-r) and one-against-one (1-a-1) [8].

6.3.3 EXPERIMENT CONTENT AND RESULT ANALYSIS

This chapter uses three real hyperspectral data to verify the performance of the proposed method. The N-FINDR algorithm is used as the endmember extraction algorithm in this chapter. We select COI's spectral endmembers to form EOI. To get a fair comparison, the LSSVM method is used as a classification tool in the MTC model and as a spectral unmixing tool in the traditional subpixel mapping model. In all experiments, the number of training samples in the LSSVM method is selected as 10% of each class, and the remaining number of each class is used as the test sample. The flowchart of the experiment is shown in Figure 6.4.

In the first set of experiments, the T-MAP-SR algorithm is compared with the traditional MAP-SR algorithm, which further demonstrates the advantages of the T-MAP-SR algorithm due to the application of the EOI operator. Three sets of experiments compare the proposed MTC model with the traditional subpixel mapping model. The MAP-SR algorithm and the T-MAP-SR algorithm are used as the super-resolution reconstruction methods of the MTC model, and the bilinear interpolation algorithm (Bilinear) [9], The bicubic interpolation algorithm (Bicubic) [10] and the MAP algorithm (MAP) [11] are used as the traditional subpixel sharpening methods of the SPM model. In addition to the mapping accuracy (%) of each class, PCC (%), Kappa, and average accuracy (AA [%]) are selected as the accuracy evaluation indicators for the experiments in this chapter.

6.3.3.1 Experiment 1

The first experimental data are the hyperspectral remote sensing image collected by the AVIRIS sensor in the Indiana agriculture and forestry in northwest Indiana in 1992. It contains 220 spectral bands, covering 16 classes of agricultural and forestry landscapes. An area with the size of 144 × 144 pixels is used as the study area. In this experiment, the traditional MAP-SR algorithm and T-MAP-SR algorithm will carry out a comprehensive comparison, including super-resolution reconstruction error, processing time, whether to protect COI and anti-noise performance, and so on, so as to reflect the special effects of the EOI operator. The original image is shown in Figure 6.5(a). Since classes 2, 3, 10, 11, and 14 have a larger number of pixels, they are selected as COI and constitute an EOI operator. Figure 6.5(b) can be used as a reference image. At the same time, the original image is downsampled through the scale to obtain a simulated low-resolution image. The scale of this experiment is set to 2, and each mixed pixel will contain 2 × 2 subpixels. The low-resolution image is shown in Figure 6.5(c).

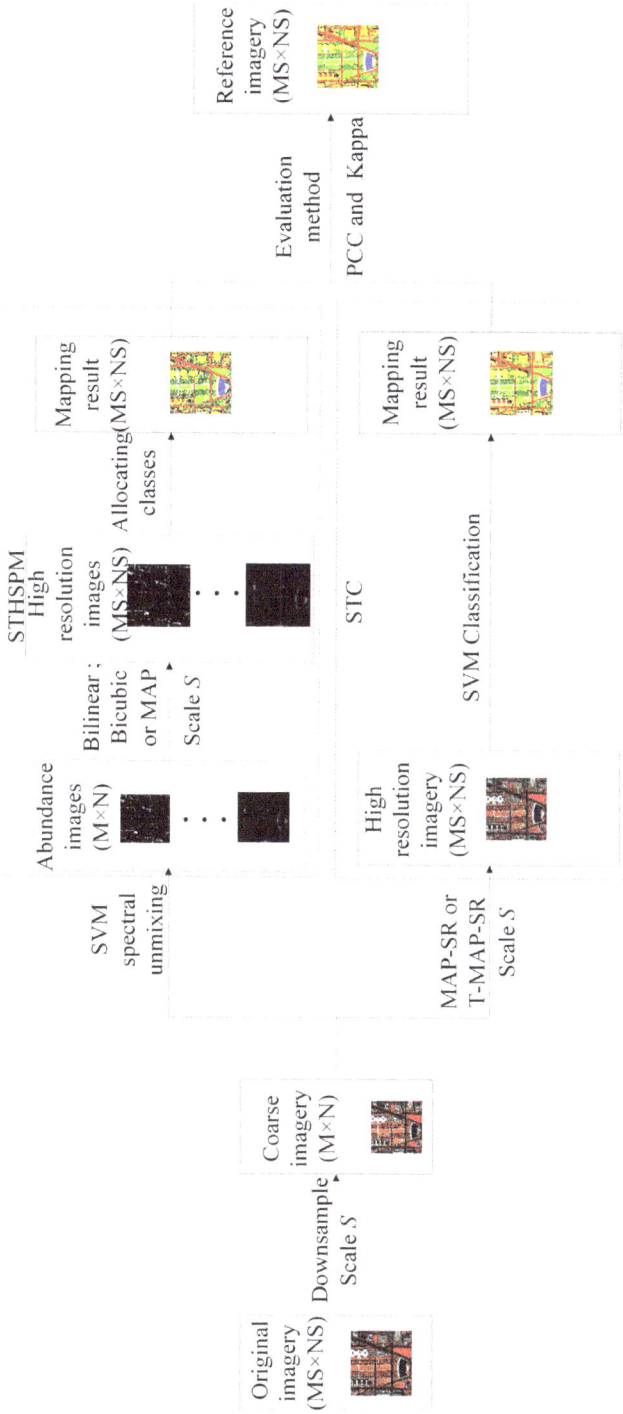

FIGURE 6.4 Flowchart of the experiment.

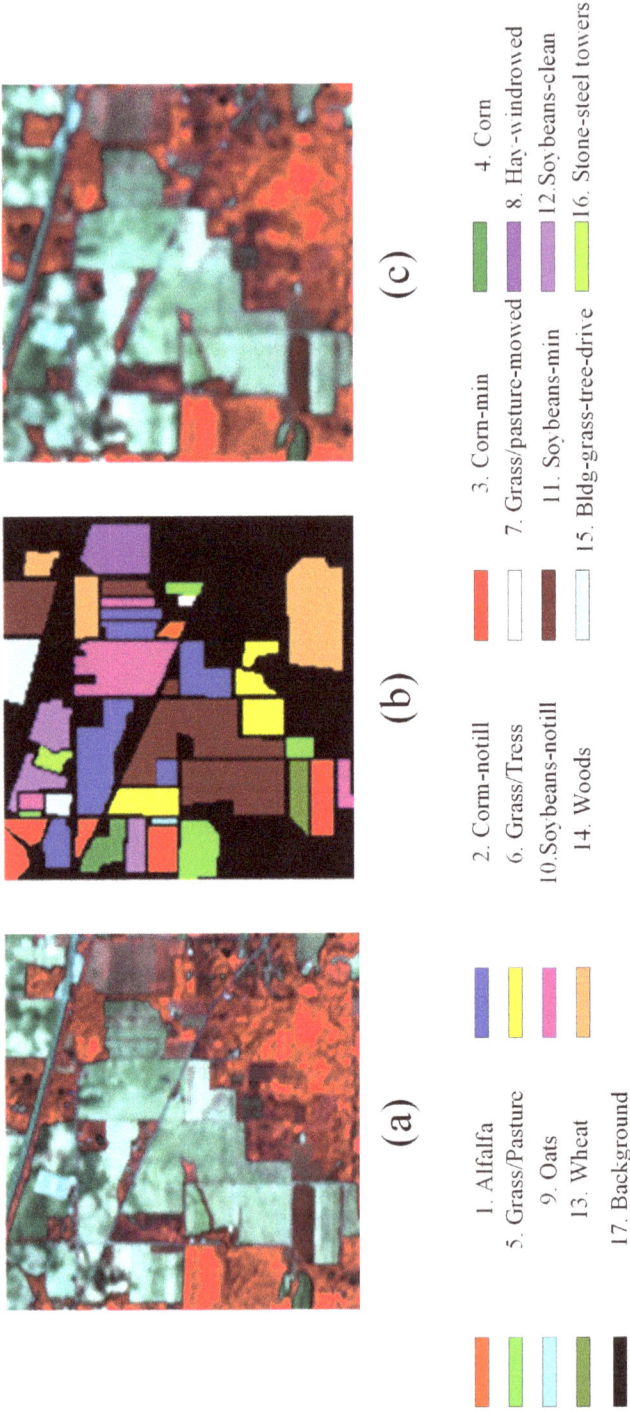

FIGURE 6.5 Indian agriculture and forestry data set: (a) false color composite image (RGB: 57, 27, 36): (b) reference image: and (c) low-resolution image (S = 2).

1. Alfalfa   2. Corn-notill   3. Corn-min   4. Corn

5. Grass/Pasture   6. Grass/Tress   7. Grass/pasture-mowed   8. Hay-windrowed

9. Oats   10. Soybeans-notill   11. Soybeans-min   12. Soybeans-clean

13. Wheat   14. Woods   15. Bldg-grass-tree-drive   16. Stone-steel towers

17. Background

The results of the MAP-SR algorithm and the T-MAP-SR algorithm are shown in Figures 6.6(a)–(b). The relative error of super-resolution reconstruction is defined as the ratio of the absolute error of all reconstructed pixels to all the total pixels in the original image. In this chapter, if the relative error is less than 0.01%, the iteration stops. In this case, when the relative error between the high-resolution image from the MAP-SR algorithm and the reference image is 2.25%, the number of iterations is 150, and the processing time is 4749 seconds. As a comparison, when the number of iterations in the T-MAP-SR algorithm is 116, the relative error between the high-resolution image from the T-MAP-SR algorithm and the reference image is only 2.22%. In addition, when the number of iterations of the T-MAP-SR algorithm reaches 150, the processing time is 275 seconds. Therefore, compared with the MAP-SR algorithm, the calculation speed of the T-MAP-SR algorithm is increased by more than 17 times, and the relative error of the super-resolution reconstruction can be lower. The results show that the T-MAP-SR algorithm has the advantages of lower calculation and better reconstruction effect. Table 6.1 shows the relative errors

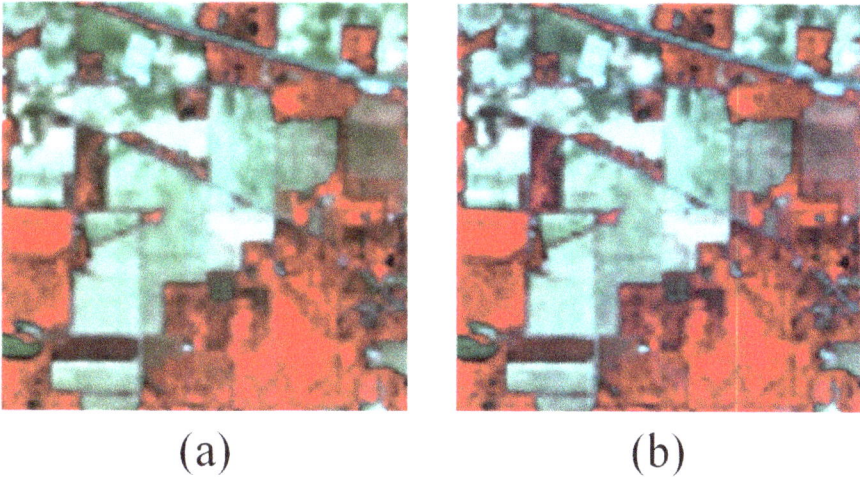

(a)   (b)

FIGURE 6.6   (a) MAP-SR. (b) T-MAP-SR.

TABLE 6.1

Different Types of Super-Resolution Reconstruction Errors

Land Cover Class	MAP-SR (%)	T-MAP-SR (%)
Class 2	1.97	1.67
Class 3	2.18	2.10
Class 10	1.82	1.47
Class 11	1.91	1.55
Class 14	2.47	2.10
Other classes	2.27	2.39

of super-resolution reconstruction for different classes in the MAP-SR algorithm and the T-MAP-SR algorithm. As shown in the table, the relative reconstruction error of COI in the T-MAP-SR algorithm is lower than that of the MAP-SR algorithm, while the reconstruction error of non-COI is slightly increased. This result shows that T-MAP-SR can protect COI at the cost of a slight reduction in the accuracy of non-COI reconstruction. In addition, when adding Gaussian noise with an average value of 267.5 (1/10 of the average value of all pixels in the image) and a variance of 1 to the original image, the relative reconstruction error in the MAP-SR algorithm increases by 0.26%, and in the T-MAP-SR algorithm, the relative reconstruction error only increased by 0.08%. This is because each pixel in the T-MAP-SR algorithm is processed in a least squares environment. At this time, the noise can be automatically and effectively filtered, so the T-MAP-SR algorithm has better resistance of noise

Next, the traditional SPM model and MTC model are used to obtain the subpixel mapping results. Figure 6.7 shows the mapping results of the five methods. Due to the limitations of the existing spectral unmixing technology, there are many unmixing errors in the traditional SPM model, and these errors affect the final mapping result. However, the MTC model not only makes full use of the original image supervision information, but it also avoids the spectral unmixing step. From the visual point of view, it can be found that the image produced by the MTC model is closer to the reference image than the traditional SPM model. For example, the class boundaries in Figure 6.7(d)–(e) are smoother than those in Figure 6.7(a)–(c).

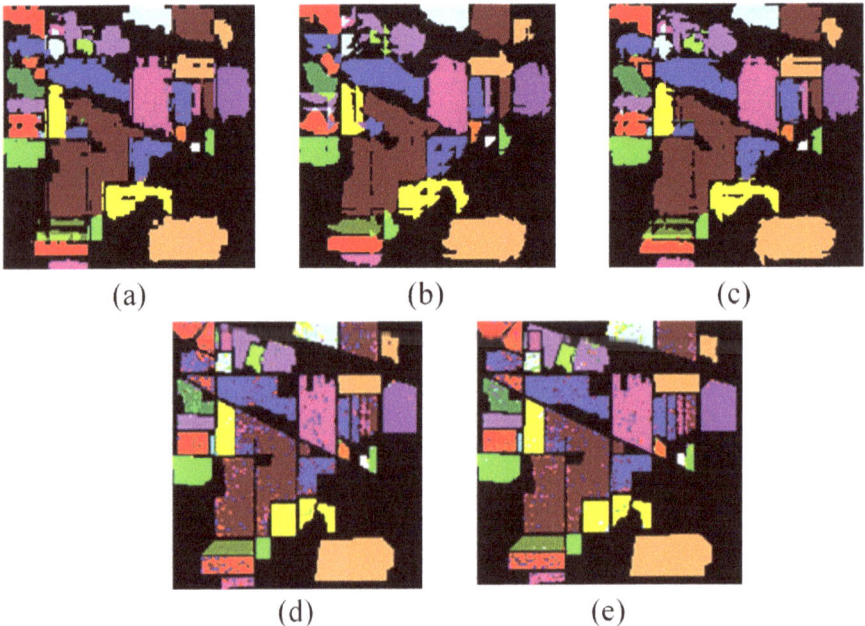

FIGURE 6.7   Mapping results in Experiment 1 ($S = 2$): (a) bilinear, (b) bicubic, (c) MAP, (d) MAP-SR, and (e) T-MAP-SR.

TABLE 6.2

Mapping Accuracy of the Five Methods in Experiment 1 ($S = 2$)

Land Cover Class	Bilinear	Bicubic	MAP	MAP-SR	T-MAP-SR
Class 2 (%)	80.47	81.66	84.31	85.43	88.15
Class 3 (%)	77.82	79.26	83.33	83.93	87.65
Class 10 (%)	78.10	81.61	83.16	84.17	86.26
Class 11 (%)	89.83	91.13	90.08	92.18	93.60
Class 14 (%)	92.74	94.67	93.12	97.85	99.77
AA (%)	72.39	73.27	85.93	92.83	93.53
PCC (%)	83.91	84.50	80.72	90.92	92.58

Table 6.2 shows the mapping accuracy (%), average accuracy (%), and PCC (%) of COI. As shown in Table 6.2, the accuracy of the MTC model is higher than that of the traditional subpixel mapping model. In the traditional subpixel mapping model, the PCC (%) of bilinear is 83.91%, the PCC (%) of bicubic is 84.50%, and the PCC (%) of MAP is 85.93%. In the MTC model, the PCC (%) of MAP-SR increased by about 5%, and the PCC (%) of T-MAP-SR increased by about 7%. Due to the effect of EOI, the mapping accuracy (%) of COI is higher than that of MAP-SR. For example, the mapping accuracy (%) of class 3 in the T-MAP-SR algorithm is 87.65%, which is about 4% higher than that of MAP-SR. At the same time, it can also be noted that due to the higher (%) value, T-MAP-SR has better stability.

The mapping accuracy (%) of the 16 classes is shown in Figure 6.8. It can be observed from Figure 6.8 that the mapping accuracy of the classes with a small number of pixels in the traditional subpixel mapping model is very unsatisfactory, such as classes 7, 9, and 16. This is mainly due to the limitations of the existing spectral unmixing technology. A large number of unmixing errors make the information expressed by the abundance images of each class inaccurate, and especially for classes with a small number of pixels, the impact is more serious. As shown in Figure 6.8, the proposed model avoids the step of spectral unmixing, and the mapping accuracy of the class with a small number of pixels has been significantly improved. At the same time, due to the special effects of EOI, it can also be noted that some non-COI mapping accuracy in the T-MAP-SR algorithm has a little loss, but the reduction is very small, and even the mapping accuracy of some classes is still the same as that of MAP-SR, such as classes 7, 9, and 16. This shows that a non-COI with a small number of pixels can also obtain good results in the T-MAP-SR algorithm.

6.3.3.2 Experiment 2

The second set of experimental data selects the Washington, DC, hyperspectral image data set with a size of 240×240 pixels as the research area, as shown in Figure 6.9(a). The reference image containing seven classes is shown in Figure 6.9(b). To simulate in real cases, the coarse image was generated by degrading Figure 6.9(a) with $S = 2$, as shown in Figure 6.9(c). In this experiment, because road, tree, grass, and roof

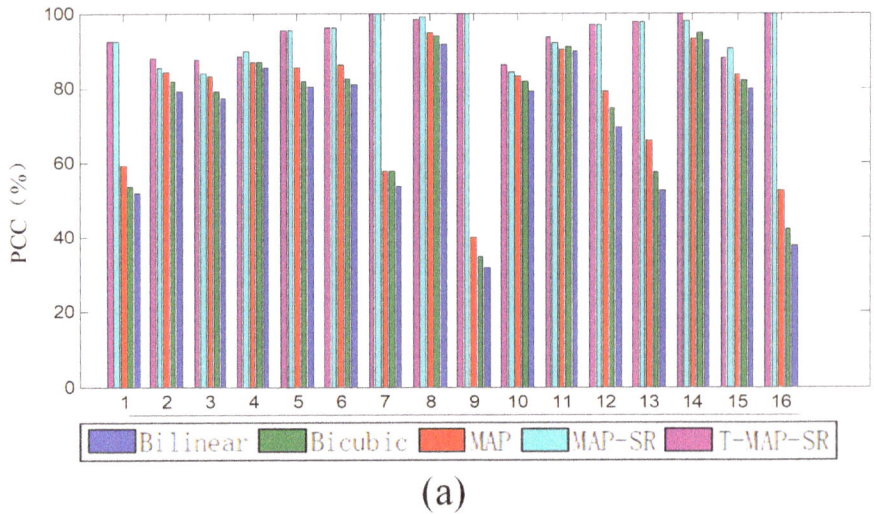

(a)

FIGURE 6.8　Mapping accuracy (%) of 16 classes.

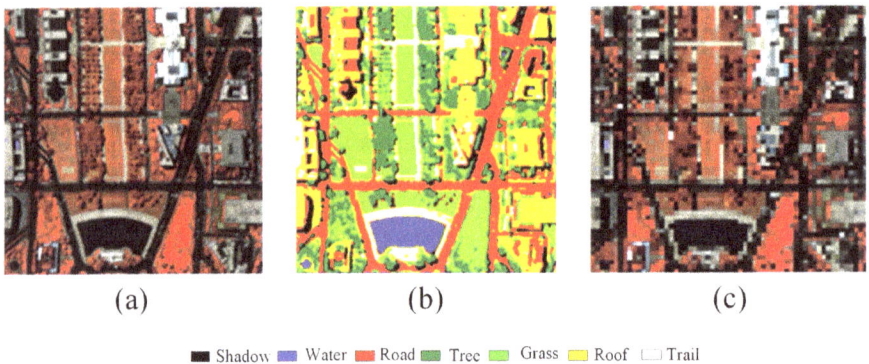

(a)　　　　　　　　　　(b)　　　　　　　　　　(c)

■ Shadow ■ Water ■ Road ■ Tree ☐ Grass ☐ Roof ☐ Trail

FIGURE 6.9　Washington, DC, data set: (a) false color composite image (RGB: 65, 52, 36), (b) reference image, and (c) low-resolution image ($S = 2$).

have more pixels, they are selected as COI and formed as EOI. Figure 6.10(a)–(b) shows the results of the MAP-SR algorithm and T-MAP-SR algorithm, respectively. It can be observed from intuitive vision that the high-resolution image obtained by the T-MAP-SR algorithm is closer to the original fine image.

The subpixel mapping results of the five methods are shown in Figure 6.11. Visual comparison shows that the MTC model can make full use of the supervision information of the original image. There are many broken hole-shaped patches in the results of the traditional subpixel mapping model. For example, in Figure 6.11(a)–(c), there are some tiny features on the narrow road between grass that cannot be well reconstructed. These erroneous mapping phenomena are significantly reduced in the MTC model. Due to the effect of EOI, the T-MAP-SR algorithm

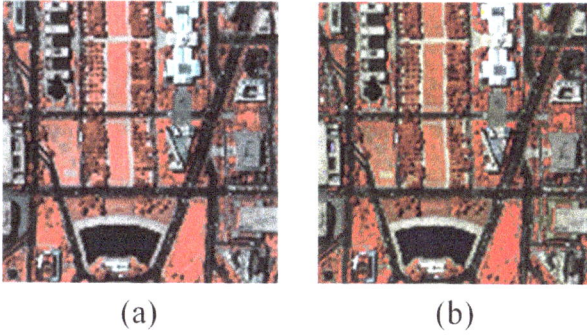

FIGURE 6.10   High-resolution image ($S = 2$): (a) MAP-SR and (b) T-MAP-SR.

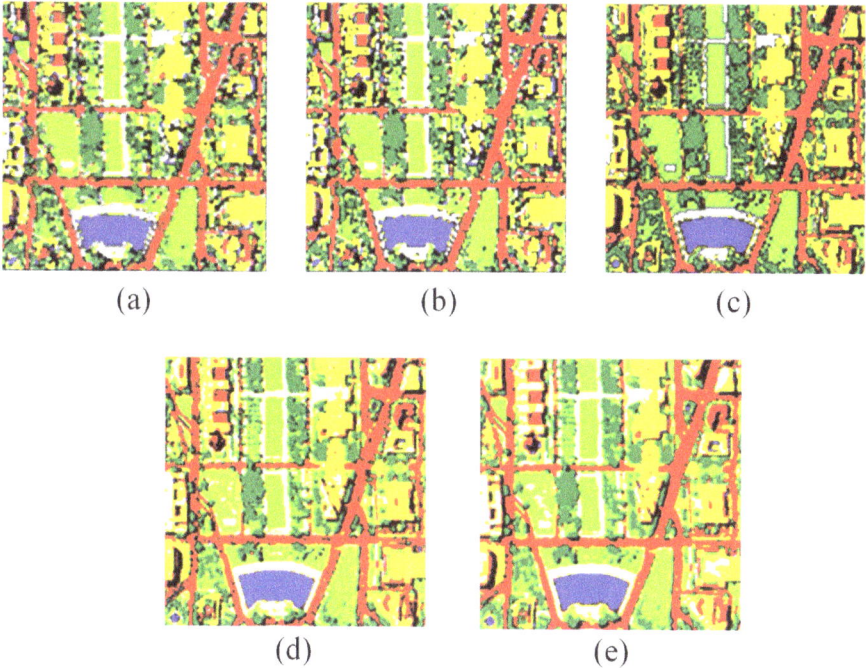

FIGURE 6.11   Mapping results in Experiment 2 ($S = 2$): (a) bilinear, (b) bicubic, (c) MAP, (d) MAP-SR, and (e) T-MAP-SR.

is closer to the reference image than the MAP-SR algorithm, especially for COI boundary protection.

In addition to intuitive visual comparison, the mapping accuracy (%) of each class, PCC (%) and AA (%) are used to evaluate the performance of the five

TABLE 6.3
Mapping Accuracy of Five Methods in Experiment 2 ($S = 2$)

Land Cover Class	Bilinear	Bicubic	MAP	MAP-SR	T-MAP-SR
Background (%)	73.44	75.03	77.50	79.42	78.77
Water (%)	85.56	88.97	90.49	95.54	95.15
Road (%)	70.55	72.74	75.73	80.81	88.75
Tree (%)	72.45	75.45	77.36	91.04	97.47
Grass (%)	74.70	78.60	82.19	86.51	88.86
Roof (%)	70.67	72.98	75.09	83.43	85.23
Trail (%)	73.88	75.58	77.98	88.35	87.16
AA (%)	74.46	77.05	79.48	86.44	88.77
PCC (%)	76.82	77.47	78.06	85.27	88.51

methods in experiment 2. As shown in Table 6.3, by observing the mapping accuracy (%) of each class in the table, it can be found that the accuracy of the MTC model is better than that of the traditional subpixel mapping model. For PCC (%), in the traditional subpixel mapping model, the PCC (%) of the bilinear algorithm is about 76.82%, the PCC (%) of the bicubic algorithm is about 77.47%, and the PCC (%) of the MAP algorithm is about 78.06%. In the MTC model, the PCC (%) of the MAP-SR algorithm increased by about 7%, and the PCC (%) of the T-MAP-SR algorithm increased by about 10%. Due to the combination of EOI, in the T-MAP-SR algorithm, the road mapping accuracy (%) is 88.75%, which is more than 8% higher than the MAP-SR algorithm, and the tree mapping accuracy (%) is 97.47%, which means the MAP-SR algorithm has improved by 6%. Although background, water, and trail are the non-COIs, the PCC (%) of background, water, and trail has some loss, but it only reduces by 0.65%, 0.39%, and 1.19%, respectively. At the same time, the PCC (%) value of the T-MAP-SR algorithm is 3.24% higher than that of the MAP-SR algorithm. AA (%) also shows that the T-MAP-SR algorithm has better stability.

The performance of subpixel mapping results will be affected by the scale S. Similarly, the other two scales $S = 4$ and $S = 6$ are used to test the performance of the five methods. Figure 6.12 shows the PCC (%) and Kappa of various methods under three scales. It can be noted that with the increase of S, the values of PCC (%) and Kappa of the five methods all decrease. However, consistent with the results in Table 6.3, the MTC model can produce higher PCC (%) and Kappa values than the traditional subpixel mapping model. At the same time, as shown in Figure 6.12, the T-MAP-SR algorithm is more accurate than the MAP-SR algorithm, which further verifies the effectiveness of the combined EOI.

6.3.3.3 Experiment 3

The Pavia city center data set with a research area of 400×400 pixels was selected as the research data for this experiment, as shown in Figure 6.13(a). The reference image is shown in Figure 6.13(b). Figure 6.13(c) is a simulated low-resolution image

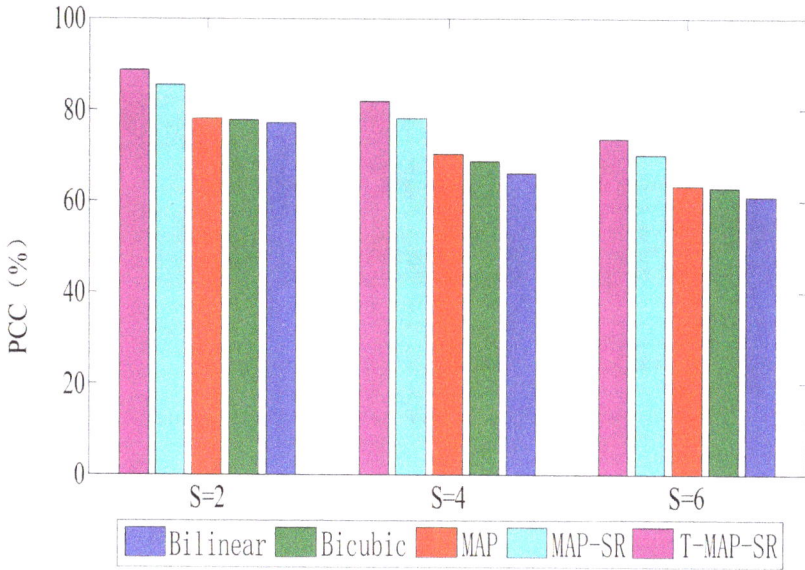

(a)

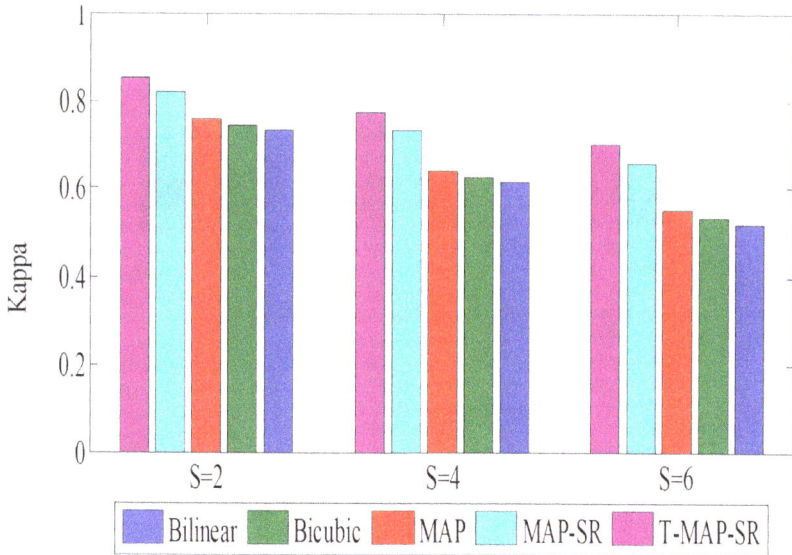

(b)

FIGURE 6.12 (a), (b) The relationship between PCC (%) and Kappa of five methods and scale S.

($S = 2$). In this experiment, lake, road, tree, grass, and roof are selected as COIs due to the large number of pixels. Through visual observation, the high-resolution image produced by the T-MAP-SR algorithm in Figure 6.14(b) is closer to the original image than the high-resolution image produced by the MAP-SR algorithm in Figure 6.14(a).

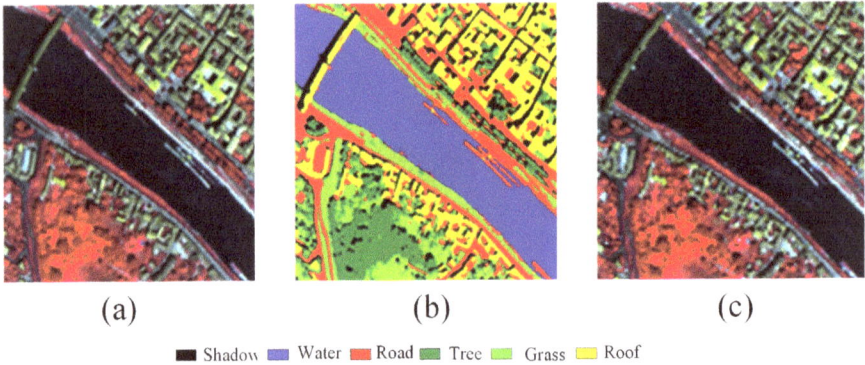

■ Shadow ■ Water ■ Road ■ Tree ■ Grass ■ Roof

FIGURE 6.13 Pavia city center data set: (a) false color composite image (RGB: 102, 56, 31), (b) reference image, and (c) low-resolution image ($S = 2$).

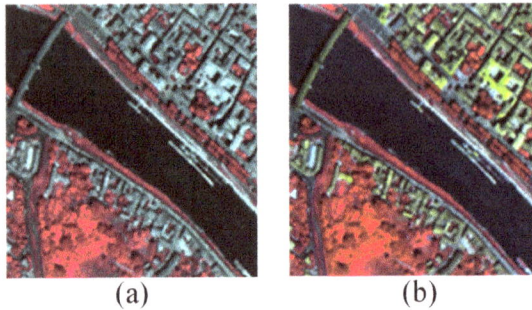

FIGURE 6.14 High-resolution image ($S = 2$): (a) MAP-SR and (b) T-MAP-SR.

Figure 6.15 shows the subpixel mapping results of the traditional subpixel mapping model and MTC model. From the results of the traditional subpixel mapping model, it can be observed that there are many disconnected or tapered patches due to the conflict of spatial features between various classes. In the MTC model, these patches have disappeared. For example, the road is more continuous, and the edge of the grass is smoother. In the MTC model, the result of the T-MAP-SR algorithm is closer to the reference image Figure 6.13(b), because the T-MAP-SR algorithm combines the EOI.

Table 6.4 shows the mapping accuracy (%) of each class, PCC (%), and AA (%). Similar to the conclusions drawn in previous experiments, the mapping accuracy (%) of each class in the MTC model is higher than that of the traditional subpixel mapping model. For the mapping accuracy of COI, the T-MAP-SR algorithm can get higher accuracy. For example, the mapping accuracy of tree is 94.68%, which is about 7% higher than the MAP-SR algorithm. The mapping accuracy of road is

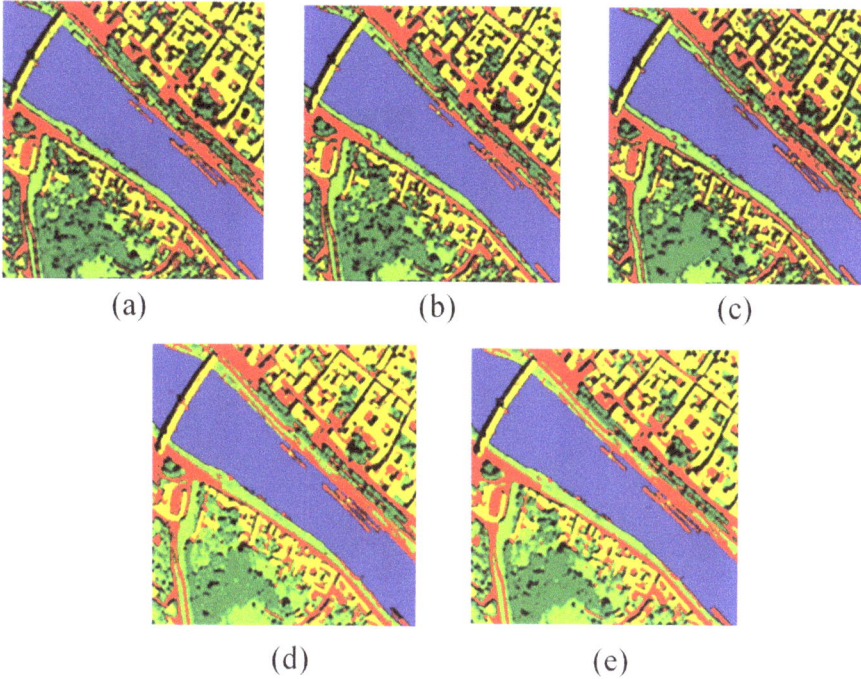

FIGURE 6.15 Mapping results in experiment 2 ($S = 2$): (a) bilinear, (b) bicubic, (c) MAP, (d) MAP-SR, and (e) T-MAP-SR.

TABLE 6.4

Mapping Accuracy of Five Methods in Experiment 3 ($S = 2$)

Land Cover Class	Bilinear	Bicubic	MAP	MAP-SR	T-MAP-SR
Background (%)	77.59	82.36	82.94	85.65	84.28
Water (%)	95.84	96.29	95.77	96.62	97.81
Road (%)	71.69	74.51	73.23	91.09	93.36
Tree (%)	74.28	75.63	77.84	87.99	94.68
Grass (%)	69.23	71.40	71.71	88.98	90.34
Roof (%)	79.75	82.07	80.81	96.84	97.69
AA (%)	78.06	80.37	80.38	91.20	93.02
PCC (%)	80.45	82.40	82.64	92.39	94.48

93.36%, which is about 2% higher than the MAP-SR algorithm. This is because the EOI protects the COI, which significantly improves the mapping accuracy of the COI. Of course, as a cost, the accuracy of the non-COI positioning is reduced.

Figure 6.16(a)–(b) shows the PCC (%) and Kappa values of the five methods under three scales S. Similar to the previous experimental results, the values of PCC (%) and Kappa in the MTC model are significantly higher than those in the traditional

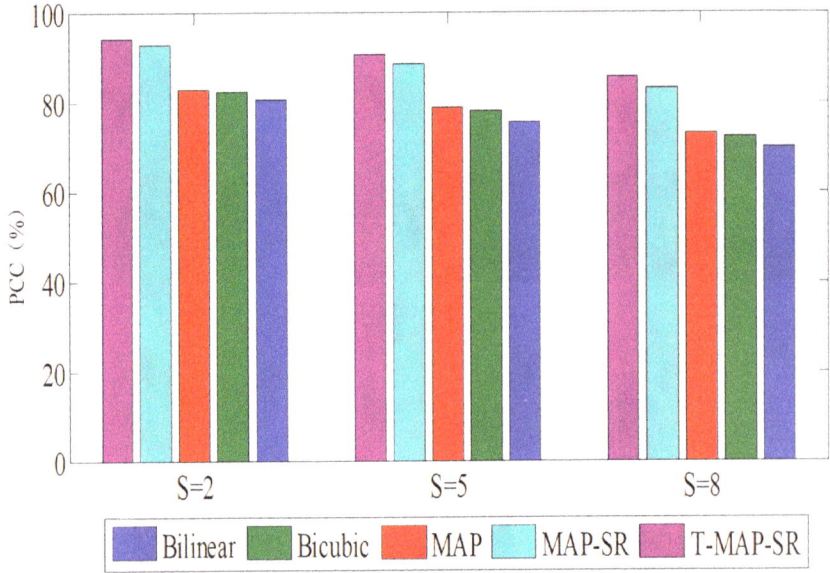

(a)

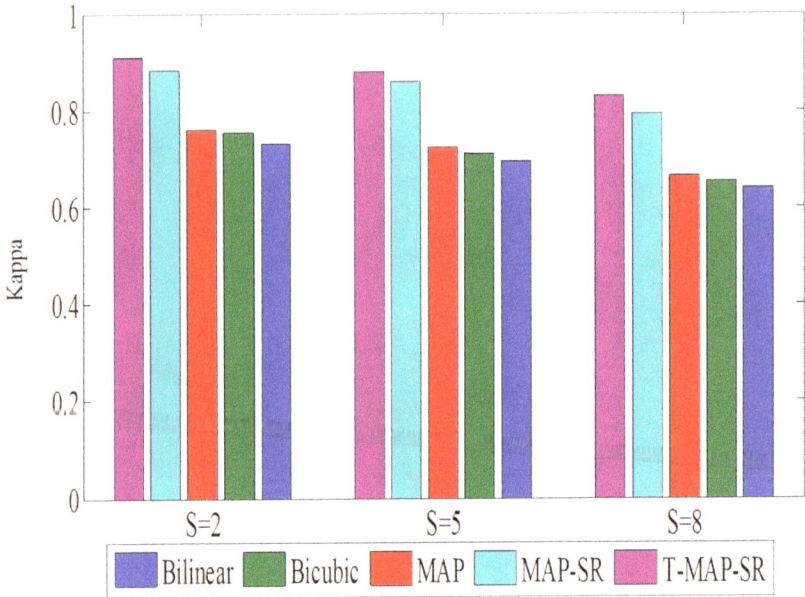

(b)

FIGURE 6.16   (a), (b) The relationship between PCC (%) and Kappa of the five methods and scale *S*.

subpixel mapping model. In addition, the results of the T-MAP-SR algorithm are more accurate than the other four methods.

6.3.3.4 McNamar Test

The McNamar test is used to count the accuracy of different methods. The difference between the results of the two methods is determined by the following formula:

$$Z = \frac{Z_{12} - Z_{21}}{\sqrt{Z_{12} + Z_{21}}} \tag{6-40}$$

where Z_{12} are the number of pixels that are correctly classified by method 1 but incorrectly classified by method 2, and Z_{21}, vice versa. Using 95% as the degree of confidence level, the difference between two results is considered to be significant if $|Z| > 1.96$.

Table 6.5 and Table 6.6 show the Z value statistics of experiment 2 ($S = 2$) and experiment 3 ($S = 5$). The numerical observations in the observation table can be obtained, and the accuracy of the results in the MTC model is more significant than that in the traditional subpixel mapping model. The Z value between MAP-SR and T-MAP-SR is much greater than zero, which indicates that the T-MAP-SR algorithm has better performance.

6.3.3.5 EOI Analysis

We selected the COI of spectral endmembers as the EOI. So the number of the COI growth will lead to the number of the EOI growth. Due to the EOI, an important issue is that the improvement of a mapping result of the COI is as a little loss of the

TABLE 6.5
McNamar Test of Five Methods in Experiment 2 ($S = 2$)

Method	Bilinear	Bicubic	MAP	MAP-SR	T-MAP-SR
Bilinear		2.0825	6.9947	54.0902	60.8301
Bicubic			3.5196	50.8100	57.6547
MAP				48.1487	53.3338
MAP-SR					6.5969
T-MAP-SR					

TABLE 6.6
McNamar Test of Five Methods in Experiment 3 ($S = 5$)

Method	Bilinear	Bicubic	MAP	MAP-SR	T-MAP-SR
Bilinear		7.8439	10.5961	39.1038	45.1272
Bicubic			2.9714	29.0144	34.0588
MAP				27.4689	33.5940
MAP-SR					5.7594
T-MAP-SR					

non-COI treatment effect. So the option of the EOI has a great effect on the performance of the T-MAP-SR algorithm. We can come to the conclusion by the previous experiments that the classes with small numbers of pixels as the non-COI also derive a good mapping result. So the classes with the higher number of pixels should be first selected as the COI. When successively adding a class into the COI according to the number of pixels from large to small order, namely, increasing the number of the EOI, we utilized the T-MAP-SR model to derive the PCC from the AVIRIS and HYDICE data sets. The PCC is shown in Figure 6.17.

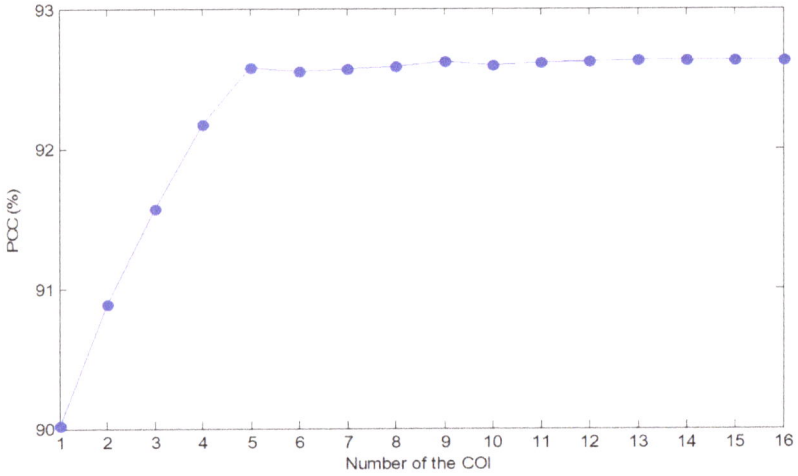

(a)

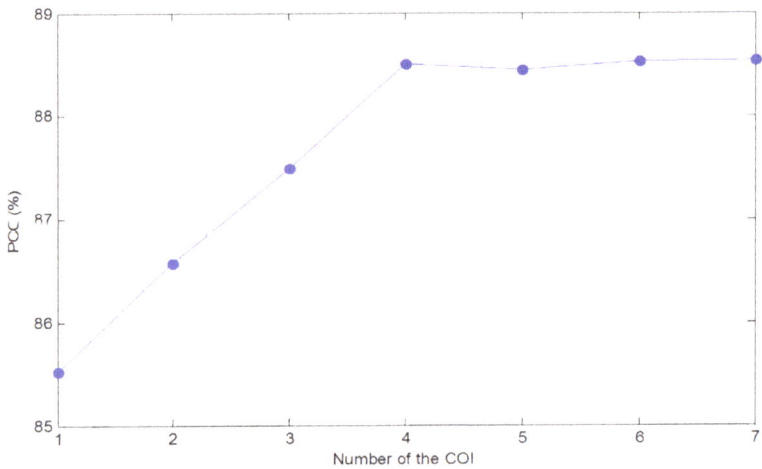

(b)

FIGURE 6.17 The relationship between PCC (%) and the number of COIs: (a) Indian agriculture and forestry data set and (b) Washington, DC, data set.

At the beginning, a substantial increase in the PCC can be observed from Figure 6.17. This is because we added the class with the higher number of pixels into the COI, the mapping accuracy of the COI derived rapid increase, and then the PCC was significantly improved. When we continued to add the class with the smaller number of pixels into the COI, the growth rate of the PCC was very slow and small. This is because the number of pixels of the class is small, it has little influence on the mapping accuracy of the COI, and the PCC cannot derive a significant increase. Sometimes it even leads to the mapping accuracy of the non-COI being much lower, and thus there is a drop in the PCC, for example, the number of the COI is 6 in Figure 6.17(a), and the number of the COI is 5 in Figure 6.17(b). At the same time, the increasing number of the EOI would be bound to increase the processing time. So in the premise of considering the processing time, the option of the EOI should comply with the following principle that the COI that constituted the EOI should have a better contribution to the study of the land cover maps, the number of the EOI should be reasonable, and the classes that are selected as the COI should be with the increased number of pixels.

6.4 SUBPIXEL MAPPING BASED ON PANSHARPENING THEN CLASSIFICATION

6.4.1 IMPLEMENTATION STEPS

When there is a suitable panchromatic image, we can use the pansharpening instead of the super-resolution reconstruction to improve the original coarse remote sensing image resolution, producing a better subpixel mapping result. In view of this, we propose a subpixel mapping based on PTC. In PTC, the PCA pansharpening with EOI proposed in Chapter 5 is used to improve the resolution of the original image, and then the classification technique is used for the pansharpening result to obtain the final subpixel mapping result. The flowchart of the proposed PTC is shown in Figure 6.18, which includes three steps:

Step 1. EOI is used to map the original high-dimensional remote sensing data to the low-dimensional transform space.

Step 2. The original remote sensing image and the high-resolution panchromatic image in the low-dimensional transformation space are fused by PCA-based pansharpening technology (see formulas [5.1] and [5.2]) to generate a high-resolution pansharpening result.

Step 3. LSSVM classification is directly performed through the improved image to generate a subpixel mapping result.

From the comparison of the flowchart in Figure 6.3 and the flowchart in Figure 6.18, it can be seen that because the PTC uses a panchromatic image, which is auxiliary data, by adding the more spatial-spectral information in the reconstruction process, the reconstructed image obtained by PTC is better than the reconstructed image obtained by MTC. Because the latter image is more accurate, a better subpixel mapping result can be obtained.

Input

Original coarse
image
($M \times N$)

Output

Mapping
result
($MS \times NS$)

PCA
pansharpening

Improved image
($MS \times NS$)

LSSVM

Sclae S

Input

High
resolution
panchromatic
image
($MS \times NS$)

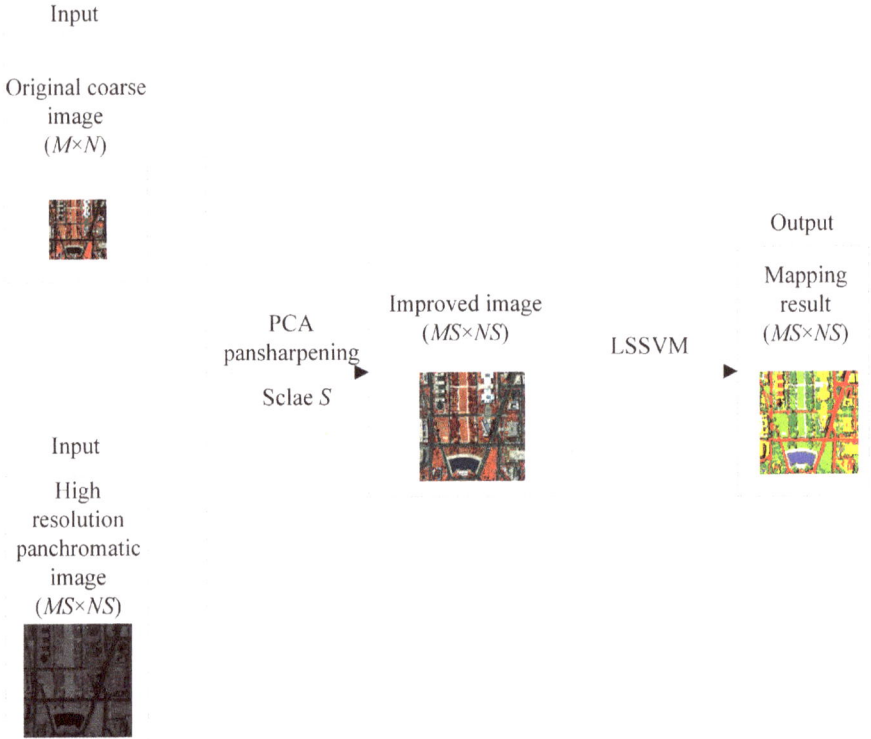

FIGURE 6.18   The framework of the PTC model.

6.4.2   Experiment Content and Result Analysis

To make a fair evaluation of the PTC and the previously proposed MTC, we still chose the same experimental conditions as in section 6.3.3, namely, the N-FINDR algorithm is selected as the endmember extraction method, and the LSSVM classification algorithm is used as the classification algorithm of the two methods. The training samples number of each class is selected as 10%, and the remaining number of each class is used as the test sample. In addition, the test data are the Washington, DC, data set and the Pavia city center data set in Section 6.3.3, and the coarse input image is simulated by downsampling the original fine image with $S = 2$. At the same time, in order to avoid the influence of a panchromatic image acquisition error on the final subpixel mapping result, only considering the influence of pansharpening on the subpixel mapping result and the spectral response of the IKONOS satellite are still used to generate a suitable panchromatic image.

6.4.2.1   Experiment 1

Experiment 1 selects the Washington, DC, data set with a size of 240×240 pixels, as shown in Figure 6.9(a). The reference image is shown in Figure 6.9(b). The simulated coarse image is shown in Figure 6.9(c). The panchromatic image derived

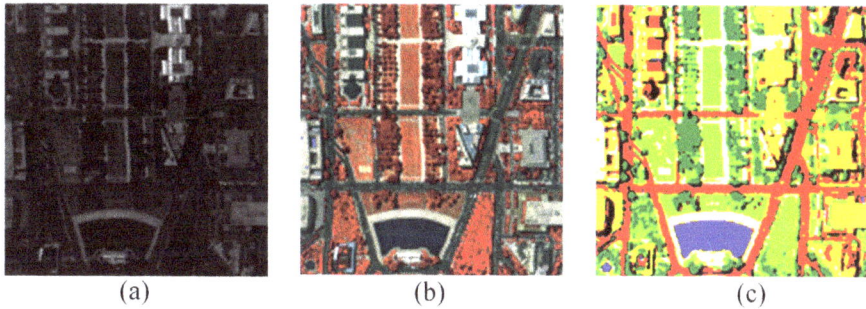

FIGURE 6.19 (a) Panchromatic image. (b) Panchromatic sharpening result. (c) PTC result.

TABLE 6.7
Different Classes of Super-Resolution Reconstruction Errors (%)

Land Cover Class	T-MAP-SR (%)	PCA With EOI (%)
Background	3.27	2.66
Water	4.18	3.50
Road	2.58	1.81
Tree	2.96	2.14
Grass	2.31	1.47
Roof	1.27	0.82
Trail	1.46	0.51

from the spectral response of the IKONOS satellite is shown in Figure 6.19(a). The PCA pansharpening technology based on EOI is used to fuse the coarse remote sensing image shown in Figure 6.9(c) and the panchromatic image shown in Figure 6.19(a) to generate the pansharpening result shown in Figure 6.19(b). Through visual comparison, it can be found that the pansharpening result shown in Figure 6.19(b) is closer to the original fine image shown in Figure 6.9(a) than the result obtained by T-MAP-SR as shown in Figure 6.10(b). In addition, the relative errors of super-resolution reconstruction are used to quantitatively evaluate the two reconstruction results, as shown in Table 6.7. Because the pansharpening technology provides the more spatial-spectral information into the coarse remote sensing image, the relative error of each class of reconstruction in the pansharpening result is lower than that of each class of reconstruction error in T-MAP-SR.

Next, the LSSVM classification is directly performed on the pansharpening result shown in Figure 6.19(b) to obtain the PTC result shown in Figure 6.19(c). Compared with the MTC result shown in Figure 6.11(e), it can be found that the result obtained by PTC is closer to the reference image shown in Figure 6.9(b). Table 6.8 shows the quantitative evaluation of MTC results and PTC results. Because the more spatial-spectral information is used in PTC, the mapping accuracy of roof is 88.75%, which is about 3.2% higher than MTC, and the mapping accuracy of road is 90.31%, which

TABLE 6.8
Mapping Accuracy of MTC and PTC in Washington, DC, Data Set (S = 2)

Land Cover Class	MTC	PTC
Background (%)	78.77	80.13
Water (%)	95.15	95.54
Road (%)	88.75	90.31
Tree (%)	97.47	98.04
Grass (%)	88.86	89.51
Roof (%)	85.23	88.43
Trail (%)	87.16	90.35
AA (%)	88.77	90.33
PCC (%)	88.51	89.62

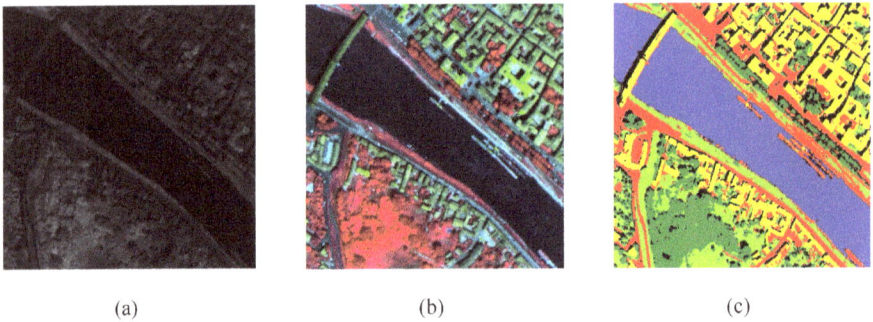

(a)   (b)   (c)

FIGURE 6.20   (a) Panchromatic image. (b) Pansharpening result. (c) PTC result.

is about 1.6% higher than MTC. At the same time, PTC produces the highest PCC (%) and AA (%).

6.4.2.2   Experiment 2

As shown in Figure 6.13(a), the Pavia city center data set selects a size of 400 × 400 pixels as the test area. Figure 6.9(b) shows the reference image. Figure 6.9(c) shows the simulated coarse remote sensing image. Figure 6.20(a) shows a panchromatic image generated by the spectral response of the IKONOS satellite. The PCA pansharpening technology based on EOI is used to merge Figure 6.9(c) and Figure 6.20(a) to generate a pansharpening result, which is as shown in Figure 6.20(b). Figure 6.20(c) shows the PTC results obtained by using the LSSVM classification algorithm to classify the pansharpening result. Compared with the MTC result in Figure 6.15(e), the PTC result is closer to the reference image. At the same time, Table 6.9 shows the quantitative evaluation of MTC result and PTC result. The mapping accuracy (%) of the tree in PTC is 97.04%, which is about 2.4% higher than that

TABLE 6.9

Mapping Accuracy of MTC and PTC in Pavia City Center Data Set ($S = 2$)

Land Cover Class	MTC	PTC
Background (%)	84.28	86.13
Water (%)	97.81	98.54
Road (%)	93.36	95.31
Tree (%)	94.68	97.04
Grass (%)	90.34	92.51
Roof (%)	97.69	98.43
AA (%)	93.02	94.66
PCC (%)	94.48	95.92

of MTC, and the mapping accuracy (%) of the road in PTC is 95.31%, which is about 2% higher than that of MTC. This is because using pansharpening technology can provide the more spatial-spectral information. In addition, PTC can obtain the higher PCC (%) and AA (%).

6.4.2.3   Comparison of PTC and STHSRM-PAN

Through the study of the soft-then-hard subpixel mapping algorithm based on the pansharpening technology (STHSRM-PAN) in Section 5.2, it can be seen that the pansharpening technology can also be used in the soft-then-hard subpixel mapping type. Although STHSRM-PAN and PTC belong to different types, the comparison between STHSRM-PAN and PTC is still worth studying. In PTC, LSSVM classification algorithm, the number of training samples is selected according to 20%, 10%, and 3% of each class, and the rest are used as test samples. We named the PTC with 20% training samples as PTC1, the PTC with 10% training samples as PTC2, and the PTC with 3% training samples as PTC3. Experiments 1 and 2 compare the STHSRM-PAN method with the three PTC methods. Figure 6.21 shows the test result. When there is enough supervision information (i.e., training samples), the mapping result of PTC is better than that of STHSRM-PAN. On the contrary, in the absence of enough supervision information, STHSRM-PAN can obtain a higher PCC (%) than PTC. Therefore, the subpixel mapping method based on the reconstruction then classification is not always more accurate than the traditional subpixel mapping methods.

6.5   SUMMARY

This chapter proposes a new model to obtain a subpixel mapping result, namely, subpixel mapping based on reconstruction then classification. A subpixel mapping based on MAP super-resolution reconstruction then classification (MTC) is proposed. The MTC model makes full use of the supervision information of the original image and effectively avoids the spectral unmixing step. In the MTC model, the high-resolution image is first obtained from the original coarse remote sensing image by the MAP

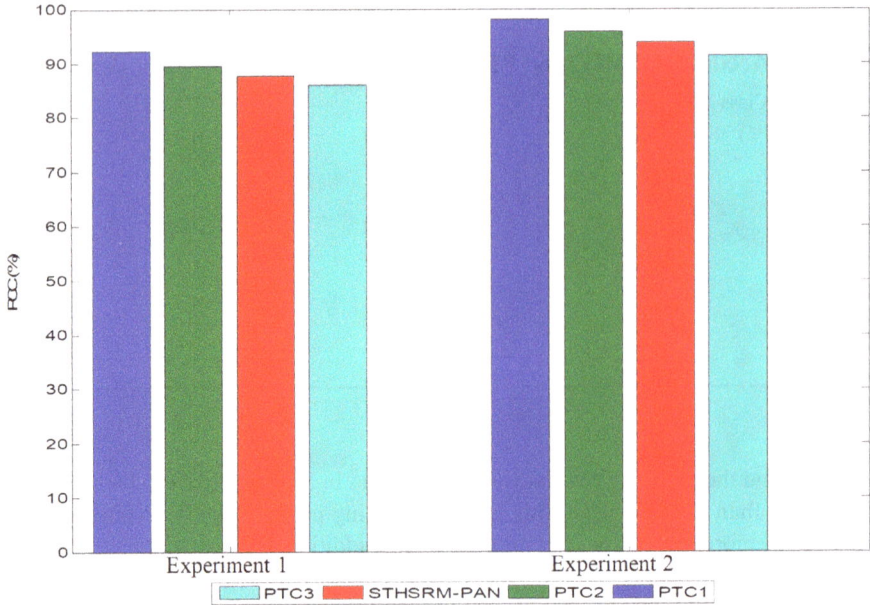

FIGURE 6.21   PCC (%) of PTC and STHSRM-PAN.

super-resolution reconstruction algorithm, and then the high-resolution image is classified by the LSSVM classification algorithm to obtain the subpixel mapping result. In particular, the MAP super-resolution reconstruction algorithm combined with EOI, namely, T-MAP-SR, can reduce super-resolution reconstruction errors, reduce calculation time, protect COI, and have good anti-noise performance. Experimental results show that when some conditions are met, the mapping accuracy of the proposed MTC model is higher than that of the traditional subpixel mapping model. Furthermore, a subpixel mapping based on pansharpening then classification is proposed. When a suitable panchromatic image exists, the pansharpening technique can be used in the reconstruction process instead of super-resolution, improving the final subpixel mapping result by adding the more spatial-spectral information.

REFERENCES

[1] Su H, Zhou J, Zhang Z. A review of super-resolution image reconstruction methods[J]. Acta Automatica Sinica, 2013, 39(8): 1202–1213.
[2] Wang L, Zhao Y. MAP-based super-resolution method for hyperspectral images[J]. Spectroscopy and Spectral Analysis, 2010, 30(4): 1044–1048.
[3] Hao S. Research on Spatial-Spectral Collaborative Classification for Hyperspectral Image [D]. Harbin: Harbin Engineering University, 2015.
[4] Yang Y. Research on Image Super-resolution Reconstruction Algorithm[D]. Beijing: University of Science and Technology of China, 2013.
[5] Zhao Y. Research on Super-Resolution Method of Hyperspectral Image Based on MAP[D]. Harbin: Harbin Engineering University, 2010.

[6]   Zhang Y, Atkinson P M, Li X, Ling F, Wang Q, Du Y. Learning-based spatial-temporal superresolution mapping of forest cover with MODIS images[J]. IEEE Transactions on Geoscience and Remote Sensing, 2017, 55(1): 600–614.

[7]   Wang L, Zhao C. Hyperspectral Image Processing Technology[M]. Beijing: Engineering Press, 2013.

[8]   Wei F. Research on the Method of Hyperspectral Image Band Selection[D]. Harbin: Harbin Engineering University, 2013.

[9]   Wang L, Wang Z, Dou Z, Wang Y. Edge-directed interpolation-based sub-pixel mapping[J]. Remote Sensing Letters, 2013, 12(4): 1195–1203.

[10] Wang Q, Shi W. Utilizing multiple subpixel shifted images in subpixel mapping with image interpolation[J]. IEEE Geoscience and Remote Sensing Letters, 2014, 11(4): 798–802.

[11] Xu X, Zhong Y, Zhang L, Zhang H. Sub-pixel mapping based on a MAP model with multiple shifted hyperspectral imagery[J]. IEEE Journal of Selected Topics in Applied Earth Observations and Remote Sensing, 2013, 6(2): 580–593.

7 Application of Subpixel Mapping Technology in Remote Sensing Imaging

7.1 INTRODUCTION

The use of subpixel mapping technology can obtain more accurate distribution information of land cover classes from remote sensing imaging. The demand for accurate distribution information is increasing in the military and civilian fields. The subpixel mapping technology has been successfully applied to thematic mapping of land cover classes, extraction of river and lake boundaries, detection of coastlines, calculation of the landscape index, and change detection of land cover classes [1–2]. In addition, subpixel mapping technology has been widely used in crop changes, forest coverage detection, drinking water quality investigation, environmental detection, climate change, and natural disaster analysis [3–5]. Specifically for specific targets, such as fire areas [6], forest coverage areas [7], flooded areas [8–10], and urban construction areas [11], many subpixel mapping models have been successfully established. However, many subpixel mapping models based on these specific targets often only use space information to obtain the final subpixel mapping results, while this space information is not very accurate. In addition, the unique spectral information of remote sensing images is not fully utilized, which will affect the final mapping accuracy.

In view of this content, this chapter introduces various normalized differential target class indexes into the subpixel mapping model, so that the spectral information of the specific target class in the remote sensing image can be more fully utilized, thereby improving the final accuracy of the subpixel mapping result. In addition, the more accurate space information described in Section 3.4 is introduced into subpixel mapping for a specific target. This chapter aims to propose three subpixel mapping methods for flood-inundation area, urban building area, and burned area, respectively. Three subpixel mapping methods are improving subpixel flood-inundation mapping for multispectral remote sensing image by supplying more spectral information (SRFIM-MSI), subpixel mapping for urban building by using spatial-spectral information from spaceborne multispectral remote sensing image (SMUB), and multispectral image subpixel burned-area mapping based on space-temperature information (STI). Experiments show that the three subpixel mapping methods can make full use of the spectral information of flood-inundation area, urban building area, and burned area. Especially, STI uses the accurate space information based on objects to improve the mapping accuracy of a burned area.

DOI: 10.1201/9781003279082-7

7.2 IMPROVING FLOOD SUBPIXEL MAPPING FOR MULTISPECTRAL IMAGE BY SUPPLYING MORE SPECTRAL INFORMATION

Due to the changes in global climate and land use, the severity and frequency of floods all over the world have increased significantly, and mapping the flood-inundation area is very important for flood research. At present, various multispectral remote sensing images have been widely used to map flood-inundation areas. However, while pursuing rapid acquisition of multispectral images, the spatial resolution of multispectral images is sometimes coarse. In addition, the severe weather during the flood also brings great difficulties to the acquisition of multispectral images, resulting in low spatial resolution of multispectral images. These problems usually lead to the widespread existence of mixed pixels, and subpixel mapping is an effective way to solve this problem.

To make the subpixel mapping method more effective for flood-inundation mapping (subpixel flood-inundation mapping, SRFIM), Li et al. used the algorithms such as genetic algorithm [8], particle swarm optimization algorithm [9], and neural network [10] to improve the existing SRFIM algorithm. However, the existing SRFIM algorithm usually uses the spatial correlation to obtain the flood-inundation area mapping result, and the flood spectral information of the multispectral remote sensing image is usually not fully utilized, which will affect the final mapping result.

To solve this problem, this section proposes an improved SRFIM called SRFIM-MSI. The normalized difference water index (NDWI) is a quantitative water spectral index. NDWI has been widely used in water body extraction by enhancing water body information while suppressing land information in a remote sensing image [12]. This algorithm uses the spectral information of the green band and short-wave infrared band to calculate NDWI. A new spectral item composed of NDWI is added to the traditional SRFIM. The experimental results show that when compared with the traditional SRFIM method, the proposed SRFIM-MSI method can obtain a more accurate flood-inundation area mapping result.

7.2.1 EXISTING SRFIM

The existing SRFIM method maximizes its spatial correlation to obtain the spatial distribution of subpixels of flood-inundation area in the mixed pixels. Figure 7.1 shows an example to illustrate how SRFIM utilizes the theory of spatial correlation. Remote sensing images of flood inundation are divided into two classes: inundation and non-inundation. Figure 7.1(a) shows the spectral unmixing results of the flood-inundation class, where the score value in the figure represents the proportion of the flood-inundation class in the mixed pixel. In Figure 7.1(a), there are 3×3 mixed pixels, and each mixed pixel is marked with the proportion of flood-inundation class. The scale $S = 4$ indicates the ratio between the mixed pixel and its subpixels. When the spectral unmixing result is upsampled with the scale S, a mixed pixel is divided into four subpixels. Then 25% indicates that 4×4 subpixels in the central mixed pixel belong to the flood-inundation class. Figure 7.1(b)–(d) describes three possible subpixel mapping results. Considering the principle of spatial correlation, the most

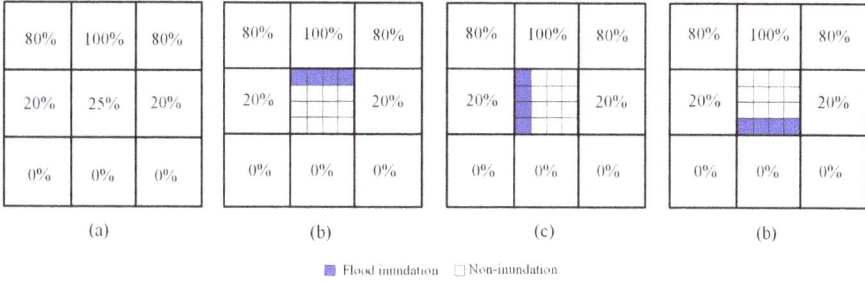

80%	100%	80%
20%	25%	20%
0%	0%	0%

(a)

80%	100%	80%
20%		20%
0%	0%	0%

(b)

80%	100%	80%
20%		20%
0%	0%	0%

(c)

80%	100%	80%
20%		20%
0%	0%	0%

(b)

■ Flood inundation   ☐ Non-inundation

FIGURE 7.1   Spatial correlation: (a) spectrum unmixing result of flood inundation, (b) distribution possibility 1, (c) distribution possibility 2, and (d) distribution possibility 3.

likely SRFIM result is assumed to be the result with the greatest spatial correlation. Therefore, Figure 7.1(b) is considered to be the optimal distribution result.

According to the principle of spatial correlation, SRFIM is defined as a maximum combinatorial optimization problem. Flood-inundation spatial correlation index ($FISDI_i$) is used to calculate spatial correlation between the central subpixel p_i ($I = 1$, $2, \ldots, MS^2$, M is the number of mixed pixels, MS^2 is the number of subpixels) and the number of adjacent mixed pixels. The mathematical model is given by the following formula:

$$FISDI_i = o_i \cdot \sum_{C=1}^{M_n} w_{iC} \cdot L(P_C) + (1-o_i) \cdot \left[\sum_{C=1}^{M_n} w_{iC} \cdot (1-L(P_C)) \right] \qquad (7.1)$$

where o_i is the binary label of each subpixel (the flood-inundation label is 1, and the non-inundation label is 0). $L(P_c)$ is the proportion of flood-inundation classes in the C adjacent mixed pixel P_c, and M_n is the number of adjacent mixed pixels. The number of adjacent pixels is considered as eight here.

Also, w_{iC} is the weight of the spatial correlation between the central subpixel p_i and the adjacent mixed pixel P_C, which can be described by the following formula:

$$w_{iC} = \exp(-d(p_i, P_C)^2 / r) \qquad (7.2)$$

where $d(p_i, P_C)$ is the Euclidean distance from the central subpixel p_i to the adjacent mixed pixel P_C, and r is the non-linear parameter of the exponential model.

The $FISDI_i$ of all subpixels constitutes a spatial term $E^{spa}(i)$, which is calculated as follows:

$$E^{spa}(i) = \text{Max} \sum_{i=1}^{MS^2} FISDI_i \qquad (7.3)$$

The evaluation criteria of SRFIM can be considered as that a higher $E^{spa}(i)$ value will lead to a better subpixel mapping result being obtained. Therefore, when SRFIM allocates flood-inundation labels to subpixels, it needs to satisfy the $E^{spa}(i)$ maximization.

7.2.2 SRFIM-MSI

This section proposes the SRFIM-MSI algorithm that can utilize the more spectral information. Compared with the traditional SRFIM, SRFIM-MSI adds a new spectral item $E^{spe}(i)$, whose purpose is to minimize the difference between the observed NDWI value ($NDWI^{obs}$) and the simulated NDWI value ($NDWI^{sim}$) of the multispectral remote sensing image. Here, we use the green band (Green) and short-wave infrared band (SWIR) in the multispectral remote sensing image to calculate in formula (7.4):

$$NDWI^{obs} = \sum_{1}^{M} \frac{\rho_{Green} - \rho_{SWIR}}{\rho_{Green} + \rho_{SWIR}} \tag{7.4}$$

where ρ_{Green} and ρ_{SWIR} are the reflectivity of each mixed pixel in their respective bands, and M is the number of mixed pixels.

Suppose r^{f}_{Green} and r^{f}_{SWIR} are the reflectivity of the flood-inundation area in the Green band and the SWIR band, and r^{n}_{Green} and r^{n}_{SWIR} are the reflectivity of the non-inundated area in the Green band and the SWIR band, respectively. For each mixed pixel in the green band and short-wave infrared band, the proportion of flood-inundation area a^{f}_{Green} and a^{f}_{SWIR} can be calculated by dividing the number of flood-inundation subpixels by the total sub-pixel number. The proportion of non-inundation area are then $1 - a^{f}_{Green}$ and $1 - a^{f}_{SWIR}$. The reflectance of each mixed pixel is regarded as the linear summation of the spectral reflectance of all subpixels in it, and the simulated reflectance R^{Sim}_{Green} and R^{Sim}_{SWIR} of each mixed pixel in the green band and short-wave infrared band are calculated as follows:

$$R^{sim}_{Green} = \left(r^{f}_{Green} \cdot a^{f}_{Green} \right) + \left[r^{n}_{Green} \cdot \left(1 - a^{f}_{Green} \right) \right] \tag{7.5}$$

$$R^{sim}_{SWIR} = \left(r^{f}_{SWIR} \cdot a^{f}_{SWIR} \right) + \left[r^{n}_{SWIR} \cdot \left(1 - a^{f}_{SWIR} \right) \right] \tag{7.6}$$

$NDWI^{sim}$ can be expressed as

$$NDWI^{sim} = \sum_{1}^{M} \frac{R^{sim}_{Green} - R^{sim}_{SWIR}}{R^{sim}_{Green} + R^{sim}_{SWIR}} \tag{7.7}$$

The spectral term $E^{spe}(i)$ is expressed as

$$E^{spe}(i) = \text{Min}\left(NDWI^{obe} - NDWI^{sim} \right)^{2} \tag{7.8}$$

At the same time, $E^{spa}(i)$ is transformed into the following form as

$$E^{spa}(i) = -\text{Min}\sum_{i=1}^{MS^{2}} FISDI_{i} \tag{7.9}$$

In other words, the goal of the proposed SRFIM-MSI is to minimize the objective function $E(i)$, which includes the spatial term $E^{spa}(i)$ and the spectral term $E^{spe}(i)$ integrated by the trade-off parameter λ.

$$\text{Min } E(i) = (1 - w)E^{spa}(i) + wE^{spe}(i) \tag{7.10}$$

The particle swarm optimization (PSO) algorithm is used to optimize the objective function. First, the inundation label or the non-inundation label is randomly assigned to all subpixels. Then, the labels of these subpixels are iteratively updated until the minimum value is reached. In each iteration, the submerged tag is converted to a non-inundation label or the non-inundation label is converted to an inundation label. If the value decreases, the conversion is accepted. If the value increases, the conversion will be rejected. When the converted label is less than 0.1%, PSO will terminate.

7.2.3 Experiment Content and Result Analysis

To quantitatively estimate the performance of SRFIM-MSI, a simulated coarse multispectral image is obtained by downsampling the original high-resolution multispectral image with a scale S. This is because under such conditions, land cover classes at subpixel-level can be obtained, which is convenient for evaluating the influence of image registration errors on the method. The spectral unmixing method based on least squares support vector machine (LSSVM) is selected to obtain the abundance image of each class. We compare the proposed SRFIM-MSI with the other three SRFIM methods: HNN (HNNA) with an anisotropic spatial correlation model [13], radial basis function (RBF) interpolation [14], SRFIM based on discrete particle swarm optimization (DPSO) [9]. PCC (%) and Kappa are used to evaluate the performance of the four SRFIM methods.

7.2.3.1 Test Data

Typhoon NARI followed strong seasonal rains and caused massive floods along the Mekong and Tonle Sap rivers in Cambodia. The floods affected more than 500,000 people and destroyed more than 250,000 acres of rice fields. Therefore, accurate acquisition of fine spatial resolution images of the flood-inundation area is of great significance for disaster relief. On October 24, 2013, two Landsat 8 OLI data sets were collected in the flood-inundation area. The geographic coordinates of the center of data 1 are 12°07′N and 104°37′E, and the geographic coordinates of the center of data 2 are 11°47′N and 105°02′E. The two data sets have six bands including red, green, blue, near-infrared, short-wave infrared 1, and short-wave infrared 2.

As shown in Figure 7.2(a), the size of data 1 is 500 × 500 pixels. As shown in Figure 7.3(a), the data 2 has a more complex flood distribution and a larger area, with a size of 1000 × 1000 pixels. As shown in Figure 7.2(b) and Figure 7.3(b), the two data sets are downsampled to generate the simulated coarse images with $S = 6$. The reference images of the two data sets are shown in Figure 7.2(c) and Figure 7.3(c), respectively. The trade-off parameter λ for both data sets is chosen to be 0.6.

7.2.3.2 Results and Analysis

As shown in Figure 7.2(b) and Figure 7.3(b), it is difficult to accurately obtain the spatial distribution information of flood-inundation area in coarse remote sensing images. To solve this problem, SRFIM is used to obtain accurate spatial distribution of the flood-inundation area. Observing from the four SRFIM results of data 1 and

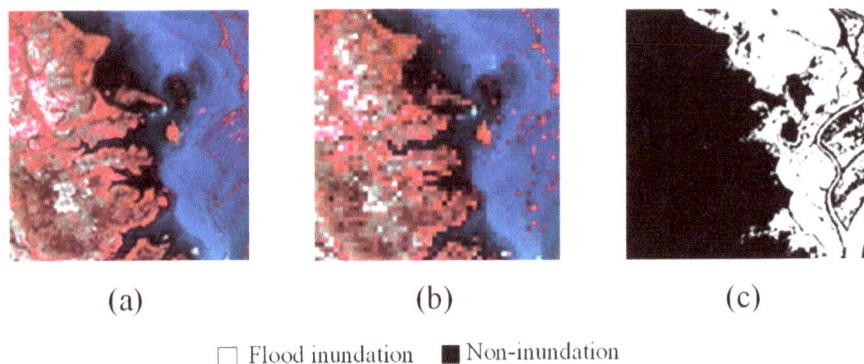

☐ Flood inundation ■ Non-inundation

FIGURE 7.2   Data set 1: (a) false color composite image (RGB: short-wave infrared 1, near-infrared, blue); (b) low-resolution image ($S = 6$); and (c) reference image.

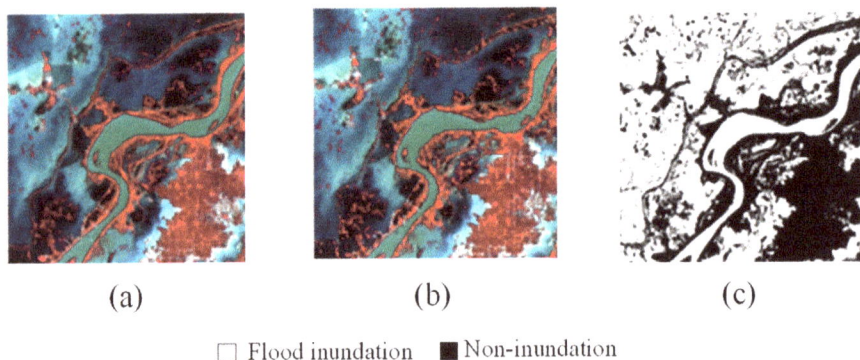

☐ Flood inundation ■ Non-inundation

FIGURE 7.3   Data set 2: (a) false color composite image (RGB: short-wave infrared 1, near-infrared, blue); (b) low-resolution image ($S = 6$); and (c) reference image.

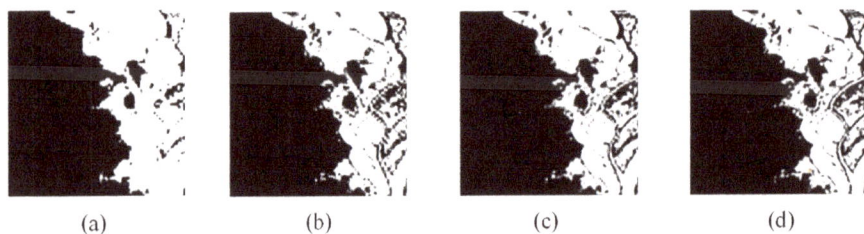

FIGURE 7.4   Mapping result in data set 1: (a) HNNA, (b) RBF, (c) DPSO, and (d) SRFIM-MSI.

data 2 shown in Figure 7.4 and Figure 7.5, we can find that SRFIM can ideally display the spatial distribution of the flood-inundation area; the mapping result obtained by the proposed SRFIM-MSI is closer to the reference image. Through intuitive visual comparison, the proposed SRFIM-MSI method is superior to the other three SRFIM

(a)   (b)   (c)   (d)

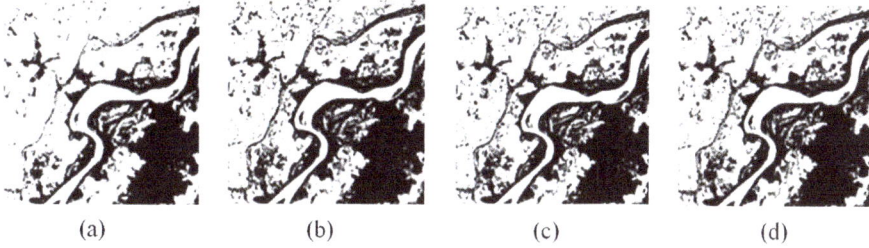

FIGURE 7.5   Mapping result in data set 2: (a) HNNA, (b) RBF, (c) DPSO, and (d) SRFIM-MSI.

TABLE 7.1
Evaluation Indicators of the Four Methods in Data Set 1

Method	PCC (%)	Kappa
HNNA	90.08	0.8137
RBF	91.56	0.8386
DPSO	94.65	0.8782
SRFIM-MSI	96.81	0.9137

TABLE 7.2
Evaluation Indicators of the Four Methods in Data Set 2

Method	PCC(%)	Kappa
HNNA	85.51	0.7482
RBF	87.39	0.7828
DPSO	90.83	0.8323
SRFIM-MSI	93.95	0.8916

methods. Since more spectral information is provided, SRFIM-MSI can generate the smoother flood-inundation area.

In addition, PCC (%) and Kappa are used to evaluate the performance of the four SRFIM methods on two data sets. According to Table 7.1 and Table 7.2, SRFIM-MSI can obtain the highest PCC (%) and Kappa. Compared with the DPSO method, the PCC (%) and Kappa of SRFIM-MSI increased by about 2.2% and 0.036 in data 1 and increased by about 3.1% and 0.059 in data 2.

7.2.3.3   Discussion

To study the influence of scale S on the proposed method, data 2 is downsampled with $S = 3$, 6 and 10 to generate the simulated coarse image. The relationship between the three scales of the four methods and the PCC (%) of the four methods is shown in Figure 7.6. Similar to the results in Table 7.1 and Table 7.2, SRFIM-MSI still achieved the highest PCC (%).

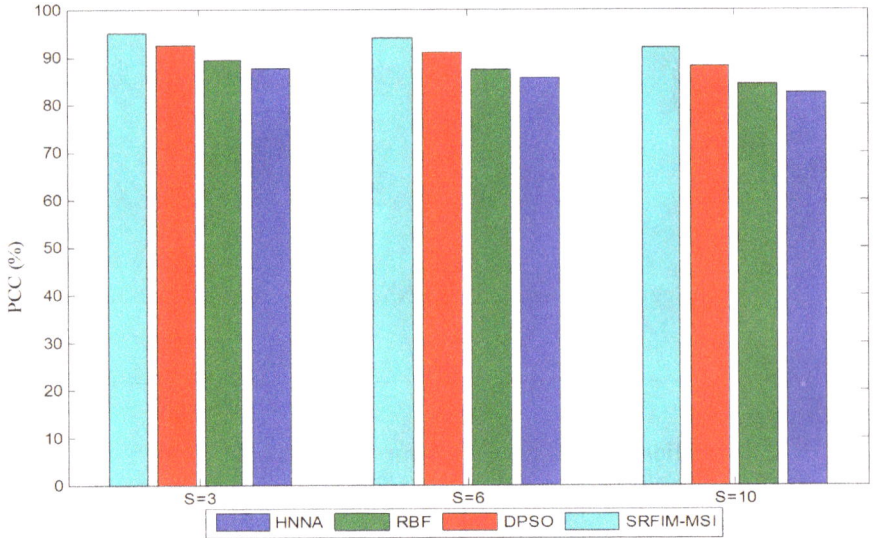

FIGURE 7.6   The relationship between PCC (%) and scale S of the four methods.

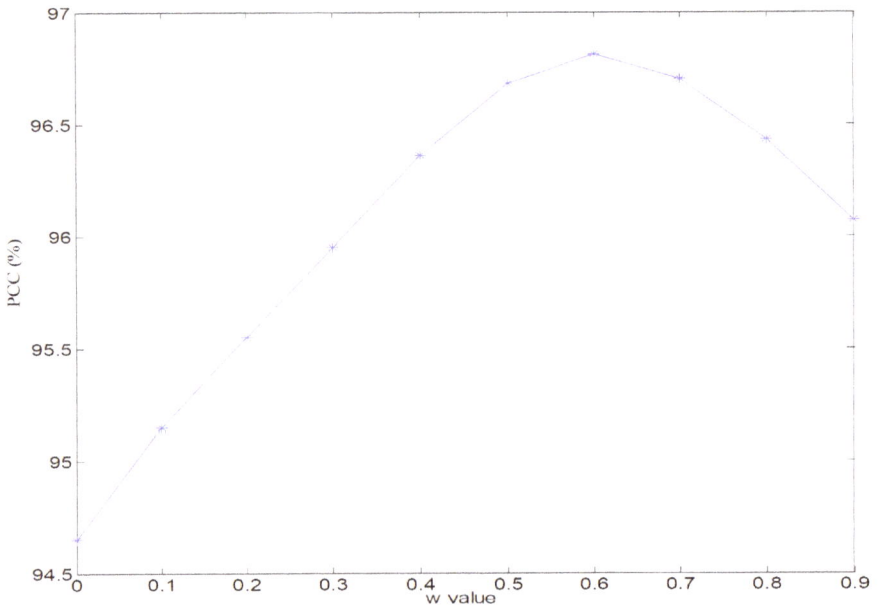

FIGURE 7.7   The relationship between PCC (%) and weight parameter λ of the SRFIM-MSI method.

To study the impact of the trade-off parameter λ of SRFIM-MSI on its performance, data 1 set ($S = 6$) was tested, the test range is [0, 0.9], and the interval is 0.1. As shown in Figure 7.7, the spectral term does not play any role when $\lambda = 0$. With the increase of w, PCC (%) is also observed to increase. This is because more spectral information is provided through the spectral item $E^{spa}(i)$. When $w = 0.6$, the PCC (%) value is the highest. However, when it continues to increase, the contribution of the spatial term $E^{spa}(i)$ to formula (7.10) decreases, which leads to a decrease in spatial information and affects the final mapping result.

7.3 SUBPIXEL MAPPING OF URBAN BUILDINGS BASED IN MULTISPECTRAL IMAGE WITH SPATIAL-SPECTRAL INFORMATION

Many basic data used in urban environmental science are obtained by classifying very high-resolution (VHR) multispectral images, such as urban tree map [15], urban vegetation classification [16], urban structure analysis [17], urban object extraction [18], and so on. In addition, the urban building mapping [19–20] obtained from the classification results of VHR images is also one of the basic data, which is mainly used in population monitoring, urban ecosystems, and air pollution. However, due to the influence of the external environment and the limitations of the sensor, the resolution of the multispectral image is sometimes coarse, resulting in a large number of mixed pixels. Since a mixed pixel contains multiple land cover classes, the traditional classification technique can only assign one land cover class to one pixel, and it often cannot effectively deal with mixed pixels. Therefore, the urban building mapping obtained by using the classification technique on coarse multispectral images is often not ideal.

In view of this, it is more effective to use subpixel mapping technology to obtain urban building mapping from coarse multispectral images than to use classification technology. Ling et al. proposed a subpixel mapping model based on urban building by using prior shape information [21]. However, this method cannot effectively map irregularly shaped urban buildings in long-distance satellite-borne multispectral images, and this method also requires sufficient prior information. However, the spatial-spectral information of the multispectral image, especially the spectral information of urban building, has not been fully utilized. Drawing lessons from the previous section, the subpixel mapping model for a flood-inundation area is proposed. This section proposes subpixel mapping for urban building by using spatial-spectral information from a spaceborne multispectral remote sensing image (SMUB). SMUB is a kind of subpixel mapping that does not require prior shape information and can effectively map irregularly shaped urban building in spaceborne multispectral images. Experimental results prove that the proposed SMUB can use more spatial-spectral information to obtain better subpixel mapping of urban buildings.

7.3.1 Spaceborne Multispectral Remote Sensing Image

The SMUB model is the same as the proposed SRFIM-MSI model in the previous section, and it also contains spatial and spectral terms. The spatial term $E^{spa}(i)$ is also the maximum spatial attraction model that is defined according to the principle of spatial correlation, as shown in formula (7.11).

$$E^{spa}(i) = \text{Max} \sum_{i=1}^{MS^2} \left\{ o_i \cdot \sum_{C=1}^{M_n} w_{iC} \cdot L(P_C) + (1 - o_i) \cdot \left[\sum_{C=1}^{M_n} w_{iC} \cdot (1 - L(P_C)) \right] \right\} \quad (7.11)$$

where p_i ($i = 1, 2, \ldots, MS^2$, M is the number of mixed pixels, MS^2 is the number of subpixels) is the central subpixel; $L(P_C)$ is the adjacent Cth mixed pixel P_C in the proportion of urban building class; M_n is the number of adjacent mixed pixels, eight adjacent mixed pixels are considered here; o_i is the binary label of each subpixel (1 means the urban building label, 0 means the background label); w_{iC} is the weight of the spatial correlation between the central subpixel p_i and the adjacent pixels P_C.

$$w_{iC} = \exp(-d(p_i, P_C)^2 / r) \tag{7.12}$$

where $d(p_i, P_C)$ is the Euclidean distance from the central subpixel p_i to the adjacent mixed pixel P_C, and r is the non-linear parameter of the exponential model.

The spectral term $E^{\mathrm{spe}}(i)$ is similar to the spectral term formula of the SRFIM-MSI. The difference is that the normalized difference built-up index (NDBI) is used to represent the spectral information of urban building instead of the normalized difference water index. The specific calculation formulas are (7.13) and (7.14):

$$NDBI = \sum_1^M \frac{o_B - o_A}{o_B + o_A} \tag{7.13}$$

$$E^{\mathrm{spe}}(i) = \mathrm{Min}\left(NDBI^{\mathrm{obs}} - NDBI^{\mathrm{sim}}\right)^2 \tag{7.14}$$

where o_A and o_B are the reflectivity of each mixed pixel in band A and band B, and M is the number of mixed pixels. The basis for selecting two bands in the multispectral image is that the reflectivity of the urban building area has a sharp increase from the A band to the B band [22]. Then $NDBI^{\mathrm{obs}}$ is the observed $NDBI$ value, and $NDBI^{\mathrm{sim}}$ is the simulated $NDBI$ value; E is obtained by integrating the spatial term $E^{\mathrm{spa}}(i)$ and the spectral term $E^{\mathrm{spe}}(i)$ through the parameter w ($0 \leq w < 1$), and the objective function is minimized as

$$\mathrm{Min}\, E = wE^{\mathrm{spe}}(i) - (1 - w)E^{\mathrm{spa}}(i) \tag{7.15}$$

Finally, the PSO algorithm is still used for optimization processing, and the ideal subpixel mapping result is obtained. Because of the integration of spatial term $E^{\mathrm{spa}}(i)$ and spectral term $E^{\mathrm{spe}}(i)$, the spatial-spectral information of remote sensing image can be fully utilized, and the mapping accuracy of urban building is improved.

7.3.2 EXPERIMENT CONTENT AND RESULT ANALYSIS

The performance of this method was tested by using two Landsat 8 OLI data. In order to obtain the simulated coarse image with the same spatial resolution in each band, we selected six bands including blue, green, red, near-infrared, short-wave infrared 1, and short-wave infrared 2 to form the test image. The spatial resolution of six bands is 30 meters. In order to understand the land cover classes at the subpixel scale and how the impact of image registration errors is directly evaluated, the test image is performed by $S \times S$ mean filtering to obtain the simulated coarse remote sensing image. Since the reflectivity of the urban building area increases sharply from the short-wave infrared band 1 to the near-infrared band in the Landsat 8 OLI data [23], the two bands are used for calculation of *NDBI* here. The abundance image of urban building is obtained by spectral unmixing based on least square support vector machine (LSSVM).

The proposed SMUB method is compared with pixel-swapping algorithm (PSA) [24], hybrid spatial attraction models (HSAM) [25], and Hopefield neural network with anisotropic spatial dependence model (HNNA) [26]. At the same time, PCC (%) and Kappa are used to evaluate the four subpixel mapping methods.

7.3.2.1   Experiment 1

In April 2013, the multispectral image was obtained near Ulysses, Kansas, as shown in Figure 7.8(a). The urban building area in this image is smaller than the background area. The center of the image is located at 36°34' north latitude and 101°21' west longitude. By downsampling the test image, the simulated coarse image is obtained, as shown in Figure 7.8(b). Figure 7.8(c) shows the reference image. Due to the low resolution of coarse images, it is difficult to obtain information on the spatial distribution of the urban building area. Therefore, subpixel mapping technology is applied to the simulated coarse image to generate the fine urban building mapping. The weight parameter w is set to 0.5.

The subpixel mapping results of the four methods are shown in Figure 7.9(a)–(d). Due to the lack of abundant spatial-spectral information, there are many disconnected

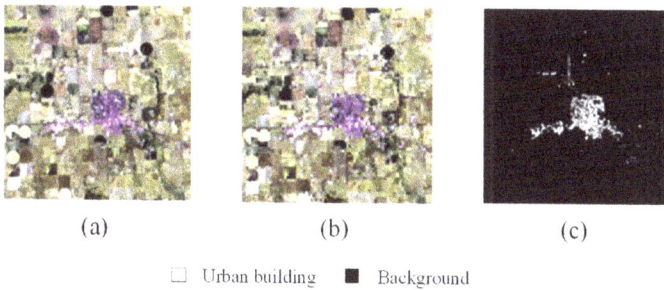

(a)   (b)   (c)

☐ Urban building   ■ Background

FIGURE 7.8   Data set 1: (a) false color composite image (RGB: short-wave infrared 2, short-wave infrared 1, red); (b) low-resolution image ($S = 5$); and (c) reference image.

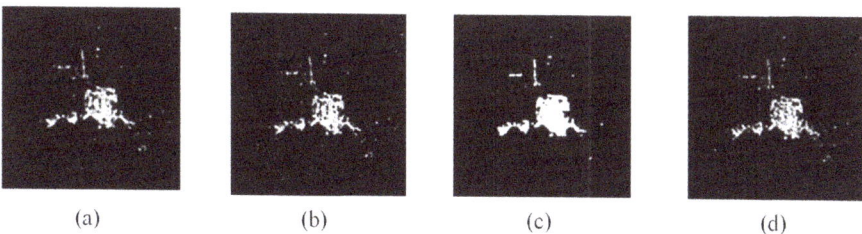

(a)   (b)   (c)   (d)

FIGURE 7.9   Mapping results: (a) PSA, (b) HSAM, (c) HNNA, and (d) SMUB.

patches and obvious burrs in Figure 7.9(a)–(b). As shown in Figure 7.9(c), since the HNNA constraint remains in an excessive state, the boundary is too smooth, and the target with a small area almost disappears. Because the proposed SMUB can more fully utilize the spatial-spectral information, the mapping result of Figure 7.9(d) is closer to the reference image than the mapping results of the other three methods.

Table 7.3 shows the mapping accuracy (%) of each class, PCC (%), and Kappa of the four subpixel mapping methods. Through the inspection of the mapping accuracy (Urban building [%]) of the urban building class in Table 7.3, the urban building (%) of SMUB is about 1.3% higher than that of HNNA. In addition, due to the use of more spatial-spectral information, SMUB could obtain the highest PCC (%) value of 98.85% and Kappa value of 0.6653.

7.3.2.2 Experiment 2

In experiment 2, we tested the performance of the proposed method in a research area with more urban buildings. The testing multispectral image was collected in Rome, Italy, in June 2014, as shown in Figure 7.10(a). The testing image is downsampled with $S = 5$ to generate a simulated coarse image, as shown in Figure 7.10(b). The reference image is shown in Figure 7.10(c). The weight parameter w is chosen to be 0.6.

TABLE 7.3
Evaluation Indicators of the Four Methods in Data Set 1

Method	Urban Building (%)	Background (%)	PCC (%)	Kappa
PSA	64.87	97.80	96.67	0.6367
HSAM	65.18	97.81	96.69	0.6398
HNNA	66.21	98.12	97.03	0.6421
SMUB	67.50	98.89	98.85	0.6653

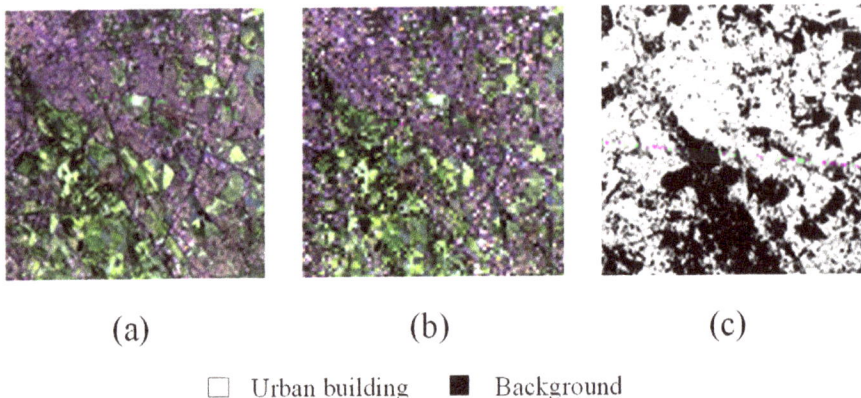

(a)   (b)   (c)

☐ Urban building   ■ Background

FIGURE 7.10 Data set 2: (a) false color composite image (RGB: short-wave infrared 2, short-wave infrared 1, red), (b) low-resolution image ($S = 5$); and (c) reference image.

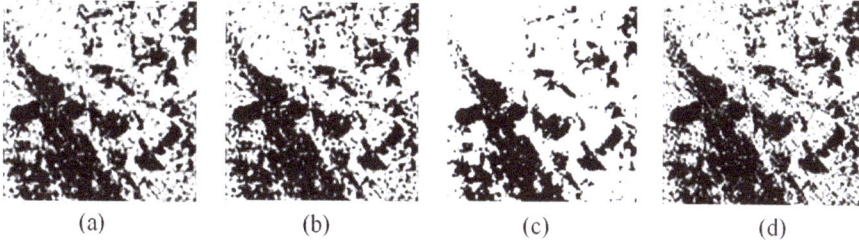

FIGURE 7.11   Mapping results: (a) PSA, (b) HSAM, (c) HNNA, and (d) SMUB.

TABLE 7.4
Evaluation Indicators of the Four Methods in Data Set 2

	Urban Building (%)	Background (%)	PCC (%)	Kappa
PSA	82.69	79.77	80.94	0.6246
HSAM	83.11	80.29	81.41	0.6340
HNNA	83.78	80.95	81.95	0.6433
SMUB	85.13	81.57	83.54	0.6770

The results of the four subpixel mapping methods are shown in Figure 7.11(a)–(d). From the perspective of intuitive visual effects, the SMUB shown in Figure 7.11(d) is closer to the reference image than the other three methods shown in Figure 7.11(a)–(c). Table 7.4 shows the mapping accuracy (%) of each class, PCC (%), and Kappa of the four subpixel mapping methods, which are consistent with the results of experiment 1. Each evaluation index of the proposed SMUB is higher than that of PSA, HSAM, and HNNA.

7.3.2.3   Discussion

The performance of the subpixel mapping method is affected by the scale S. Therefore, we use the multispectral image of experiment 1 to test the four subpixel mapping methods at the other two scales $S = 2$ and $S = 8$. Figure 7.12 shows urban building (%) for all three scales. We can notice that urban building (%) decreases as the scale S increases among the four methods. This is because a higher scale S means that the simulated image is coarser, it is more difficult for the subpixel mapping technology. But consistent with the results in Table 7.4, SMUB still gets the highest urban building (%).

To study the influence of the weight parameter w on the experimental results, the image of experiment 2 ($S = 5$) tested 10 combinations in the range of [0, 0.9] at 0.1 intervals. As shown in Figure 7.13, when $w = 0$, the spectral term $E^{spa}(i)$ has no effect. As w increases, urban building (%) also increases. When $w = 0.6$, urban building (%) can get the highest value. However, when w continues to increase, the spatial term reduces the influence on formula (7.16). Due to the reduction of spatial information, the urban building (%) is reduced.

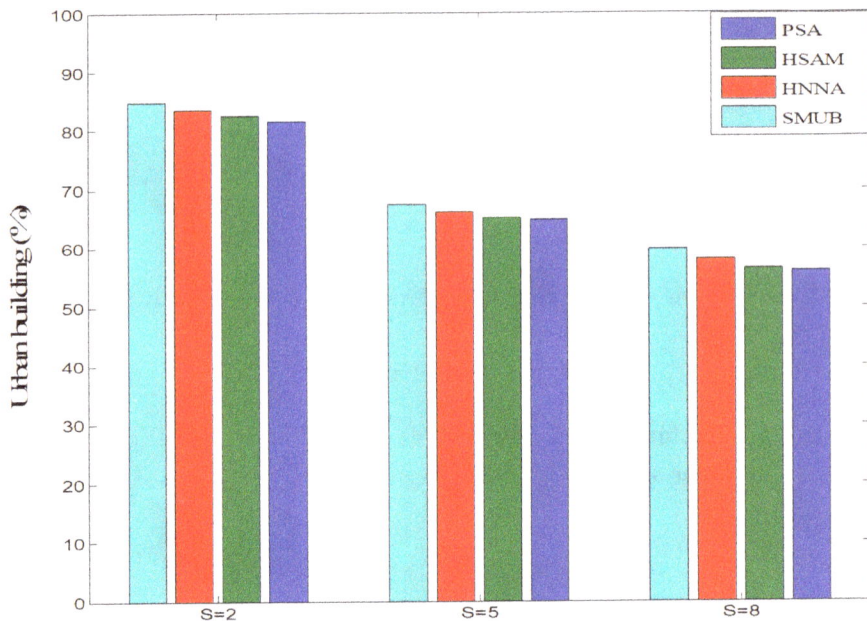

FIGURE 7.12   The relationship between urban building (%) and scale S of the four methods.

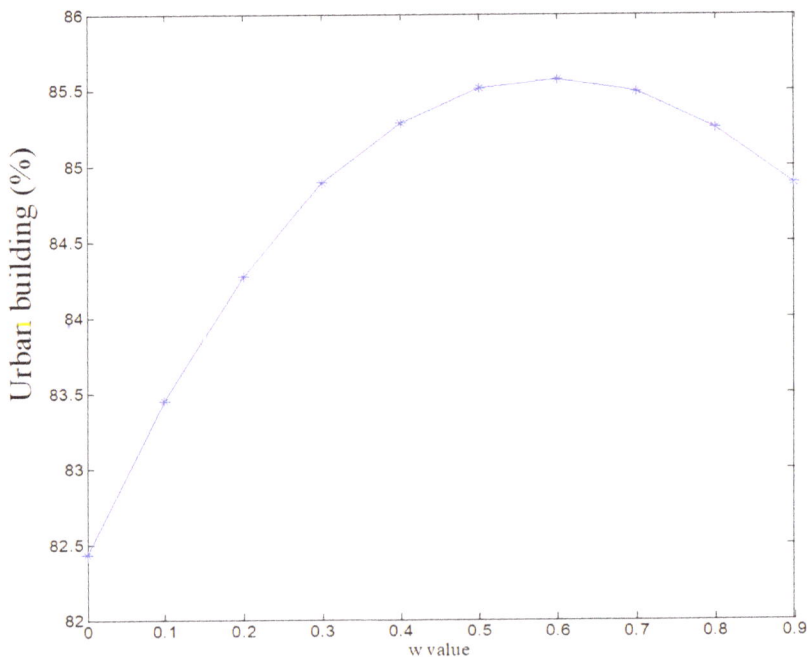

FIGURE 7.13   The relationship between urban building (%) of the SRFIM-MSI method and the weight parameter w.

7.4 MULTISPECTRAL SUBPIXEL BURNED-AREA MAPPING BASED ON SPACE-TEMPERATURE INFORMATION

Multispectral imaging (MI) provides important information for burned-area mapping. Due to the severe conditions of burned areas and the limitations of sensors, the resolution of collected multispectral images is sometimes very rough, hindering the accurate determination of burned areas. Subpixel mapping has been proposed for mapping burned areas in rough images to solve this problem, allowing subpixel burned-area mapping (SRBAM). However, the existing SRBAM methods do not use sufficiently accurate space information and detailed temperature information. To improve the mapping accuracy of burned areas, an improved SRBAM method utilizing space-temperature information (STI) is proposed here. STI contains two parts, a space part and a temperature part. We utilized the random-walker algorithm (RWA) to characterize the space part, which encompassed accurate object space information, while the temperature part with rich temperature information was derived by calculating the normalized burn ratio (NBR). The two parts were then merged to produce an objective function with space-temperature information. The particle swarm optimization (PSO) algorithm was employed to handle the objective function and derive the burned-area mapping results. The data set of the Landsat 8 Operational Land Imager (OLI) from Denali National Park, Alaska, was used for testing and showed that the STI method is superior to the traditional SRBAM method.

7.4.1 Space Part

We introduce the space part T^{spa} with the object space information to obtain more accurate space information. Figure 7.14 shows the process of producing the space

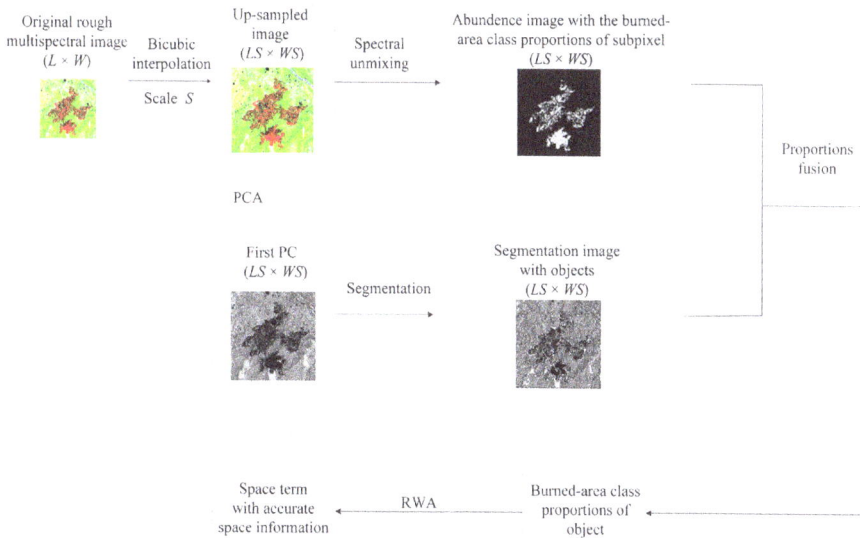

FIGURE 7.14 The flowchart of producing space part.

part. The rough multispectral image is upsampled by bicubic interpolation. A fractional image with burned-area classes of subpixel proportions is received by unmixing the upsampled image.

The first principal component (PC) is then extracted from the upsampled image through principal component analysis (PCA). Because there is a lot of space information contained in this first PC, we segment it to obtain objects using a multiresolution segmentation method. Then Q is defined as the segmentation scale parameter, which determines the object size and the condition of merger termination. The segmentation method is given by

$$H = \lambda \times H^{\text{spectral}} + (1 - \lambda) \times H^{\text{shape}} \tag{7.16}$$

where H represents regional difference, and λ is a free parameter to balance the shape difference H^{shape} and spectral difference H^{spectral}.

The shape difference H^{shape} is calculated as

$$H^{\text{shape}} = \lambda^{\text{shape}} \times A / \sqrt{N} + (1 - \lambda^{\text{shape}}) \times A / R \tag{7.17}$$

where A is the actual frontier length of the object region, R is the rectangular frontier length of the object region, N is the subpixel number in the object region, $A / \sqrt{N}$ and A/R represent the smoothness and compactness of the object region, respectively, and λ^{shape} is the free parameter.

The spectral difference H^{spectral} is defined as

$$H^{\text{spectral}} = \sum_{b=1}^{B} \lambda_b^{\text{spectral}} \times D_b \tag{7.18}$$

where b represents a spectral band ($b = 1, 2, \ldots, B$; with B being total band numbers), D_b is the bth band spectral value standard deviation in the object region, and $\lambda_b^{\text{spectral}}$ is the free parameter here.

Among adjacent object regions, we merge the two objects with the minimum difference. When H is larger than Q, we terminate the merging process and extract the final objects.

Third, the space part T^{spa} with the object space information is derived by RWA. The M objects U_m ($m - 1, 2, \ldots, M$) are derived by segmenting the upsampled image, where object O_m comprises N_m subpixels. The burned-area class proportion $L(p_i)$ of subpixel p_i ($i = 1, 2, \ldots, N_m$) is from the spectral unmixing of the upsampled image, and we average the burned-area class proportions of subpixels to generate the burned-area class proportion $G(O_m)$ of object O_m:

$$G(O_m) = \sum_{i=1}^{N_m} L(p_i) \bigg/ N_m \tag{7.19}$$

The space part $T(i)^{\text{spa}}$ corresponding to the ith subpixel is then obtained by RWA, as given in Equation (7.20):

$$T(i)^{\text{spa}} = \beta T^{\text{among}}(\mathbf{G}) + (1 - \beta) T^{\text{within}}(\mathbf{G}) \tag{7.20}$$

where T^{among} (**G**) is the space information among objects, and T^{within} (**G**) represents the space information within each object, $\mathbf{G} = \left[G\left(O_1 \right), G\left(O_2 \right), ..., G\left(O_m \right) \right]$ is a column vector, and β is set to 0.5.

T^{among} (**G**) is given by

$$T^{among}\left(\mathbf{G}\right) = \mathbf{G}^T \mathbf{L} \mathbf{G} \tag{7.21}$$

where **L** is the Laplacian matrix:

$$\mathbf{L} = \begin{cases} \sum -v_{mb} & \text{if } m = b \\ -v_{mb} & \text{if } m \text{ and } b \text{ are adjacent objects} \\ 0 & \text{otherwise} \end{cases} \tag{7.22}$$

where $v_{mb} = \exp(-\varepsilon(\hat{v}_m - \hat{v}_b)^2)$. The free parameter ε is set to 0.6, and $\hat{v}_m$ is the mth object O_m spectral value:

$$\hat{v}_m = \sum_{i=1}^{N_m} v_i \Big/ N_m \tag{7.23}$$

where v_i is the spectral value of the ith subpixel in object O_m. T^{within} (**G**) is defined as

$$T^{within}\left(\mathbf{G}\right) = \left(\mathbf{1} - \mathbf{G}\right)^T \overline{\Lambda} \left(\mathbf{1} - \mathbf{G}\right) + \left(\mathbf{G} - \mathbf{1}\right)^T \Lambda \left(\mathbf{G} - \mathbf{1}\right) \tag{7.24}$$

where the diagonal values in the diagonal matrix $\overline{\Lambda}$ are the background class proportions, and the diagonal values in the diagonal matrix Λ represent the burned-area class proportion.

We minimize the space part T^{spa} for all land cover classes. The minimize formula is given by

$$x\left(p_i\right) = \begin{cases} 1, & \text{if subpixel } p_i \text{ belongs to burnt-area class} \\ 0, & \text{otherwise} \end{cases} \tag{7.25}$$

$$T^{spa} = \text{Min} \sum_{i=1}^{N_m} x\left(p_i\right) \times T\left(i\right)^{spa} \tag{7.26}$$

7.4.2 Temperature Part

A new temperature part T^{tem} is proposed to fully utilize the temperature information. The T^{tem} aims to minimize the difference in spectrum between the observed NBR value (NBR^{obe}) and the simulated NBR value (NBR^{sim}). The near-infrared (NIR) band and short-wave infrared band 1 (SWIR1) were used to calculate NBR^{obe} here [27]:

$$NBR^{obe} = \sum_1^K \frac{\rho_{NIR}^{obe} - \rho_{SWIR1}^{obe}}{\rho_{NIR}^{obe} + \rho_{SWIR1}^{obe}} \tag{7.27}$$

where the observed reflectance of both NIR band ρ_{NIR}^{obe} and SWIR1 band ρ_{SWIR1}^{obe} are obtained directly from the original MI, and K is the number of mixed pixels.

Suppose $r_{\text{NIR}}^{\text{bur}}$ and $r_{\text{SWIR1}}^{\text{bur}}$ are the reflectance of the burned area in the NIR and SWIR1 bands, and $r_{\text{NIR}}^{\text{non}}$ and $r_{\text{SWIR1}}^{\text{non}}$ are the corresponding reflectance of the background. For each mixed pixel in these bands, the ratio of burned-area subpixel numbers to total subpixel numbers is the proportion of burned area $a_{\text{NIR}}^{\text{bur}}$ or $a_{\text{SWIR1}}^{\text{bur}}$. The proportion of background in the two bands is then $1-a_{\text{NIR}}^{\text{bur}}$, and $1-a_{\text{SWIR1}}^{\text{bur}}$, respectively. We built a linear mixture including all subpixel spectra to consider the reflectance of each mixed pixel. Then each mixed-pixel simulated reflectance in the NIR band $\rho_{\text{NIR}}^{\text{sim}}$ and SWIR1 band $\rho_{\text{SWIR1}}^{\text{sim}}$ are calculated using Equations (7.28) and (7.29), respectively:

$$\rho_{\text{NIR}}^{\text{sim}} = \left(r_{\text{NIR}}^{\text{bur}} \times a_{\text{NIR}}^{\text{bur}}\right) + \left[r_{\text{NIR}}^{\text{non}} \times \left(1-a_{\text{NIR}}^{\text{bur}}\right)\right] \qquad (7.28)$$

$$\rho_{\text{SWIR1}}^{\text{sim}} = \left(r_{\text{SWIR1}}^{\text{bur}} \times a_{\text{SWIR1}}^{\text{bur}}\right) + \left[r_{\text{SWIR1}}^{\text{non}} \times \left(1-a_{\text{SWIR1}}^{\text{bur}}\right)\right] \qquad (7.29)$$

NBR^{sim} is given by

$$\text{NBR}^{\text{sim}} = \sum_{1}^{K} \frac{\rho_{\text{NIR}}^{\text{sim}} - \rho_{\text{SWIR1}}^{\text{sim}}}{\rho_{\text{NIR}}^{\text{sim}} + \rho_{\text{SWIR1}}^{\text{sim}}} \qquad (7.30)$$

The temperature part T^{tem} is then obtained by minimizing the difference between NBR^{obe} and NBR^{sim}:

$$T^{\text{tem}} = \text{Min}\left(\text{NBR}^{\text{obe}} - \text{NBR}^{\text{sim}}\right)^2 \qquad (7.31)$$

7.4.3 IMPLEMENTATION OF STI

To improve the burned-area mapping result, STI is proposed as shown in Figure 7.15. It includes the following three steps:

Step 1. Bicubic interpolation, segmentation, and RWA are utilized to obtain the space part T^{spa} with more accurate space information. At the same time, the temperature part T^{tem}, which contains rich temperature information, is obtained by calculating the NBR.

Step 2. We merge the space part T^{spa} and the temperature part T^{tem} through a trade-off parameter θ to produce the objective function T with space-temperature information. The aim of the proposed STI is to minimize T:

$$\text{Min } T = \left(1-\theta\right)T^{\text{spa}} + \theta T^{\text{tem}} \qquad (7.32)$$

Step 3. To optimize the objective function, PSO is employed. First, we randomly assign a burned-area or background label to all subpixels. Second, the labels of these subpixels are iteratively changed until the minimum value of T is derived. During each iteration, the burned-area label is changed to the background label, or vice versa. If T increases, the change is rejected, otherwise it is accepted. When less than 0.1% of labels are changed, the PSO is terminated.

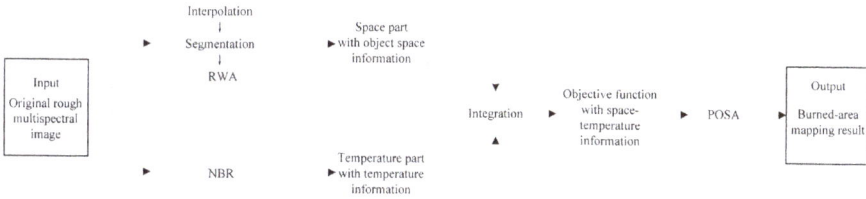

FIGURE 7.15   The flowchart of STI.

7.4.4   Experiment Content and Result Analysis

7.4.4.1   Experimental Settings

Lightning ignited the Castle Rocks fire in the deep backcountry of Denali National Park, Alaska, in July 2013. More than 12,900 acres were burned in 2 months. The fire caused great losses to the local ecosystem and economy. Therefore, it is very important to obtain fine space distribution of burned areas for firefighting and disaster relief. The experimental data set used was an image of this area obtained by Landsat 8 OLI on 26 August 2013, which can be downloaded from the US Geological Survey (USGS) website: https://earthexplorer.usgs.gov/. The image has a size of 2968 × 2052 pixels, 30 m space resolution, and is centered at 64°31'N, 152°52'W. As shown in Figure 7.1a, the five visible main burned areas are marked in red due to the false color of the image. For quantitative evaluation, we need a reference image derived from Figure 7.16(a) by a classification algorithm based on least squares support vector (LSSVM). There are two class labels (burned area and background) in the reference image shown in Figure 7.16(b). To highlight the burned area, the label is marked in red, and the background label is marked in black.

Fine images of five visible main burned areas from the experimental data set are shown in Figure 7.16. The test sizes of the five burned areas are 720 × 720 pixels, 300 × 300 pixels, 720 × 720 pixels, 400 × 400 pixels, and 500 × 500 pixels. A flow-chart of the experimental process is shown in Figure 7.17. We used the most common experimental process of subpixel mapping to conduct the experiments. The five visible main burned areas were downsampled via an S × S mean filter to produce the coarse multispectral image. Here, scale S was set to 8, namely 8 × 8 pixels in the original fine image were merged into one mixed pixel in the simulated coarse image. In this case, we can directly evaluate the impact of the error of image registration on subpixel mapping. In addition, quantitative evaluation can be carried out more reasonably in this way; a reference image can be derived from the classification result of the fine image, which is compared with the subpixel mapping result from the simulated coarse image. Coarse images of the five burned areas are shown in Figure 7.18. Although the false color coarse image can highlight the burned area, it is difficult to obtain more accurate distribution and boundary information of the burned area due to the coarse resolution. In addition, it is difficult for the classification technology to handle the mixed pixels, because one mixed pixel contains more than one land cover class. To solve this problem, subpixel mapping is utilized to handle the mixed pixels to produce accurate burned-area mapping. Least squares linear mixture model (LSLMM) was applied on the coarse images to derive the fractional images as inputs.

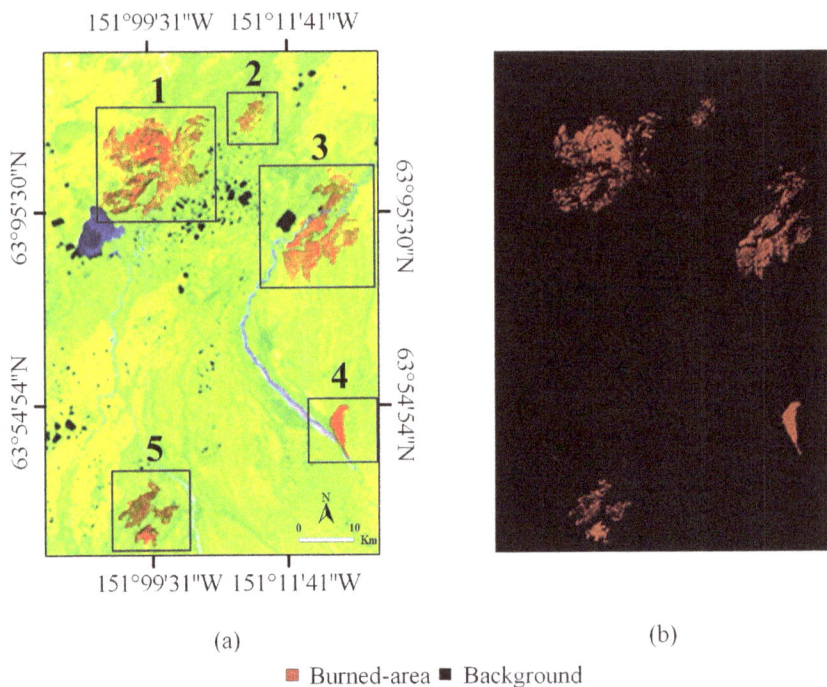

(a)   (b)

■ Burned-area ■ Background

FIGURE 7.16   (a) False color image (RGB: short-wave infrared 2 band, near-infrared band, and blue band for red, green, and blue, respectively). (b) Reference image.

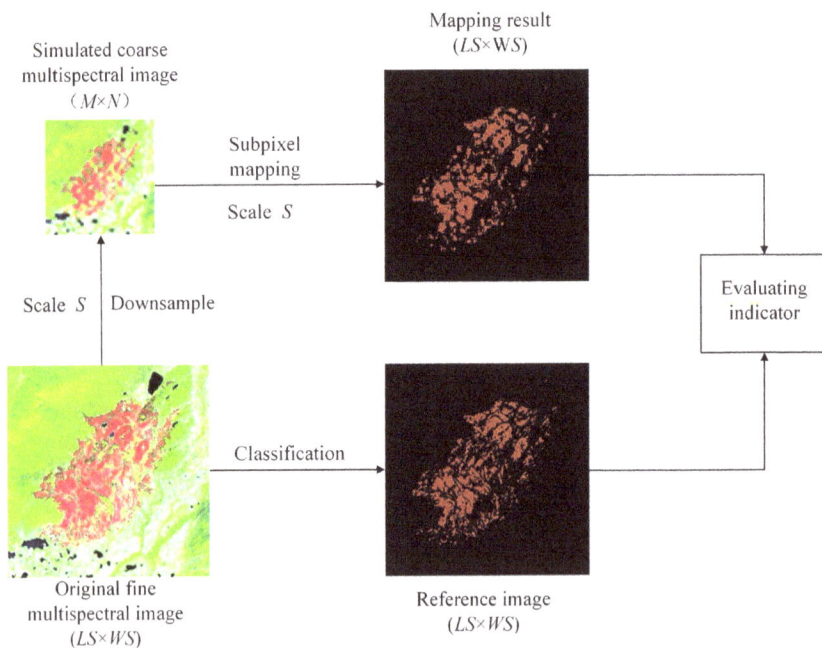

FIGURE 7.17   The flowchart of experimental process.

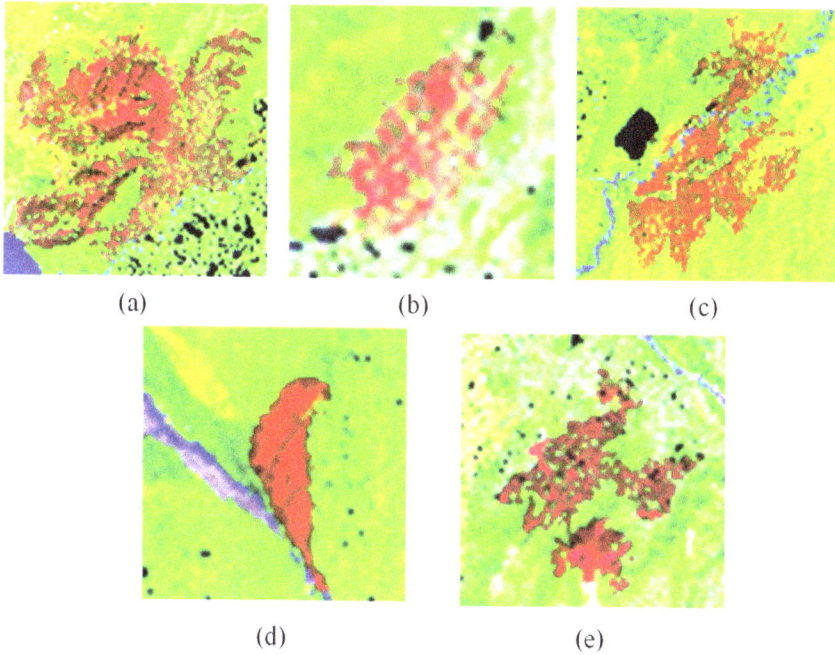

FIGURE 7.18   Coarse image of five burned areas: (a) area 1, (b) area 2, (c) area 3, (d) area 4, and (e) area 5.

In the segmentation method, the selected λ, λ^{shope}, and $\lambda_b^{\text{spectral}}$ were set to 0.5, 0.4, and 1, respectively, according to multiple tests. The trade-off parameter θ was set to 0.4, 0.5, 0.4, 0.6, and 0.4 in the five test areas, while the segmentation scale parameter Q was set to 15, 10, 15, 20, and 15. All experiments were performed using MATLAB 2018a.

We tested four subpixel mapping methods: hybrid spatial attraction model (HSAM) [25], object-scale spatial subpixel mapping (OSRM) [28], SRBAM [6], and the proposed STI. The ratio of the number of correct mapping subpixels belonging to burned areas derived from each subpixel mapping result to the total number of subpixels belonging to burned areas derived from the reference image is defined as burned area (%). The ratio of the number of correct mapping subpixels derived from each subpixel mapping result belonging to background to the total number of subpixels belonging to background derived from the reference image is defined as background (%). The four methods were evaluated using the accuracy of each class (burned area [%] and background [%]), overall accuracy (PCC [%]), and kappa coefficient (Kappa).

7.4.4.2   Results Analysis

First, visual comparison was analyzed. The burned-area mapping results of the subpixel mapping method in the five test areas are given in Figures 7.19–7.23. The detailed area is marked in a rectangular white frame. When we compare the reference images with the four experimental results, we find that STI outperforms the other three subpixel mapping methods, and the results from STI are more similar to the reference images. For burned areas with complex distribution, such as areas 1, 3, 4, and 5, there are many

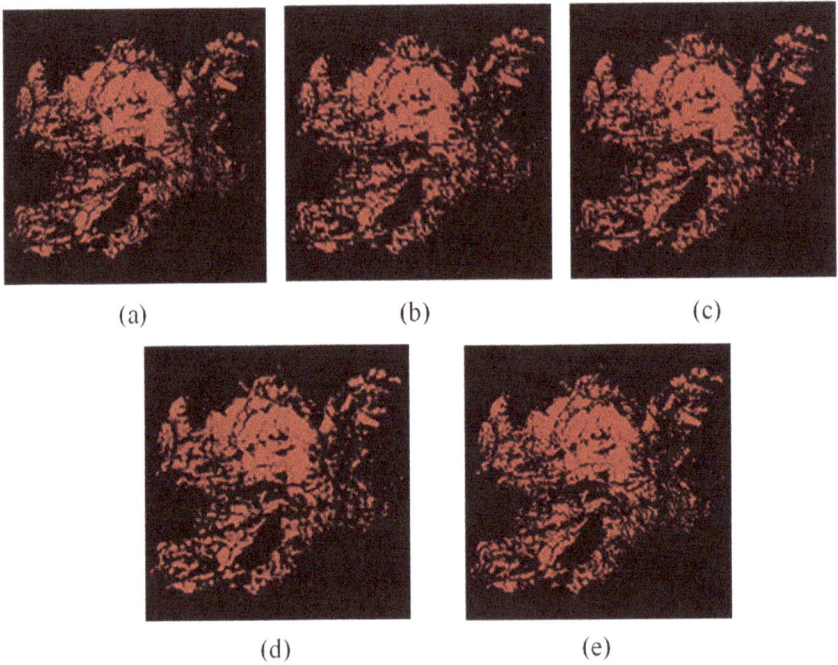

FIGURE 7.19   Burned-area mapping results in area 1: (a) reference image, (b) HSAM, (c) OSRM, (d) SRBAM, and (e) STI.

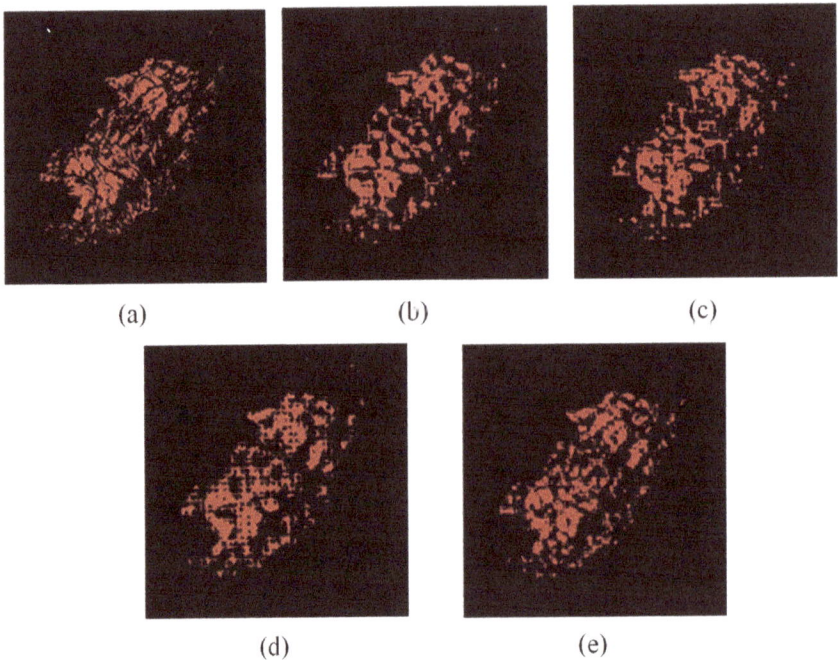

FIGURE 7.20   Burned-area mapping results in area 2: (a) reference image, (b) HSAM, (c) OSRM, (d) SRBAM, and (e) STI.

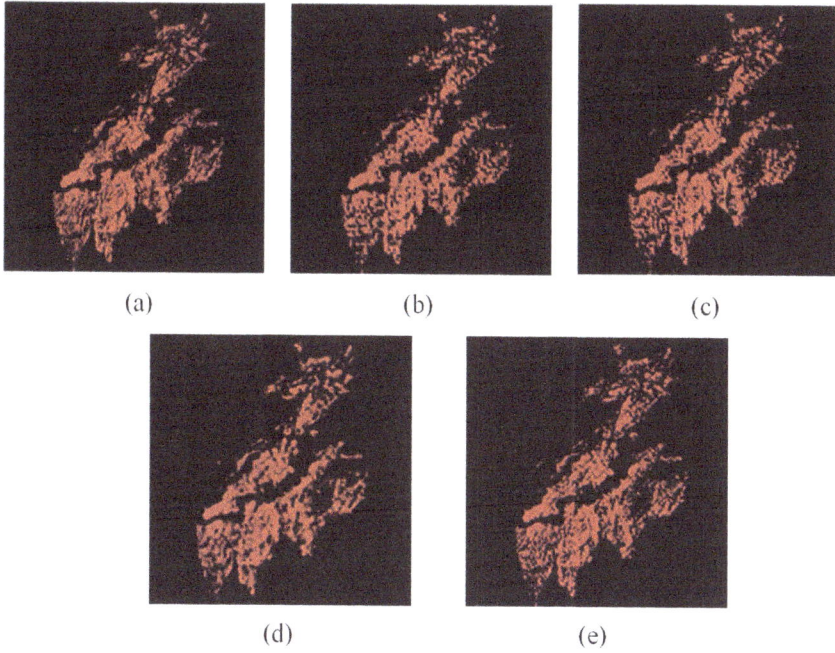

FIGURE 7.21 Burned-area mapping results in area 3: (a) reference image, (b) HSAM, (c) OSRM, (d) SRBAM, and (e) STI.

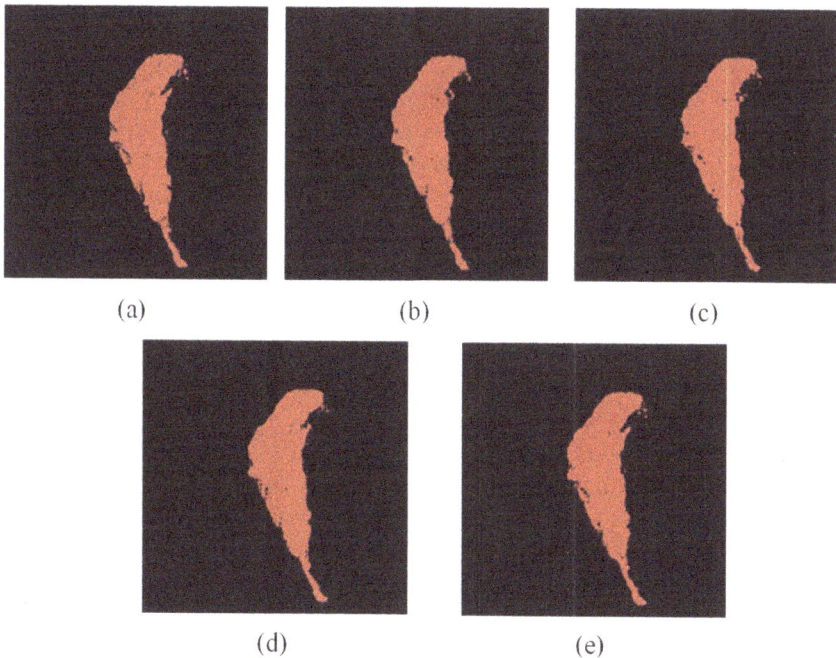

FIGURE 7.22 Burned-area mapping results in area 4: (a) reference image, (b) HSAM, (c) OSRM, (d) SRBAM, and (e) STI.

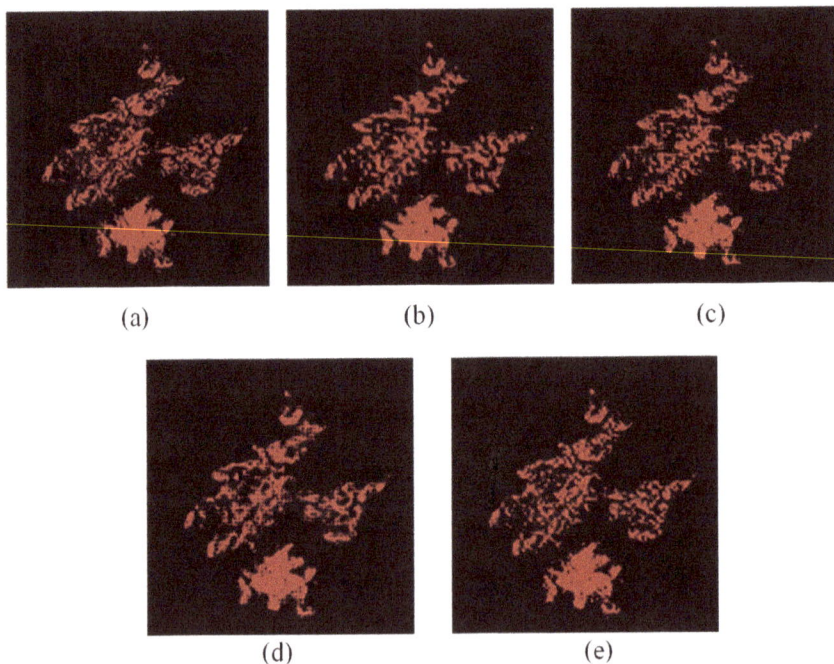

FIGURE 7.23 Burned-area mapping results in area 5: (a) reference image, (b) HSAM, (c) OSRM, (d) SRBAM, and (e) STI.

disconnected patches, and some small areas have disappeared in the results of HSAM, OSRM, and SRBAM. For some areas with simple distribution, such as area 2, there are many obvious burrs in the boundary of burned area in the results of HSAM, OSRM, and SRBAM. There are two reasons for these phenomena: First, the space information is not accurate enough. HSAM and SRBAM only consider pixel-level space information, which is rougher than object-level space information. Although OSRM utilizes object-level space information, it only calculates space information among object regions and does not consider space information within object regions. Since the proposed STI utilizes object-level space information among and within object regions through RWA, the space information is more accurate in STI than in the other three methods. In addition, STI is better able to make full use of temperature information than the other methods, so it obtains better burned-area mapping results.

Second, we analyzed the accuracy evaluation index. The performance of the four subpixel mapping methods was evaluated by burned area (%), background (%), PCC (%), and Kappa. Checking the evaluating indicator in Table 7.5, the burned area (%) of STI is higher than that of the other three methods. Compared with SRBAM, the burned-area (%) of STI is increased by 3.29%, 4.73%, 2.85%, 0.63%, and 3.55% in the five test areas, respectively. With the aid of space-temperature information, the proposed STI produces the highest PCC (%) and Kappa.

Third, we tested the performance of subpixel mapping by different scales S, which represent the simulated rough images with different resolution as inputs. The scales S set to different values verify that the STI still has the best performance for inputs with

TABLE 7.5
Evaluating Indicator of Four Methods

Area 1

Land Cover Class	HSAM	OSRM	SRBAM	STI
Burned area (%)	76.30	77.66	79.84	83.13
Background (%)	93.10	93.50	94.13	95.09
OA (%)	89.31	89.93	90.91	92.39
Kappa	0.6940	0.7116	0.7397	0.7622

Area 2

Land Cover Class	HSAM	OSRM	SRBAM	STI
Burned area (%)	56.50	59.73	63.66	68.39
Background (%)	95.62	95.94	96.34	96.82
OA (%)	92.04	92.63	93.35	94.21
Kappa	0.5212	0.5567	0.6000	0.6321

Area 3

Land Cover Class	HSAM	OSRM	SRBAM	STI
Burned area (%)	72.18	73.98	77.02	79.87
Background (%)	95.52	95.81	96.09	96.76
OA (%)	92.28	92.78	93.39	94.42
Kappa	0.6770	0.6979	0.7112	0.7463

Area 4

Land Cover Class	HSAM	OSRM	SRBAM	STI
Burned area (%)	94.23	95.35	95.41	96.04
Background (%)	98.54	98.59	98.60	99.23
OA (%)	98.18	98.26	98.47	99.01
Kappa	0.9448	0.9494	0.9531	0.9596

Area 5

Land Cover Class	HSAM	OSRM	SRBAM	STI
Burned area (%)	71.60	73.14	76.27	79.82
Background (%)	96.41	96.61	97.01	97.45
OA (%)	93.63	93.98	94.68	95.48
Kappa	0.6801	0.6975	0.7328	0.7627

different resolution. HSAM, OSRM, SRBAM, and STI are tested for the other two scales (5 and 10) in the five test areas. The burned area (%) of these methods in relation to $S = 5$ and $S = 10$, shown in Figure 7.24. We find that as the value of S increases, the burned area (%) of the four methods decreases. This is because as S is larger, the input image becomes rougher, bringing greater challenge to subpixel mapping. The experimental results show that STI still obtains the highest burned area (%) with different scales S.

Fourth, the influence of parameter θ selection on the proposed method was studied. Five test areas ($S = 8$) were rerun for ten combinations of θ from 0 to 0.9 with an interval of 0.1. The results are shown in Figure 7.25. There is no contribution

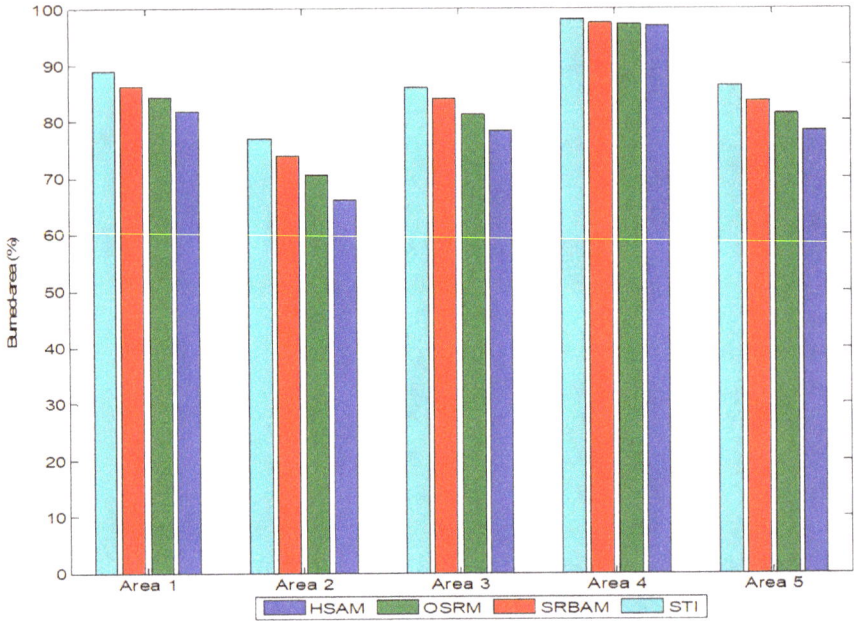

(a)

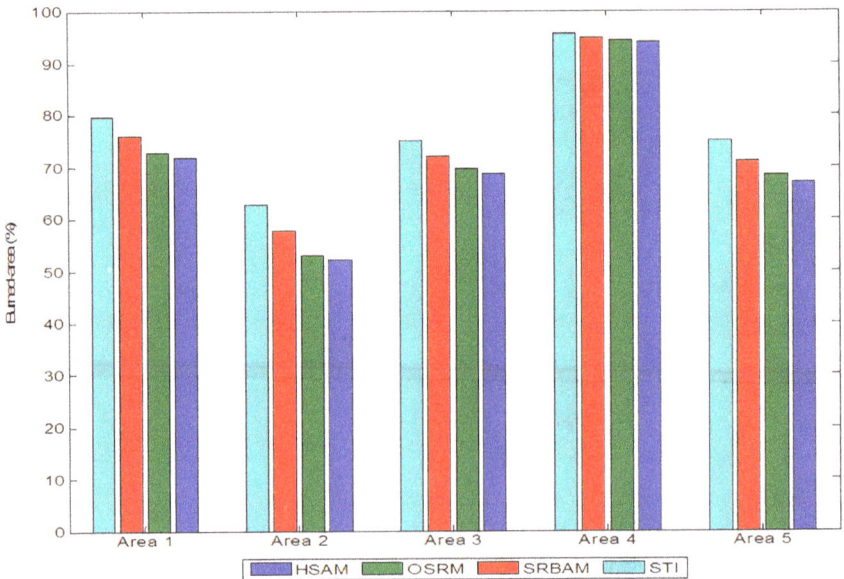

(b)

FIGURE 7.24   Burned area (%) derived using the four methods tested for different values of
S: (a) S = 5 and (b) S = 10.

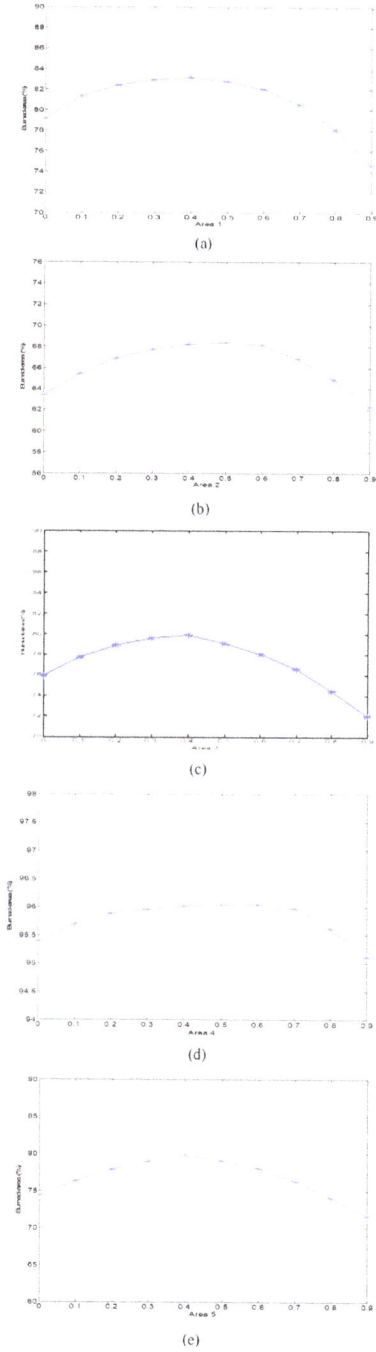

FIGURE 7.25   Burned area (%) derived using the four methods tested for different values of weight parameter θ: (a) area 1, (b) area 2, (c) area 3, (d) area 4, and (e) area 5.

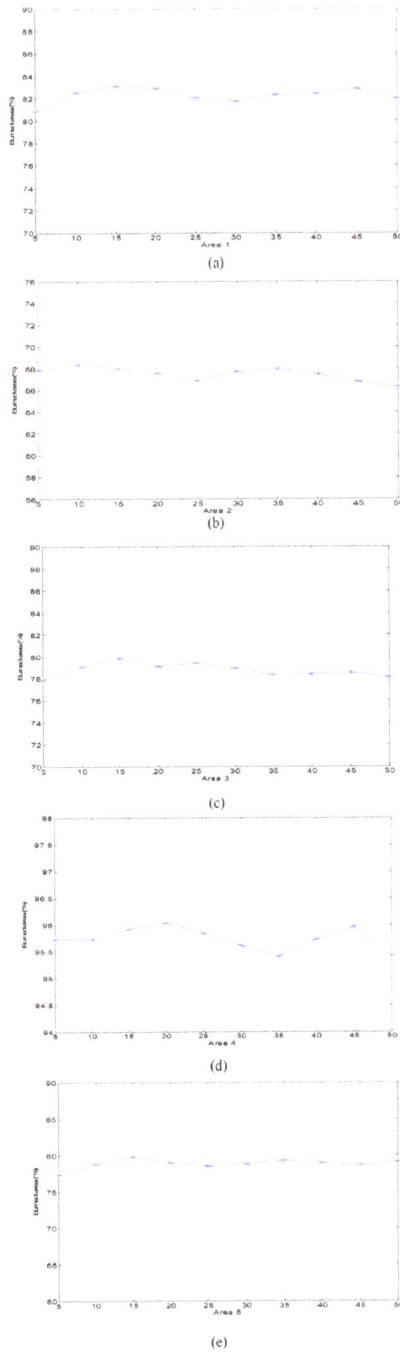

FIGURE 7.26   Burned area (%) derived using the four methods tested for different values of segmentation scale parameter Q: (a) area 1, (b) area 2, (c) area 3, (d) area 4, and (e) area 5.

from the temperature part T^{tem} when $\theta = 0$. At this time, only the space part T^{spa} is working, so the value of burned area (%) is low. As θ increases, the burned area (%) increases. This is because the use of temperature information from the temperature part T^{tem} increase as θ increases. When $\theta = 0.4$, $\theta = 0.5$, $\theta = 0.4$, $\theta = 0.6$, and $\theta = 0.4$ in the five test areas, the burned area (%) has its highest value. At this time, the contributions of the space term T^{spa} and the temperature term T^{tem} reach a state of the best balance. However, when θ increases, the space term T^{spa} reduces its contribution to formula (7.33). The burned-area mapping accuracy is affected due to the decreased space information from the space term T^{spa}.

Fifth, the impact of segmentation scale parameter Q on the proposed method was studied. In the proposed STI, the space part T^{spa} is obtained by calculating the class proportion of objects through RWA. Therefore, the step of segmentation that produces the object is very important for STI, and the quality of objects is decided by segmentation scale parameter Q in the segmentation method. Because Q determines the object size and the condition of merger termination, we study the optimal selection of Q in this experiment. Ten Q values from 5 to 50 with an interval of 5 are tested in five test areas ($S = 8$). As shown in Figure 7.26, it turns out that the selection of Q has an impact on the final mapping accuracy. When the value of Q is not properly selected, the burned area (%) is low. This is because an inappropriate Q results in producing low-quality objects, which affects the accuracy of space information in space part T^{spa}. After many experiments, it is noted that the best Q of five test areas is 15, 10, 15, 20, and 15. In the future, an adaptive method for the selection of Q is worth studying.

Finally, we analyzed the operation time (s). Figure 7.27 shows the operation time (s) of the four subpixel mapping methods in the five test areas ($S = 8$). The results

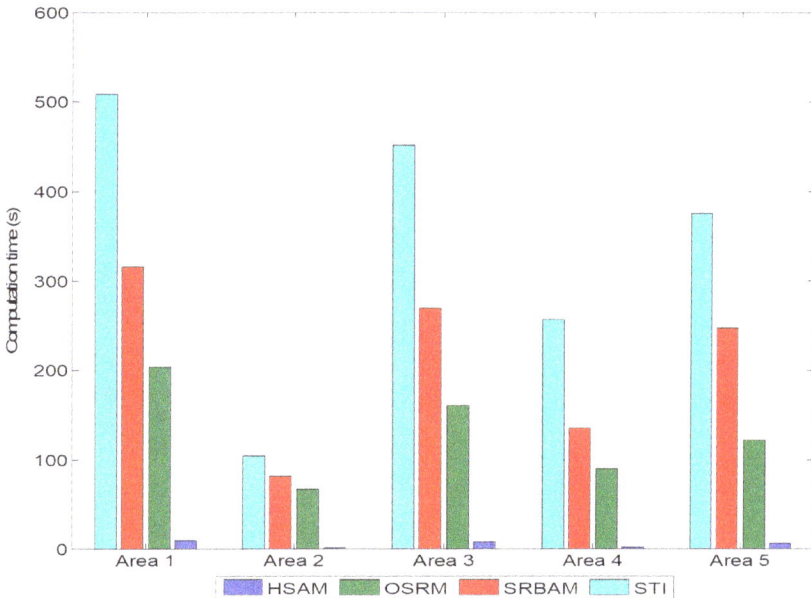

FIGURE 7.27   Operation time (s) in relation to five subpixel mapping methods.

show that STI takes the most time. This is because there is more complex processing in the proposed STI. Although STI takes more computation time than the other subpixel mapping methods, it shows improved performance.

7.5 SUMMARY

This chapter mainly introduces the application of subpixel mapping technology in practice. In SRFIM-MS, the spectral information from the green band and the SWIR band is utilized to calculate *NDWI*. A new spectral term, which is constituted by aiming minimize the difference of spectral indicator between $NDWI^{obe}$ and $NDWI^{sim}$, is added to the traditional SRFIM. The more spectral information is supplied to improve the SRFIM results by adding the new spectral term. Two Landsat 8 OLI data sets are conducted in the experiments for validation of the proposed SRFIM-MSI method. Both visual and quantitative assessments show that the SRFIM-MSI can noticeably increase the mapping accuracy of flood inundation. The experimental results of the SRFIM-MSI are visually more continuous and smoother. The PCC and Kappa of the SRFIM-MSI are significantly higher than those of conventional SRFIM methods.

In SMUBA, the spatial-spectral information of satellite multispectral imagery is more fully utilized by combining the spatial term and spectral term. Experiments using data from two Landsat 8 OLI satellites were performed. The experimental results of the SMUBA are visually more continuous and smoother. The quantitative assessments show that SMUBA can increase the mapping accuracy of urban built-up areas noticeably. In the future, the performance of the proposed method in a full-scene imagery with heterogeneous and randomly distributed features will be investigated.

STI is proposed to improve burned-area mapping by fully utilizing the space-temperature information of burned areas. The space part and temperature part are proposed in STI. RWA is used to compute the segmented objects to obtain the space part with accurate and comprehensive space information. At the same time, the temperature part with full temperature information is obtained by calculating the difference between NBR^{obe} and NBR^{sim}. An objective part with the space-temperature information is derived by integrating the space part and the temperature part. Finally, PSOA is utilized to optimize the objective part to produce burned-area mapping results. Due to space-temperature information, the proposed STI obtains better burned-area mapping results than the existing SRBAM. Experiments on Landsat 8 OLI images of burned areas in Denali National Park, Alaska, show STI produce the highest PCC (%) and kappa, achieving 92.39%, 94.21%, 94.42%, 99.01%, and 95.48% in the five tested areas.

REFERENCES

[1] Ling F, Wu S, Xiao F, Wu K, Li X. A review of remote sensing image sub-pixel positioning research[J]. Chinese Journal of Image and Image, 2011, 16(8): 1335–1345.

[2] Tatem A J, Lewis H G, Atkinson P M, Nixon M S. Increasing the spatial resolution of agricultural land cover maps using a Hopfield neural network[J]. International Journal of Geographical Information Science, 2003, 17(7): 647–672.

[3] Thornton M W, Atkinson P M, Holland D A. Sub-pixel mapping of rural land cover objects from fine spatial resolution satellite sensor imagery using super-resolution pixel-swapping[J]. International Journal of Remote Sensing, 2006, 27(3): 473–491.

[4] Zhang H, Shi J, Liu S. Research on sub-pixel mapping algorithm of lakes[J]. Advances in Water Science, 2006, 17(3): 376–382.

[5] Foody G M, Muslim A M, Atkinson P M. Super-resolution mapping of the waterline from remotely sensed data[J]. International Journal of Remote Sensing, 2005, 26(24): 5381–5392.

[6] Ling F, Du Y, Zhang Y, Li X, Xiao F. Burned-area mapping at the subpixel scale with MODIS images[J]. IEEE Geoscience and Remote Sensing Letters, 2015, 12(9): 1963–1967.

[7] Zhang Y, Atkinson P M, Li X, Ling F, Wang Q, Du Y. Learning-based spatial: Temporal superresolution mapping of forest cover with MODIS images[J]. IEEE Transactions on Geoscience and Remote Sensing, 2017, 55(1): 600–614.

[8] Li L, Chen Y, Xu T, Liu R, Shi K, Huang C. Super-resolution mapping of wetland inundation from remote sensing imagery based on integration of back-propagation neural network and genetic algorithm. Remote Sensing of Environment, 2015, 164: 142–154.

[9] Li L, Chen Y, Xu T, Liu R, Huang C. Sub-pixel flood inundation mapping from multispectral remotely sensed images based on discrete particle swarm optimization[J]. ISPRS Journal of Photogrammetry and Remote Sensing, 2015, 101: 10–21.

[10] Li L, Xu T, Chen Y. Improved urban flooding mapping from remote sensing images using generalized regression neural network-based super-resolution algorithm[J]. Remote Sensing, 2016, 8(8): 625.

[11] Ling F, Li X, Xiao F, Fang S, Du Y. Object-based subpixel mapping of buildings incorporating the prior shape information from remotely sensed imagery[J]. International Journal of Applied Earth Observation and Geoinformation, 2012, 18(1): 283–292.

[12] Xie H, Luo X, Xu X, Pan H, Tong X. Automated subpixel surface water mapping from heterogeneous urban environments using Landsat 8 OLI Imagery[J]. Remote Sensing, 2016, 8(7): 584.

[13] Li X, Du Y, Ling F, Feng Q, Fu B. Superresolution mapping of remotely sensed image based on hopfield neural network with anisotropic spatial dependence model[J]. IEEE Geoscience and Remote Sensing Letters, 2014, 11(7): 1265–1269.

[14] Wang Q, Shi W, Atkinson P M. Sub-pixel mapping of remote sensing images based on radial basis function interpolation[J]. ISPRS Journal of Photogrammetry and Remote Sensing, 2014, 92(1): 1–15.

[15] Pu R, Landry S. A comparative analysis of high spatial resolution IKONOS and WorldView-2 imagery for mapping urban tree species[J]. Remote Sensing of Environment. 2012, 124(9): 516–533.

[16] Tigges J, Lakes T, Hostert P. Urban vegetation classification: benefits of multitemporal RapidEye satellite data[J]. Remote Sensing of Environment, 2013, 136(5): 66–75.

[17] Zhang X, Du S. A linear Dirichlet mixture model for decomposing scenes: Application to analyzing urban functional zonings[J]. Remote Sensing of Environment, 2015, 169: 37–49.

[18] Sebari I, He D C. Automatic fuzzy object-based analysis of VHSR images for urban objects extraction[J]. ISPRS Journal of Photogrammetry and Remote Sensing, 2013, 79 (5): 171–184.

[19] Du S, Zhang F, Zhang X. Semantic classification of urban buildings combining VHR image and GIS data: An improved random forest approach[J]. ISPRS Journal of Photogrammetry and Remote Sensing, 2015, 105: 107–109.

[20] Hu T, Huang X, Li J, Zhang L. A novel co-training approach for urban land cover mapping with unclear Landsat time series imagery[J]. Remote Sensing of Environment, 2018, 217: 144–157.

[21] Ling F, Li X, Xiao F, Fang S, Du Y. Object-based sub-pixel mapping of buildings incorporating the prior shape information from remotely sensed imagery[J]. International Journal of Applied Earth Observation and Geoinformation, 2012, 18: 283–292.

[22] Zha Y, Gao J, Ni S. Use of normalized difference built-up index in automatically mapping urban areas from TM imagery[J]. International Journal of Remote Sensing, 2003, 24(3): 583–594.

[23] Sekertekin A, Abdikan S, Marangoz A M. The acquisition of impervious surface area from LANDSAT 8 satellite sensor data using urban indices: A comparative analysis[J]. Environmental Montitoring and Assessment, 2018, 190(7), 381.

[24] Atkinson P M. Sub-pixel target mapping from soft-classified remotely sensed imagery[J]. Photogrammetric Engineering and Remote Sensing, 2005, 71(7): 839–846.

[25] Ling F, Li X, Du Y, Xiao F. Sub-pixel mapping of remotely sensed imagery with hybrid intra- and inter-pixel dependence[J]. International Journal of Remote Sensing, 2013, 34(1): 341–357.

[26] Li X, Du Y, Ling F, Feng Q, Fu B. Superresolution mapping of remotely sensed image based on Hopfield neural network with anisotropic spatial dependence model[J]. IEEE Geoscience and Remote Sensing Letters, 2014, 11(7): 1265–1269.

[27] Holden Z A, Smith A M S, Morgan P, Rollins M G, Gessler P E. Evaluation of novel thermally enhanced spectral indices for mapping fire perimeters and comparisons with fire atlas data. International Journal of Remote Sensing, 2005, 26: 4801–4808.

[28] Chen Y, Ge Y, Heuvelink G B M, An R, Chen Y. Object-based superresolution land-cover mapping from remotely sensed imagery. IEEE Transactions on Geoscience and Remote Sensing, 2018, 56: 328–340.

Appendix
Abbreviations

Subpixel mapping (SPM)
Remote sensing (RS)
Aero imaging spectrometer (AIS)
National Aeronautics and Space Administration (NASA)
Airborne visible/infrared imaging spectrometer (AVIRIS)
Hyperspectral Digital Imagery Collection Experiment (HYDICE)
Spatially Enhanced Broadband Array Spectrograph System (SEBASS)
Fluorescence Line Imager (FLI)
Compact Airborne Spectrographic Imager (CASI)
Shortwave Infrared Airborne Spectrographic Imager (SASI)
Thermal Airborne Spectrographic Imager (TASI)
Reflective Optics Imaging Spectrometer (ROSIS)
Hyperspectral Mapper (HyMap)
Coastal Ocean Imaging Spectrometer (COIS)
Naval EarthMap Observer (NEMO)
Pushbroom hyperspectral imaging (PHI)
Operational modular imaging spectrometer (OMIS)
China Moderate Resolution Imaging Spectroradiometer (CMODIS)
Pixel purity index (PPI)
Iterative error analysis (IEA)
Pixel-swapping algorithm (PSA)
Hopfield neural network (HNN)
Subpixel-pixel spatial attraction model (SPSAM)
Backpropagation (BP)
Linear optimization technique (LOT)
Units of subpixel (UOS)
Highest soft attribute values first (HAVF)
Markov random field (MRF)
Light detection and ranging (LiDAR)
Multiple subpixel shifted images (MSIs)
Maximum a posteriori (MAP)
Subpixel mapping based on spatial-spectral correlation (SSC)
Linear spectral unmixing model (LSMM)
Non-linear spectral unmixing model (NSMM)

Indicator cokriging (ICK)

Error mapping pixels (EMPs)

Percentage of correctly classified (PCC)

Kappa coefficient (Kappa)

Subpixel mapping based on Hopfield neural network with more prior information (I-HNN)

Subpixel mapping based on extended random walker (SPMERW)

Subpixel mapping based on spatial-spectral correlation (SSC)

Bilinear interpolation (BI)

Bicubic interpolation (BIC)

Support vector machine (SVM)

Reflective Optics System Imaging Spectrometer (ROSIS)

Least squares linear mixture model (LSLMM)

New-style subpixel mapping based on bicubic interpolation (NBIC)

Subpixel mapping of the hybrid spatial attraction model (HSAM)

HNN with anisotropic spatial correlation (HNNA)

SPM based on object spatial correlation (OSPM)

Extended random walk (ERW)

Principal component analysis (PCA)

Least squares support vector machine (LSSVM)

Subpixel method based on dual-path bicubic interpolation (DPBIC)

Subpixel method based on the spatial correlation of pixels and subpixels (PSSD)

Mixed spatial attraction model (MSAM)

Kullback-Leibler distance (KLD)

Kullback-Leibler (KL)

Probability density functions (PDFs)

Subpixel-scale spatial attraction model (SSAM)

Spatial-spectral interpolation (SSI)

Object spatial dependence (OSD)

Overall accuracy (OA)

Kennedy Space Center (KSC)

Multiple subpixel shifted images with spatial-spectral information in soft-then-hard subpixel mapping (MSI-SS)

Subpixel mapping based on the spatial attraction model with multi-scale subpixel shifted images (SAM-MSSI)

Spatiotemporal subpixel mapping by considering the point spread function effect (FCSTD)

Subpixel mapping based on bilinear interpolation of multi-shift images (MSI-BI)

Subpixel mapping based on bicubic interpolation of multiple multi-shift images (MSI-BIC)

Subpixel mapping based on bilinear interpolation of multiple multi-shift images with spatial-spectral information (MSI-SS-BI)

Subpixel mapping based on bicubic interpolation of multi-shift images with spatial-spectral information (MSI-SS-BIC)

Multi-scale subpixel shifted images (MSSIs)

Single-scale subpixel shifted images (SSIs)

Spatial attraction model based on multi-shift images (SPSAM-SSI)

Subpixel shifted images with multi-scale spatial-spectral information for subpixel mapping (SSI-MSSI)

Projection onto convex set (POCS)

Subpixel mapping based on multiple subpixel shifted images by bicubic interpolation (SSI-BIC)

Subpixel mapping based on multiple subpixel shifted images with multi-scale spatial-spectral information (SSI-MSAM)

Prior fine spectral image (PFSI)

Original coarse spectral image (OCSI)

Point spread function (PSF)

Spatiotemporal subpixel mapping model based on fine and coarse scales temporal dependence by considering point spread function effect (FCSTD)

Area-to-point kriging (ATPK)

Pixel swapping-algorithm (PSA)

Radial basis function interpolation (RBF)

National Land-Cover Database (NLCD)

Soft-then-hard subpixel mapping based on the pansharpening technique for remote sensing image (STHSRM-PAN)

Subpixel land cover mapping based on dual processing paths for hyperspectral image (DPP)

Subpixel mapping based on multi-source remote sensing fusion data for land cover classes (SPM-MRSFD)

Digital surface model (DSM)

Multi-source remote sensing fusion data (MRSFD)

Component substitution (CS)

Gram schmidt (GS)

Intensity-hue-saturation (HIS)

Endmember of interest (EOI)

Units of class (UOC)

Edge interpolation (EI)

Spatial-spectral bicubic interpolation (SSBIC)

High-accuracy surface modeling (HASM)

HNN based on fused image (HNNF)

Class of interest (COI)

Deep Laplace pyramid networks (DLPNs)

Hybrid interpolation by parallel paths (HIPP)

Iterative interpolation deconvolution (IID)

Hybrid spatial attraction model (HSAM)

HNN with panchromatic image (HNNP)

Erreur Relative Globale Adimensionnelle de Synthèse (ERGAS)

Spectral Angle Mapper (SAM)

Universal Image Quality Index (UIQI)

Principal components (PCs)

Kernel principal component analysis (KPCA)

Feature fusion based on graph method (FFG)

Subpixel mapping based on intra- and inter-pixel dependence (NSAM)

Subpixel mapping based on radial basis function interpolation by reducing point spread function effect (RBF-PSF)

Subpixel mapping based on HNN with LiDAR image (HNN-LiDAR)

Subpixel mapping based on pansharpening technology (SPM-PAN)

Band-dependent spatial detail (BDSD)

Super-resolution recovery then classification (MTC)

Pansharpening then classification (PTC)

Maximum posterior probability and convex set projection hybrid (MAP/POCS)

Maximum likelihood (ML)

Support vector machine (SVM)

MAP super-resolution reconstruction algorithm based on the transformed space (T-MAP-SR)

Existing MAP super-resolution reconstruction algorithm (MAP-SR)

Improving subpixel flood-inundation mapping for multispectral remote sensing image by supplying more spectral information (SRFIM-MSI)

Subpixel mapping for urban building by using spatial-spectral information from spaceborne multispectral remote sensing image (SMUB)

Multispectral image subpixel burned-area mapping based on space-temperature information (STI)

Subpixel flood-inundation mapping (SRFIM)

Normalized difference water index (NDWI)

Flood-inundation spatial correlation index ($FISDI_i$)

Observed NDWI value ($NDWI^{obs}$)

Simulated NDWI value ($NDWI^{sim}$)

Green band (Green)

Short-wave infrared band (SWIR)

Particle swarm optimization (PSO)

SRFIM based on discrete particle swarm optimization (DPSO)

Very high-resolution (VHR)

Multispectral imaging (MI)

Subpixel burned-area mapping (SRBAM)

Normalized burn ratio (NBR)

Operational Land Imager (OLI)

Observed NBR value (NBR^{obe})
Simulated NBR value (NBR^{sim})
Near-infrared (NIR)
Short-wave infrared band 1 (SWIR1)
US Geological Survey (USGS)
Object-scale spatial subpixel mapping (OSRM)

Content Validity

The book is divided into nine chapters. The first two chapters briefly introduce the background and significance of the research on remote sensing image subpixel mapping technology, the current situation of subpixel mapping technology research, and the basic principles of subpixel mapping technology, which are convenient for readers with different needs. Chapters 3 to 7 are mainly based on the author's research results in recent years, from subpixel mapping based on single remote sensing image, remote sensing image subpixel mapping based on multi-shift image, remote sensing image subpixel mapping based on fusion technology, five aspects of remote sensing image subpixel mapping based on reconstruction then classification, and the application of remote sensing image subpixel localization technology. These are systematically organized and explained in detail, aiming to provide readers with a more complete remote sensing image subpixel localization technology, framework and newer research methods.

This book can be used as a reference book for undergraduates and postgraduates majoring in remote sensing, surveying and mapping, signal, and information processing in colleges and universities, and can also be referred to by researchers at different levels in related fields.

Index